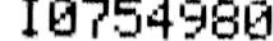

1

2

3

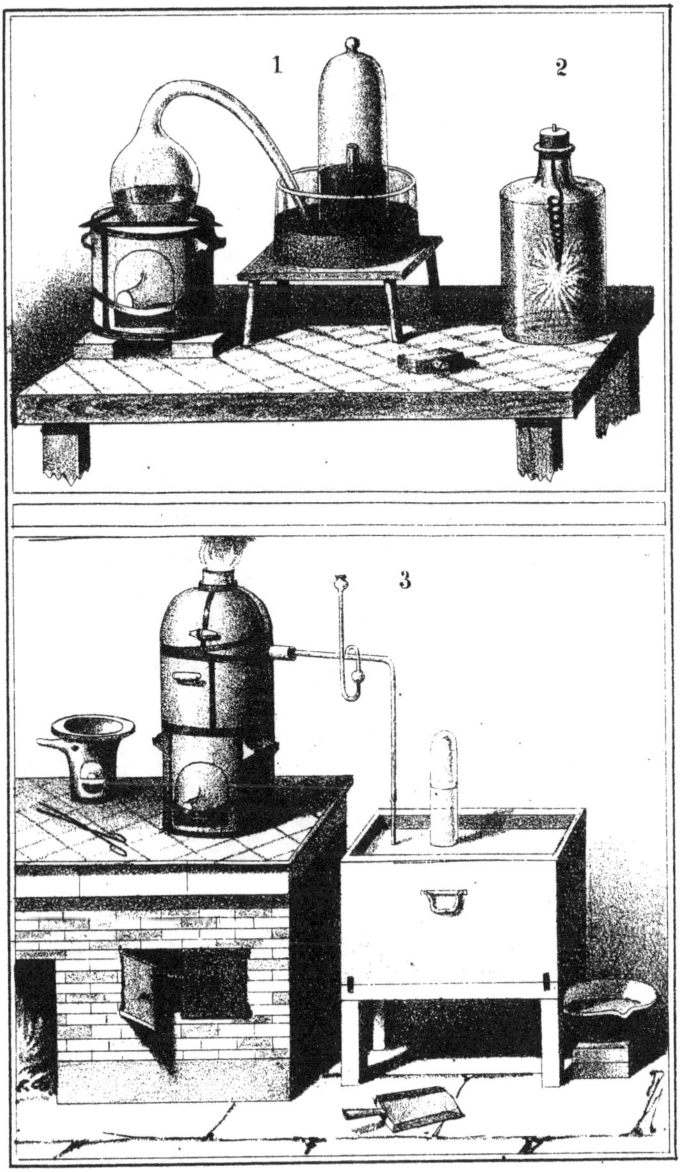

GENERAL NOTIONS OF CHEMISTRY,

BY

J. PELOUZE & E. FREMY.

TRANSLATED FROM THE FRENCH.

PHILADELPHIA.

J. B. LIPPINCOTT & CO.

GENERAL NOTIONS

OF

CHEMISTRY,

BY

J. PELOUZE AND E. FREMY.

TRANSLATED FROM THE FRENCH.

BY

EDMUND C. EVANS, M.D.

PHILADELPHIA:
LIPPINCOTT, GRAMBO & CO.
1854.

PREFACE.

THIS work is intended for persons who, unaccustomed to scientific studies, wish to acquire a general knowledge of Chemistry and its principal applications.

Among the numerous facts which compose this science, we have chosen those which recommend themselves by their importance in the arts; these we have attempted to make clear by freeing them from formulas and details purely scientific, which we have given in other works.

To render our explanations more intelligible, we have (when desirable) accompanied them with plates which faithfully represent the form and arrangement of apparatus used in laboratories and manufactories.

Our object will have been gained, if the "GENERAL NOTIONS OF CHEMISTRY" contribute to develope a taste for a science which renders such great services to the industrial arts, that ignorance of its first elements is, at this day, inadmissible.

CONTENTS.

COMBINATION OF SULPHUR WITH OXYGEN.

COMBINATIONS OF CARBON WITH OXYGEN.

COMBINATIONS OF CARBON WITH HYDROGEN.

ORGANIC CHEMISTRY.

THE ASTRINGENT PRINCIPLES OF VEGETABLES.

NEUTRAL BODIES.

NOTE BY THE TRANSLATOR.

It was thought best to give the French system of weights and measures, and not to attempt to convert them into the English system in the text. This renders it necessary here to place before the reader a table of the comparative value of the French with English weights and measures. Also, a comparative scale of Fahrenheit's Thermometer with Centigrade's, which last is in universal use in France.

The metre is the standard of linear measure. It is the ten millionth part of the distance from the Equator to the North Pole. This measure is marked on a bar of platinum, which is preserved in the archives of the Academy of Sciences. Thus in linear measure —

The Metre is equal to $39\frac{37}{100}$ English inches, or $3\frac{28}{100}$ feet.

Decimetre, or $\frac{1}{10}$ of a metre, $3\frac{98}{100}$ inches.

Centimetre, or $\frac{1}{100}$ of a metre, $\frac{39}{100}$ of an inch.

Millimetre, $\frac{1}{1000}$ of a metre, $\frac{39}{1000}$ of an inch.

Decametre is 10 times 1 metre, or $393\frac{71}{100}$ inches, or 10 yards 2 feet 7 inches.

Hectometre, 100 times 1 metre, or $3,937\frac{1}{10}$ inches, or 100 yards 1 foot 1 inch.

Chiliometre, 1000 times 1 metre, or 39,371 inches.

SUPERFICIAL OR SQUARE MEASURE.

Are is a square decametre, $3\frac{95}{100}$ English perches.

Decare is 10 ares, or $39\frac{5}{10}$ English perches.

Hectare 100 ares, or 395 perches, or 2 acres 1 rood 35 perches.

SOLID MEASURE.

Stere is a cubic metre, or $35\frac{31}{100}$ cubic feet.

Decistere is $\frac{1}{10}$th of a stere.

Decastre is 10 cubic steres.

MEASURES OF CAPACITY.

Litre is the cubic decimetre. It is the standard measure of capacity in the decimal system. It contains 61.$\frac{028}{1000}$ English cubic inches. The English imperial gallon is equal to 4.$\frac{54}{100}$ litres.

The Decalitre is 10 litres.

Hectolitre is 100 "

Decilitre is $\frac{1}{10}$th of a litre.

Centilitre is $\frac{1}{100}$ "

Millilitre is $\frac{1}{1000}$ "

By keeping in mind that a metre is the basis of all other measures both of length, of superficies and capacity, and that it is rather more than 3 feet and 3 inches, and that it is a decimal system of measures, the reader can readily follow the authors.

WEIGHTS.

The unit of weight is the *gramme*. It is the weight of the 100th part of a cubic metre of distilled water at the temperature of melting ice.

A gramme is equal to 15$\frac{434}{1000}$ troy grains.

A decigramme is equal to 1$\frac{543}{1000}$ troy grains.

A centigramme is equal to $\frac{154}{1000}$ " "

A milligramme is equal to $\frac{015}{1000}$ " "

A decagramme is 10 grammes or 154$\frac{34}{100}$ grains.

A hectogramme is 100 grammes or 1543$\frac{4}{10}$ "

A kilogramme is 1000 grammes or 15,434 "

A myriagramme is 10,000 grammes or 154,340 grains.

A kilogramme is equal to 2 lbs. 3 oz. 4·428 drachms avoirdupois weight.

THERMOMETERS.

THREE scales or methods of division of the thermometer have been adopted. They are those of Fahrenheit, mostly used here and in England; Reaumur, now seldom used; and the Centigrade scale used in this work, and most generally adopted in France and on the Continent of Europe. Properly, the point marked Zero 0° should in all thermometers mark the freezing point of water. This is the case in the Centigrade; but on Fahrenheit's scale 0° is 32° below freezing. In order to have a clear idea of the division of these scales, it must be borne in mind that the starting point in both is the freezing of water, which is a constant temperature (that is under ordinary natural circumstances); the next point to be fixed in the scale is the boiling of water; this, too, at the level of the sea in an open vessel, the water being pure, is a constant temperature. These two facts being remembered, and also that the space passed over by the mercury on the scale between these two points is, on the Centigrade scale, divided into 100 equal parts called degrees, and in Fahrenheit's scale into 180 equal parts or degrees, it will be easy to compare the degrees on Centigrade with those of Fahrenheit, by always comparing those of Centigrade above or below zero with those above or below the *freezing point of Fahrenheit*, 32° *F.*, without reference to his zero. 32° *F. and* 0° *C. mark the same temperature.* Then by the single rule of three, it is very easy to convert any degree marked on Centigrade's scale to the same temperature marked on Fahrenheit. What degree, for example, on Fahrenheit's scale corresponds with 5° C.? Say as follows: as 100 (the number of degrees between freezing and boiling of water on Centigrade's scale,) is to 5° C., so is 180° (the number of degrees between freezing and boiling of water in Fahrenheit's scale,) to the answer required, which is 9°, that is, 9° *above freezing*, (or 9° added to 32°,) or 41° F.; if we wish to compare 5° C. below 0° C., or freezing, then it would be by the same rule, 32° F. less 9°, or 23° F. Take again 40° C. above 0° or freezing, this will be found equal to 72° above freezing of Fahrenheit; that is, 72° added to 32° F., which is 104° F.; if it is 40° below 0° C. (the freezing point), it will correspond to 72° F. below freezing (32° F.), which is 40° below zero, Fahrenheit. As water when kept perfectly still may have its temperature reduced to 12° below 0° C. without freezing, and as ice or snow when melting preserves a constant temperature, viz., 0° C. or 32° F., melting ice or snow is used for fixing the zero, or freezing point of water, in making thermometers. This point is, therefore, properly speaking, the zero, or starting point.

NOMENCLATURE.

In relation to the nomenclature of the so-called *hydracids*, the following is from Regnault: — "Certain combinations of the metalloids with each other are energetic acids, which are scarcely below the most powerful *oxacids;* as, for example, the chloride of hydrogen, fluoride of hydrogen, etc., etc. It has unfortunately been thought proper to establish for these compounds a particular rule of nomenclature. It was thought that, in these new acids, hydrogen acted a part analogous to that of oxygen in the oxacids, and the name *hydracids* was given to them. This was, however, a grave error; in the oxacids, *oxygen is the electro-negative element,* while in the hydracids, *hydrogen is constantly the electro-positive element.*" (That is, when subjected to the action of a feeble voltaic pile, the oxacid is decomposed, and the *oxygen goes to the* POSITIVE POLE; but when subjected to the same influence, the *hydrogen* of the *hydracid goes to the negative pole,* and is therefore the *electro-positive element.*)

"The nomenclature of hydracids is, however, so generally employed that we are obliged to adopt it ourselves. The chloride of hydrogen takes the name of *chlorhydric acid*; the sulphuret of hydrogen takes the name of *sulphydric acid.* The names *hydrochloric acid, hydrosulphuric acid,* are often given to these same acids; but these names are more defective than the first, for they are an infraction of that general principle, according to which *the name of the compound body ought always to commence with the name of the electro-negative element.*" — REGNAULT'S COURS ELEMENTAIRE DE CHIMIE.

GENERAL NOTIONS

OF

CHEMISTRY.

INTRODUCTION.

The special object of Chemistry is the study of phenomena, which, taken together, enable us to characterise bodies.

Bodies may be ranged in two classes: the first class comprises *simple bodies;* the second, *compound bodies.*

A *simple body* is one from which but a single substance can be extracted; for example, from sulphur, in whatever way it be treated, we get nothing but sulphur.

A *compound body* is one from which two or more substances may be separated, having different properties. If oxide of mercury be heated, oxygen and mercury are obtained from it; the oxide of mercury is, then, a compound body.

It is known that bodies present themselves in three states; they may be either *solid*, *liquid*, or *gaseous*. A large number of bodies, such as sulphur and water, are known in these three states; others, as platinum among the metals, and wax among the organic bodies, are only known in the solid and liquid states. Some, as carbon, lime, and lignine, are only solid; while others, called *permanent gases*, as oxygen, hydrogen, and nitrogen, are only known at this time in the gaseous state.

Heat, cold, compression, the solvents, are frequently used to modify the state of aggregation of bodies. Many gases are liquefied by compression or cold, or more frequently by both of these united. A gas which has been liquefied by great pressure, may even be solidified.

The force which unites the molecules of solid and liquid bodies, is called *cohesion*. Cohesion is very great in solid bodies, feeble in liquids, and does not exist in gases; the molecules of a gas, on the contrary, have a tendency to separate from each other. Heat tends to destroy the force of cohesion; thus it often causes the fusion, and even volatilization, of solid bodies.

The force which unites the molecules of simple bodies to form the molecule of a compound body, is called *affinity*. It is this which, in the oxide of mercury, unites the oxygen with the mercury. Affinity plays a great part in chemical phenomena; it determines the combinations of bodies, and a great number of decompositions. All the causes which tend to destroy cohesion, such as heat, and solution in a liquid, tend equally to increase affinity; thus there is a great number of bodies which only unite together under the influence of heat, or solvents. Sometimes heat will, according to its intensity, produce different results, and destroy combinations which it at first produced.*

NOMENCLATURE.

NOMENCLATURE OF SIMPLE BODIES.

THE number of simple bodies known at present, is sixty-one.

* It may be well to explain by means of an illustration, or otherwise, what is meant by certain apparatus, or parts of apparatus, constantly made use of in experiments and processes in this work, viz.:

Many simple bodies derive their names from some of their essential properties. The word chlorine, for example, has reference to the green color of this gas; the word bromine, to the fœtid odor of this body.

Retort is a globular vessel with a long neck, employed in a variety of distillations. It may be either without a stopper, as in Fig. 6, Plate 2, or provided with a stopper called a tubulum, forming, in this last case, a tubulated retort, as in Fig. 9, Plate 3.

Receiver is a vessel intended for receiving the product of distillation, or for receiving and containing gases or other products over a pneumatic trough, or for placing over the plate of an air-pump, as a recipient. A receiver may be cylindrical, globular, or of any shape, of any suitable material. For some experiments and purposes, they must be tubulated, or have two mouths or necks; a bottle will sometimes answer. We have examples in Fig. 4, and in Figs. 7 and 9.

Matrass is an oval or globular vessel, with a long open neck, used for digestion, evaporation, &c. (Fig. 9 and Fig. 1.)

Alembic, Fig. 10, used in laboratories (as well as in the large way), for distilling water when required in large quantities. It is composed of a bottom part, or boiler, in which the liquid to be distilled is placed, (properly called *cucurbit*), and a head or dome (which is in fact the alembic) which is provided with a neck or beak communicating with a serpentine or worm, which descends through a vessel of cold water to condense the steam; this condensed water, which is the distilled water, is conducted to a vessel or receiver, from which it is taken for use. This entire apparatus is called a *distilling apparatus*.

Cupel is a small cup or vessel used in the process of refining metals, called cupellation. The cupel is porous, so as to absorb the baser vitrified metals when subjected to heat, and the precious metals remain. It is used, also, in certain experiments and chemical processes.

Lute is a composition of tenacious substances, used for stopping the junctures of vessels so closely as to prevent the escape of, or entrance of gas or air. This term is also applied to an external covering of clay, sand, or other refractory substance, to protect glass retorts, or other vessels, from the effects of heat.

It will be impossible for those ignorant of chemistry to obtain any, even the slightest, knowledge of the science, without going through the preliminary, and to some, no doubt, dry explanations of nomenclature and terms. These matters have, however, been so generalized in this work, that they are less dull than in most treatises.

Simple bodies are divided into two classes, metalloids and metals.

Metals are distinguished from metalloids by certain physical characteristics, but more especially by their essential property of forming bases in their union with oxygen; while the metalloids, in combining with oxygen, only produce neutral compounds or acids. We will explain, further on, what is meant by the words *bases, acids, neutral bodies.*

No salifiable base is known which results from the combination of a metalloid with oxygen.

The following is a list of metalloids and metals:—

METALLOIDS.

Arsenic,*	Carbon,	Iodine,	Silicon,
Nitrogen,	Chlorine,	Oxygen,	Sulphur,
Boron,	Fluorine,	Phosphorus,	Tellurium.
Bromine,	Hydrogen,	Selenium,	

METALS.

Aluminum,	Erbium,	Niobium,	Tantalum,
Antimony,	Tin,	Gold,	Terbium,
Silver,	Iron,	Osmium,	Thorium,
Barium,	Glucinium,	Palladium,	Titanium,
Bismuth,	Iridium,	Pelopium,	Tungsten,
Cadmium,	Lantanium,	Platinum,	Uranium,
Calcium,	Lithium,	Lead,	Vanadium,
Cerium,	Magnesium,	Potassium,	Yttrium,
Chromium,	Manganese,	Rhodium,	Zinc,
Cobalt,	Mercury,	Rhuthenium,	Zirconium,
Copper,	Molybdenum,	Sodium,	
Didymium,	Nickel,	Strontium,	

NOMENCLATURE OF COMPOUND BODIES.

The principle of chemical nomenclature which we owe to Guyton de Morveau, Lavoisier, Bertholet, and Fourcroy,

* *Arsenic* resembles the metals fully in its physical properties, but its combinations are so analogous with the corresponding combinations of phosphorus, that it is best to study these two bodies together. (Regnault.)

consists in designating the compound body by names indicating their composition, and sometimes even their properties.

The principal compound bodies are, acids, oxides, salts, and those binary bodies of which oxygen is not one of the elements.

ACIDS.

The name acid is given to such bodies as have the property of reddening the tincture of litmus, and of combining with bases to form salts.

Acids are divided into two principal groups: oxacids and hydracids.

Oxacids are produced by the combination of a simple body with oxygen; they are named after the following rules:—

When a simple body combines with oxygen only in a single proportion to form an oxacid, the name of this acid is composed of the name which designates the simple body with the termination *ic*.

Example: The oxacid formed by the combination of silicium with oxygen is called silicic acid.

When a simple body combines with oxygen in two proportions to form two acids, that which contains the least oxygen takes the termination *ous*, and the most oxygen takes the termination *ic*.

Example: The two acids formed by the combination of arsenic with oxygen, are termed arsenious acid, and arsenic acid; arsenious acid contains less oxygen than arsenic acid.

When in fine a simple body combines in four proportions with oxygen, the prefix *hypo*, is placed before the name of the acid terminating in *ous*, or in *ic*.

This prefix expresses a smaller quantity of oxygen than is contained in the acid terminating in *ous*, or *ic*, of which the name is not preceded by this prefix *hypo*.

Example: The acids formed by chlorine and oxygen have received the following names:—

Hypochlorous acid, chlorous acid, hypochloric acid, chloric acid.

In these compounds the proportion of oxygen goes on increasing from the hypochlorous acid to the chloric acid.

There is an acid more oxygenated than chloric acid, which is distinguished from it, and at the same time it is indicated that it contains more oxygen than chloric acid, for the same quantity of chlorine, by adding to the word chloric the prefix *per* or *hyper*. Thus it is called *perchloric* acid, or *hyperchloric* acid.

This rule has been extended to other acids, such as hyperiodic, and hypermanganic.

Hydracids. The name hydracid, is given to binary compound acids which are formed by the combination of hydrogen with a metalloid.

The names of the hydracids, are composed of the name of the simple body, which is sometimes called a *radical*, followed by the termination *hydric*.

Thus the hydracids produced by the union of hydrogen with *chlorine*, *bromine*, *iodine*, are named chlorhydric, bromhydric, iodhydric acids.

It may be remarked that hydrogen never forms but a single hydracid with the same radical.

OXIDES.

The name oxide is given to the binary compounds of oxygen which exercise no action on the colour of litmus.

Oxides are divided into two series.

The first comprises those oxides which have not the property of combining with acids to form salts; these are termed indifferent oxides.

In the second series are found those oxides which can unite with acids to form salts; these are termed *salifiable bases*.

When a simple body forms but one oxide in combining

with oxygen, the compound is designated by the collective word *oxide*, which is followed by the name of the simple body; thus the combination of zinc with oxygen is called oxide of zinc.

If the simple body is capable of uniting in several proportions with oxygen, the compounds which result are designated by adding to the term *oxide* the prefixed words, prot, sesqui, bi, and per, which express progressively increasing quantities of oxygen.

Examples: Protoxide of manganese, of iron, of copper, of tin.

Sesqui oxides of manganese, of iron, of chromium.

Binoxide of manganese, of copper, of chromium.

The name of peroxide is often given to such oxides as contain the most oxygen, and which still preserve the generic characters of oxides; thus we say *peroxide of iron, peroxide of manganese.*

SALTS.

When we make an acid act upon a base, we ordinarily see the properties of the acid and those of the base neutralize each other; thus the acid which at first reddened litmus, loses this property in proportion as we mix it with the base: in this case the acid and the base combine to form *a salt.*

A salt is then the combination of an acid with a base.

To give a name to the salt, we should have regard,

1st, to the nature of the acid; 2nd, to the salifiable base; 3rd, to the proportions in which the acid and the base have combined.

Every acid, the termination of which is in *ic*, will form a salt whose termination will be in *ate.*

Every acid whose termination is in *ous*, will form a salt whose termination will be in *ite.*

These new names terminating in *ate*, or in *ite*, will be followed by the name of the oxide which enters into the salt.

Examples: Sulphuric acid and the protoxide of iron will give the *sulphate* of the *protoxide of iron.*

Sulphurous acid and the protoxide of iron, will give the *sulphite* of the *protoxide of iron.*

Hyposulphuric acid and the protoxide of iron will give the *hyposulphate* of the *protoxide of iron.*

It often happens that to abridge the names of salts, the word oxide is suppressed; thus the salt formed by the combination of sulphuric acid and the oxide of lead is called *sulphate of lead.*

Acid and basic bodies have the property of neutralizing each other more or less exactly, and of losing more or less completely their action on coloured re-agents.

When the salt is brought as close as possible to the neutral state, its name is fixed according to the preceding rules, but if the proportion of the acid is greater than in neutral salts, the name acid salt is given to it. It is thus that the combination of sulphuric acid and potassa, which reddens litmus, would be called the *acid sulphate of potassa.*

If the base is in excess, the generic name is preceded by the preposition *sub.* Thus *subacetate of lead*, *subnitrate of bismuth;* the subsalts, are also called *basic salts.*

The combinations which water forms with simple or compound bodies are called *hydrates.*

BINARY COMPOUNDS OF WHICH OXYGEN IS NOT ONE OF THE ELEMENTS.

When a metalloid combines with a metal to form a compound which is neither acid nor basic, the combination is designated by giving to the metalloid the termination *uret*, which is followed by the name of the metal; thus the combinations of sulphur, of chlorine with iron will be sulphuret of iron, chloruret of iron (or chloride of iron.)

This nomenclature is applied also to the binary compounds

which result from the action of a hydracid on an oxide: in this case the radical of the hydracid takes the termination *uret;* thus *chlorhydric acid* in acting on the *oxide of iron*, produces the *chloruret of iron; sulphhydric acid* with the *oxide of mercury*, forms the *sulphuret of mercury.*

If the metalloid combines with the metal in many proportions, the generic name is preceded by the term *proto, sesqui, bi, trito* or *tri, quadri, penta* etc. *per;* thus, to name the different combinations of potassium with sulphur, which for the same quantities of metal contain quantities of sulphur represented by 1, 1½, 2, 3, 4, 5, would be said, protosulphuret, sesquisulphuret, bisulphuret, trisulphuret, quadrisulphuret, pentasulphuret of potassium.

Certain bases, as ammonia, combine integrally with hydracids to form true salts. In this case, the salt takes the termination *ate;* thus, the combination of *chlorhydric acid* with *ammonia* is called *chlorhydrate of ammonia.*

ALLOYS.

The combinations of metals with each other are termed *alloys.*

The alloys into which mercury enters are called *amalgams.*

Thus the alloy of mercury and silver is called *amalgam of silver.*

CRYSTALLIZATION OF BODIES.

When a solid body has lost its cohesion by the action of extraneous causes, and these causes discontinue their action, the body gradually resumes its solid state: if this change of state takes place with sufficient slowness, the body presents itself in small masses, sometimes isolated, sometimes grouped together assuming geometric forms, and terminated on all sides by plane and brilliant surfaces.

These small masses are designated by the name of *crystals.*

The crystalline forms of a body are not always apparent

to the naked eye; often these can only be distinguished by the aid of a lens or microscope.

Substances which do not assume geometric forms are called *amorphous*.

Artificial crystallization is brought about by different processes, which vary according to the properties of bodies; the chief of these methods we shall explain.

CRYSTALLIZATION BY FUSION.

A fusible body can be made to crystallize by bringing it to such a temperature as will cause it to melt, and then allowing it to cool very slowly.

The parts of the liquid in contact with the air, and those which touch the sides of the vessel in which it has been melted, cooling more rapidly, there is produced by the cooling, a crystalline layer which adheres to the sides of the vessel, and a solid crust which forms on the surface of the liquid, while the central part maintains its state of fluidity.

If we pierce this upper crust and pour out the fluid, crystals will be formed in the interior, which are larger or smaller in proportion as it has cooled slowly, and as we have been operating on a more or less considerable mass. While the melted body is cooling it should be placed where it will be free from all vibration.

It is thus that sulphur, bismuth, and a great number of metals and alloys are made to crystallize.

CRYSTALLIZATION BY VOLATILIZATION.

Solid and volatile bodies can be crystallized by volatilization; they are put into a retort of glass, or porcelain, or earthen-ware, according to the degree of their volatility. The beak of the retort is made to communicate with a receiver, properly cooled, and the retort is then brought to a temperature which will volatilize the body we wish to crystallize.

The vapours in cooling take the solid form, and give crystals which are deposited in the neck of the retort, or in the receiver.

Arsenic, certain metallic chlorides, sundry salts of ammonia, or of mercury, crystallize by volatilization.

CRYSTALLIZATION BY SOLUTION.

There are two different methods of causing bodies to crystallize by solution.

The first consists in dissolving the substance in a liquid, and in evaporating this liquid by means of heat, or else spontaneously, until the solid body is deposited. The form of the crystals is more beautiful in proportion to the slowness of the evaportion. The second method is founded on the unequal solubility of bodies in hot and cold liquids.

Suppose a body much more soluble in hot water than in cold, nitre for example: if nitre is dissolved in boiling water, until the water becomes saturated, and the liquor is then allowed to cool, it will necessarily deposit a certain quantity of nitre. If it is cooled slowly, we will obtain beautiful crystals of this salt.

It is thus that most salts are made to crystallize in laboratories, as the carbonate of soda, phosphate of soda, sulphate of copper, &c.

Leblanc has made known a method which enables us to increase the size of the crystals at will, and gives them in a state of perfect regularity. We first choose small regular crystals, obtained in a crystallization by evaporation, or by cooling a solution; these are introduced into a glass crystallizer and covered with the same liquor from which they were deposited, which is called the *mother-water;* this liquor is then left to a spontaneous evaporation.

As the liquor evaporates, it deposits on the small crystals a certain quantity of the salt which the mother-water contains; this deposit takes place in a manner so symmetrical that the crystal augments equally in all its dimensions.

We should be careful to turn over the crystals from time to time, so that they may increase equally on all their faces, and that the irregularities may be repaired.

Many causes contribute to promote the crystallization of bodies. It may be said generally that a solution crystallizes more rapidly when it is agitated with a solid body, than when it is allowed to repose quietly without agitation; however, a solution which has been agitated, always gives small crystals. Thus a syrup evaporated as is usual, gives sugar in little crystals when it is agitated, and *sugar candy*, that is to say, sugar in large crystals, when it is allowed to evaporate slowly in a stove.

When several bodies are dissolved in the same liquid, that which is first deposited is more pure and regular than the crystallization in a less dense medium. Thus the first crystals of chloride of sodium (marine salt) which are formed during the evaporation of sea-water, are more pure and more regular than those which are last produced. It happens often that a solution remains many days without giving any signs of crystallization, and will become a crystalline mass as soon as it is slightly agitated.

Solid bodies will sometimes favour crystallization, and become in some sort a *nucleus* for crystals, which form on their surface; sometimes when a liquor refuses to crystallize, small crystals of the same nature with those which ought to be deposited being thrown in, bring about, by their presence, the crystallization of the whole mass. The nature of vessels in certain cases facilitates crystallization. It is remarked that a liquor crystallizes more rapidly in rough vessels, as earthen-ware, than in glass vessels.

Vibrations exercise such an influence on the crystallization of bodies, that they not only facilitate the deposit of crystals in a liquor, but they can bring about the transformation of a solid amorphous body, into a crystalline body. It is thus that tough iron, of a good quality, which does not present to the eye any appearance of crystallization, becomes in a short time crystallized, and very brittle, when it is exposed to long continued vibrations; this is a consideration of great importance in the arts.

The form of crystals is not accidental, as might be supposed. An attentive examination has shown, that in general, and with certain exceptions, the same body always crystallizes in the same forms, and that the identity of form carries with it, if not identity of nature, at least an extreme analogy in its chemical properties.

The exterior configuration is then an important characteristic, for the distinction and classification of bodies.

Though crystalline forms may be, so to speak, innumerable, we are enabled to group them from certain characters of symmetry which determine the optical properties, and physical qualities, suitable to characterize them. These groups take the name of *crystalline systems*, and are six in number.

These preliminary notions will enable us to understand the details which will be given on the metalloids and the metals. The arrangement we observe in this work is easy to comprehend. We shall take up each simple body successively and study it in a thorough manner when it presents any thing of interest, and we will then speak of the compounds it produces with the bodies which have been previously examined, whenever these compounds are of importance in their applications.

METALLOIDS.

OXYGEN.

OXYGEN was discovered by Priestley, in 1774; and shortly after by Scheele, who isolated it without having known of the labours of Priestley.

Lavoisier was the first to study the principal properties of oxygen; he made known the part which it played in a great number of chemical phenomena, and chiefly in combustion.

This gas was first named *dephlogisticated air*, *pure air*, and *vital air;* and afterwards, at the time chemical nomenclature was made, oxygen, from two Greek words, ὀξυς, acid, and γεννάω, to produce, because it was then thought that all acids necessarily contained oxygen.

Oxygen is a permanent gas, without color, taste, or smell; its density is 1·10563. It refracts light the least of all gases. It is slightly soluble in water, which dissolves, at the ordinary temperature, one twenty-seventh of its volume of oxygen.

Oxygen, suddenly compressed in a pneumatic briquet, or piston, developes a temperature which exceeds 200° Cent., and produces a bright light. M. Thenard has shown that in this case, oxygen causes the combustion of a certain quantity of fat matter, which is used to grease the piston of the pneumatic briquet.

Oxygen is essentially proper to support combustion, which has given it the name of supporter of combustion. This property is characteristic of oxygen, and is demonstrated by the following experiment: a taper, or lighted allumet, which has been just extinguished, and on which there is the least fire, is immediately relighted, when it is plunged into a jar filled with oxygen. The protoxide of nitrogen, which is a gas resulting from the combination of nitrogen with oxygen, also inflames allumets almost extinguished, but with less rapidity than oxygen; and the combustion is less brisk than in this latter gas. All combustible bodies, such as sulphur, charcoal, &c., burn in oxygen with greater brilliancy and rapidity, than in atmospheric air.

Certain metals, even, will burn in oxygen, when their temperature is first elevated; thus, when an iron wire, having a piece of lighted tinder on its extremity, is placed in a jar filled with oxygen, the iron, Fig. 2, soon takes fire, and throws out thousands of colored sparks; in this case the iron, in uniting with the oxygen, forms the oxide of iron, which melts and penetrates to some depth into the glass of the jar. The temperature produced by the combustion of the iron in oxygen, is sufficiently elevated to cause the fusion of some globules of the iron which are found covered by the oxide.

Phosphorus inflamed, when it is placed in a jar full of oxygen, burns there with so brilliant a light, that the eyes can scarcely endure it.

The combustion of sulphur, of carbon, and of phosphorus, is shown by introducing into a glass jar, of the size of about two quarts, filled with oxygen, a small earthen or plaster cupel, supported by an iron wire fastened to a cork which is too large to enter into the mouth of the jar; the wire should be of such a length that the cupel will be suspended to within about four inches of the bottom of the jar. The combustible is then placed in the cupel, and lighted and introduced into the jar.

One of the essential characters of oxygen, is that of supporting respiration. Animals placed in this gas will live a much longer time than in the same volume of atmospheric air. Hence the name of *vital air*, which was at first given to this gas.

Electricity causes oxygen to undergo a peculiar modification, as Van Marum pointed out, in 1785. Under its influence, the affinities of this gas are more energetic than those of oxygen in its ordinary state. Thus oxygen electrified, attacks mercury and silver in presence of water and at the ordinary temperature, displaces the iodine contained in the iodides, combines directly with nitrogen to form nitric acid, causes the super-oxidation of the protoxide of lead, &c.

Electrified oxygen is odorous, its odor resembling that of phosphorus.

This curious modification of oxygen has been studied by M. Schœnbein with great care; he gave it the name of *ozone.* M. Schœnbein recognised that the oxygen which disengages itself at the positive pole of a pile, of which the poles are plunged in water, is strongly *ozoned,* and that ozone is also obtained by passing moist air over sticks of phosphorus. More lately, Messrs. Marignac and de la Rive have shown that ozone is oxygen modified by electricity. This fact is rigorously demonstrated by experiments which prove that a limited volume of very pure oxygen, exposed for several days to the influence of a series of electric sparks, becomes entirely absorbable in the cold by silver or the iodide of potassium moistened. (*Ed.* Becquerel and Fremy.)

The name of *ozone* ought henceforth to be replaced by that of *electrified* oxygen.

Preparation of oxygen. The most simple mode of preparing oxygen is to decompose, by heat, certain metallic oxides.

When the oxide of silver, or the oxide of mercury is heated, oxygen is disengaged, and silver or metallic mercury remains. These oxides are not however employed in laboratories for the preparation of oxygen, on account of their high price.

The preference is given to the peroxide of manganese, which is found in nature, and is of low price.

The apparatus in which this decomposition is made is composed of an earthen retort, into which about a pound of peroxide of manganese is introduced. The retort is placed in a reverberatory furnace; a tube provided with a safety tube to prevent absorption, connected through a cork or stopper with the neck of the retort, passes under a receiver filled with water, the retort is then heated, and gradually brought to a red heat. (Fig. 3.)

At first there is disengaged a mixture of atmospheric air and carbonic acid. The atmospheric air which comes over, was in the retort, being displaced by the disengagement of the gas; the carbonic acid comes from the carbonates which the peroxide of manganese of commerce almost always contains; these carbonates are decomposed by the heat, and produce the carbonic acid which mixes with the oxygen.

The property which carbonic acid possesses of rendering lime-water turbid, serves to show the presence of this acid in the oxygen.

The receivers of the gas which first come over are allowed to escape, and the oxygen is not collected till it relights a taper with a slight report, and when lime-water is no longer precipitated by it. To free the oxygen from the carbonic acid which it sometimes contains, it is sufficient to agitate the gas with a concentrated solution of potassa, which absorbs the carbonic acid and leaves the oxygen pure.

One-half of the oxygen which the peroxide of manganese contains, may be withdrawn from it by heating this oxide with concentrated sulphuric acid. The peroxide of manganese is an indifferent body, but there is another oxide of manganese, the protoxide, which is an energetic base; sulphuric acid causes the formation of this base, and combines with it.

Oxygen may be obtained very pure, by extricating it from the chlorate of potash, which is transformed by heat into chloride of potassium and oxygen.

We shall not press the importance of the part which oxygen plays in most chemical reactions.

Oxygen forms one of the constituent parts of atmospheric air; without it, the phenomena of vegetation and combustion could not be accomplished.

It unites itself besides to most of the bodies which we shall examine successively, and forms combinations which will enable us to complete its study.

HYDROGEN.

Hydrogen was discovered in the beginning of the seventeenth century; but it has only been well known since the year 1777, when Cavendish described its principal properties.

This gas was first called *inflammable air*, and then *hydrogen* (generator of water), from two Greek words, ὕδωρ, water, and γεννάω, to generate, because it is one of the elements of water.

Hydrogen is a permanent gas, without color, taste, or smell, when it is pure; often it exhales a slight odor of garlic, due to the presence of a carburet of hydrogen, or to traces of sulphuretted hydrogen, and of arseniuretted hydrogen. It is rendered inodorous by passing it through solutions of salts of lead, of silver, or of mercury.

Hydrogen is the lightest of all bodies. The density of the air being taken for unity at the temperature of 0°C., and under the natural pressure, 0m·76, that of hydrogen is equal to 0·06926; a litre of hydrogen weighs 0·08957 grammes.

This gas is about fourteen and a half times lighter than the air (a litre of air weighs 1·2937 g.). The lightness of this gas may be shown by means of a receiver filled with hydrogen, which is taken out of the water vertically, and then turned over. The hydrogen immediately escapes, and is replaced by atmospheric air.

If, on the contrary, the receiver is raised, keeping its orifice down towards the water, the hydrogen remains in it for some time.

Finally, if the receiver containing hydrogen is placed in communication, by its orifice, with another receiver filled with air, inverting the two receivers in such a way that the one containing the air shall be above, and the one containing the hydrogen below, it will be seen that the hydrogen has taken the place of the air, and the air that of the hydrogen.

Hydrogen easily passes through bodies which would be almost impermeable to other gases. If a sheet of paper is placed at a little distance from an orifice from which hydrogen is escaping, the gaseous current passes through the paper, without changing its direction, and it can be lighted on the opposite side of the sheet. But hydrogen does not pass through the thin pellicles of glass which are blown at the lamp. (M. Louyet.) Hydrogen, placed in a cracked glass receiver, over water or mercury, completely escapes.

Hydrogen is not a supporter of combustion, but is very combustible. Thus, a taper, plunged into hydrogen, inflames the first layers of this gas, because they are in contact with the air, but is extinguished on penetrating further into the gas.

Hydrogen is unfitted for respiration, without being poisonous. An animal dies in hydrogen only for the want of oxygen. This gas, introduced into the lungs, does not produce disorganization. Hydrogen is the most refractive of all gases. It refracts light about six and a half times more than atmospheric air.

This gas, in burning in the air, combines with oxygen, and forms water. Its flame is yellow, and has but little brilliancy, because it does not contain any solid particles.

Hydrogen is but slightly soluble in water, which only dissolves one and a half hundredths of its volume. Thus, it can be collected over water; but to obtain it pure, it must be collected over mercury, because water holds oxygen in solution, as well as nitrogen and carbonic acid, which are partly disengaged when it is traversed by a current of gas.

Action of hydrogen on oxygen. Oxygen and hydrogen do not exercise any action on each other at the ordinary temperature, but at 400° or 500° C., these two gases combine and produce water.

This combination takes place in the proportion of 2 volumes of hydrogen to 1 volume of oxygen.

The combination of hydrogen with oxygen is produced also under the influence of the electric spark, or of spongy platinum, and of several bodies in a state of minute division, which only act by their presence.

Hydrogen in combining with oxygen produces a very intense heat. The high temperature which is produced by the combustion of hydrogen is made use of to melt certain refractory bodies.

The mixture of hydrogen and oxygen, called *detonating mixture*, causes the fusion of platina which resists the heat of the forge.

To fuse platina and other refractory bodies, place two volumes of hydrogen and one volume of oxygen in separate recipients; let these two gases come out and meet on the flame of a lamp; thus will be obtained a yellow flame of but little brilliancy, but which possesses calorific properties highly developed.*

A mixture of hydrogen and oxygen enclosed in a flask, if presented to the flame of a candle, produces a violent detonation, and forms water.

The electric spark also inflames the detonating mixture.

The detonation is caused by the instantaneous condensation of vapour of water in contact with cold air. The water occupying a volume nearly 1700 times less than its vapour, a momentary vacuum is formed in the interior of the flask into which the air suddenly passes, and causes a detonation sufficiently strong to break the flask, which should be covered with a cloth when the experiment is made.

A loud explosion may be produced also by inflaming a mixture of oxygen and hydrogen contained in soap-bubbles. To make this experiment, the detonating mixture is placed

* This constitutes the oxyhydrogen blow-pipe invented by Dr. Hare. See *Ogilvie, Imperial Dictionary, art. Blow-pipe, Turner's Chemistry.*

in a bladder furnished with a stop-cork carrying a fine tube, which is plunged into soapsuds, the bladder is compressed, the gas which comes out forms bubbles, which on being lighted, cause a detonation.

A rapid succession of detonations in a glass tube gives rise to musical sounds.

This phenomenon is shown, by enclosing in a large tube open at both ends a flame of hydrogen produced by dry gas, which escapes from the extremity of a finely drawn tube. From the series of detonations which cause the column of air to vibrate, there results a sound which varies in intensity with the diameter and length of the tube.

The apparatus used for this experiment bears the name of *chemical harmonica.*

The flame of a detonating mixture, hardly visible by itself to the naked eye, acquires a brilliancy which the eye can scarcely endure, by the contact of certain solid bodies, such as platina, and particularly lime. This light has been applied to the purposes of the microscope.

The property which spongy platina possesses of inflaming hydrogen, has been applied to the construction of a briquet which bears the name of *briquet for hydrogen*, the invention of which is due to Gay Lussac.

In this apparatus, the hydrogen is produced by the reaction of zinc on sulphuric acid and water. The gas is let out by a cock, and passes through a little grating of copper containing the spongy platinum, which causes the inflammation of the hydrogen.

The gas is disengaged in a bell-glass, which contains a cylinder of zinc suspended by a wire a little above the opening; this bell plunges into a jar half full of acidulated water, the gas presses back little by little the liquid from the bell, and in a little while completely expels it, and prevents the acid reacting on the zinc when the bell is filled with hydrogen. By this ingenious arrangement, the piece of zinc is preserved from the action of the acid, when the bell is filled with hydrogen, and will answer the purpose for a long time. (Fig. 8.)

Preparation of hydrogen.—Hydrogen is extracted from water, which is composed of oxygen and hydrogen. This liquid, placed in contact with a body having a strong attraction for oxygen, is decomposed and disengages hydrogen.

The metals in general having a strong affinity for oxygen, are used in the preparation of hydrogen. Certain metals, such as potassium and sodium, decompose water in the cold. A piece of potassium, introduced into a receiver filled with mercury, which contains a small quantity of water in its upper part, disengages hydrogen, and there remains in solution, in the water, oxide of potassium (potash).

Hydrogen may be prepared by passing steam over iron heated to redness; the water is decomposed, its oxygen combines with the iron to form the magnetic oxide of iron, and hydrogen is disengaged.

The apparatus employed consists of a porcelain tube containing pieces of iron wire, and a long furnace for heating it. The porcelain tube communicates on one side with a small glass retort, into which is introduced a small quantity of water, on the other, with a tube for the disengagement of the gas, which passes under a receiver filled with water, Fig. 6.

We begin by heating the porcelain tube to a red heat, and then placing a chafing-dish of coals under the retort, cause the vapour of water to pass over the iron; the hydrogen gas soon disengages itself with great rapidity.

In the laboratory, hydrogen is always prepared by decomposing water by zinc, in presence of sulphuric acid.

Zinc alone not having sufficient affinity for oxygen to decompose water at the ordinary temperature, sulphuric acid is added to it; the water is then decomposed by the zinc, under the influence of the sulphuric acid; its oxygen

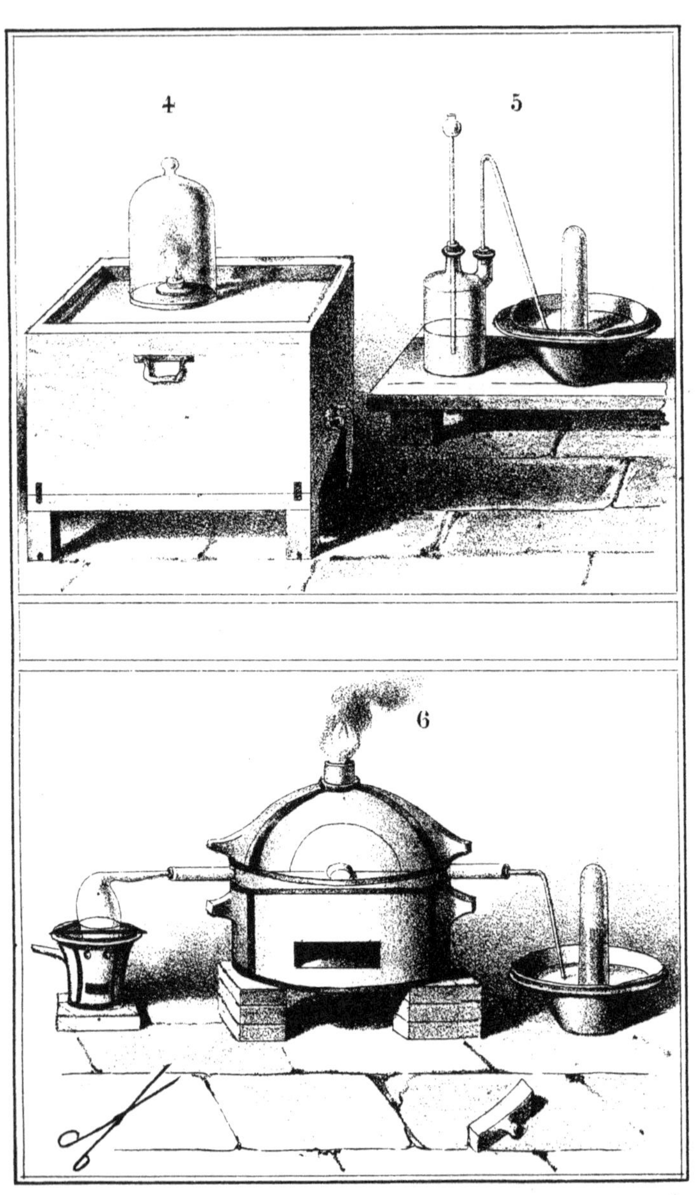
4
5
6

combines with the zinc to form oxide of zinc, which, uniting with the sulphuric acid, produces sulphate of zinc, which remains in solution in the water, and the hydrogen is disengaged.

The apparatus consists in a flask with two tubulures, into which is introduced the zinc, in grains or strips. To one of these necks is fitted a tube for collecting the gas; to the other, a tube or a funnel for pouring in water, so placed that its extremity will be in the liquid, Fig. 5.

The tube for the disengagement of the gas passes under a receiver filled with water. On pouring through the funnel a few grammes of sulphuric acid, a rapid evolution of the gas takes place, so that a large quantity may be collected in a few minutes.

Uses of hydrogen.—Hydrogen is used in chemical laboratories for reducing oxides, and restoring them to the metallic state; metals reduced by hydrogen are in general very pure. Hydrogen also serves for isolating some metals from their combinations with chlorine and sulphur.

Hydrogen is used for filling balloons. Sometimes, in this last application, it is often replaced by the gas produced in the distillation of coal.

COMBINATION OF HYDROGEN WITH OXYGEN.

WATER.

WATER, as well as air, was considered as an element until the end of the last century.

Towards the year 1781, Priestley, Watt, and Cavendish demonstrated that hydrogen, in burning in the air, produced water.

In 1789, Lavoisier demonstrated that water is composed of hydrogen and oxygen, and that these two gases form, in combining together, a quantity of water represented by the sum of their weights.

Composition of Water. Water is proved to be formed of oxygen and hydrogen:

1. By turning a jet of dry hydrogen under a bell-glass, its interior is seen to be covered with drops of water, the number of which increases as long as the combustion goes on. (Fig. 7.)

2. By placing water in contact with the metals which decompose it, whether with potassium, or at a high temperature, as with iron, tin, &c. In this decomposition, the oxygen of the water combines with the metals, and hydrogen is disengaged.

3. By decomposing water with galvanism, the oxygen appears at the positive pole, and the hydrogen at the negative. The volume of the first of these gases is sensibly one-half of that of the second.

These experiments show that water is formed of oxygen and hydrogen, but it remains to fix the relation of these two elements exactly.

To show in what proportion in volumes hydrogen combines with oxygen to form water, a mixture of pure oxygen and hydrogen is made in a graduated glass tube placed over a mercurial trough; this mixture is then passed into another tube, very thick and strong, called a *eudiometer*.

These two gases are inflamed by the action of the electric spark. In this experiment it is shown that two volumes of hydrogen combine always with one volume of oxygen to form water. If, for example, the mixture is made of exactly two volumes of hydrogen and one volume of oxygen, it disappears entirely; if it is composed of two volumes of hydrogen and two volumes of oxygen, one volume of oxygen remains after the passage of the electric spark; and if the mixture is composed of three volumes of hydrogen and one volume of oxygen, one volume of hydrogen will remain after the explosion.

Thus water is formed by the combination of two volumes

of hydrogen and one volume of oxygen. These proportions expressed in weight correspond to 88.888 of oxygen, and to 11.112 of hydrogen.

Water may be solid, liquid, or gaseous. We shall examine it under these different conditions.

SOLID WATER.

Water in solidifying may either be amorphous or regularly crystallized. The crystalline form of solid water is that of a hexahedral prism of 120°, or that of an isosceles dodecahedron. These crystals have a double refraction.

Snow often assumes the forms of stars with six points, each point or ray being a regular prism with six faces; sometimes, too, the centre of the star is occupied by a small hexagonal brilliant scale, and the rays of the star diverge from each of its angles.

In passing from the liquid to the solid state, water increases in volume. Its density becomes 0.918, that of water at +4° centigrade being 1000.

This increase of volume which water undergoes in solidifying explains,

1st. Why ice always remains on the surface of quiet water.*

2nd. Why the water contained in the cellular tissue of plants or fruits, which is solidified by a heavy frost, by its increase in volume bursts the capillary vessels, kills the

* In tumbling streams of water, solid ice is often seen on the bottom or bed of the stream, because the water being agitated at a low temperature cannot follow the law above explained; its lighter and colder particles cannot remain at the surface, but the whole mass being agitated becomes of an uniform temperature, and crystals of ice form on the rough bed of the stream, and in time a great mass of ice forms at the bottom of the stream. We can by constant agitation on a cold day, solidify the entire mass of water in a bucket, which if left to itself would merely freeze in a cake over the surface; hence, also, the necessity of agitation in making ice-cream.—*Trans.*

4

vegetables in a short time, and brings on the rapid decomposition of frozen fruits.

3rd. Why fountains and vessels filled with water become broken during a hard frost, when the water which they contain solidifies ; water-pipes which are not deeply buried also burst when the water which passes through them freezes.

4th. Why stones called *frost-stones*, which condense within them a considerable quantity of water, burst in winter in consequence of the dilatation which they undergo from the congelation of the water contained in their pores.

5th. Why water in soldifying sometimes causes the breaking of the most resisting metals and alloys.

It is thus also we can burst cannons, by filling them with water, and exposing them, after having hermetically sealed them, to a temperature which will solidify the water they contain.

Melting ice preserves a consistent temperature, which is taken for the point of departure in centigrade thermometers, and serves to fix the zero on their scale.

The point at which water freezes, presenting often great variations, is not adapted for fixing the zero of thermometers. When water is not disturbed, we can, according to Gay-Lussac, lower its temperature to —12° Cent. without solidifying it; if it is agitated it congeals instantly, and there is observed to be a disengagement of caloric which causes its temperature to rise rapidly to zero.

The congelation of water then presents two remarkable phenomena, a disengagement of heat, and an increase in volume.

Water which holds salts in solution freezes more slowly than pure water. When a saline solution suffers a partial congelation, it is the pure water which congeals in the first place, while the salts remain in the mother-water. This property is applied in cold countries to the concentration of the waters of the sea, from which ordinary salt or sea salt is extracted.

LIQUID WATER.

Water is without smell, taste, or colour, though when seen in a considerable mass, it has a green tint. When water at the temperature of zero is submitted to the action of heat, its volume diminishes till at +4° Cent.; it then increases progressively until the temperature of its ebullition which is constant. At 8° Cent., the volume of water is about the same as at zero. Its maximum of density is at 4° Cent. according to M. Despretz.

Water, considered as a solvent, is of importance in the arts, industry, and chemical analysis.

Thus among the different properties of a body, its degree of solubility or insolubility in water ranks among the foremost.

WATER IN A STATE OF VAPOUR.

The boiling point of a liquid is always the same under the same pressure. Water under a pressure of 0.760m. boils at an invariable temperature, which serves to fix the 100° Cent. of the thermometric scale.*

Water passing from 0° Cent. to the state of vapour, increases to about 1700 times its volume.

The temperature of water in ebullition varies with the pressure. Water enclosed in a sufficiently resisting case, may be raised to a very high temperature, and be prevented from boiling. This experiment is made in an apparatus called *Papin's digester.*

M. Cagniard Latour enclosed water in very thick glass tubes, freed from air and sealed with a lamp. He has proved by bringing these tubes first to a red heat, that water can be compressed in steam into a space which is but four times greater than its natural volume.

---

* The boiling point of a body is the temperature at which the tension of its vapour is in equilibrium with the pressure exercised on it; by augmenting this pressure, its boiling point necessarily rises.

The vapour of water is without smell or colour, and transparent; its density, 0.622.

Water, as is the case with all volatile bodies, gives off vapours at the lowest temperatures. This evaporation increases with the temperature. The vapour of water, when chilled, condenses and passes to the state of liquid water. This condensation is produced in the atmosphere, when it contains a quantity of vapour greater than it can keep in a state of saturation. It is thus are produced dews, white frost, fogs, rain, and snow. The vapour of water condensed in the atmosphere takes the name of fog when it is at the surface of the earth, and cloud when it is suspended at a certain height in the atmosphere. Vapour in condensing in the air forms little spheroids, which constitute *vapour in the vesicular form.*

Water, to transform itself completely into an elastic fluid, requires about five and a half times more heat than to heat it from 0° to 100° Cent.* Thus, a kilogramme of vapour of water, at 100° Cent., received into 5½ kilogrammes of water at zero, produces 6½ kilogrammes of water at 100° C. This principle is made use of in workshops, to boil large masses of water contained in vessels of wood, which would be destroyed by the direct action of the fire. Lest the steam, in condensing, be hurtful, it is made to circulate in a double bottom, or in pipes which plunge into the liquid which it is intended to heat.

Chemical properties of Water.—Water does not exercise any action on the coloured reagents. It is undecomposable by heat. Many simple bodies decompose it; some, as chlorine, combine with its hydrogen, and disengage its oxygen; others, as potassium, iron, &c., take its oxygen, and set free its hydrogen.

Water combines in definite proportions with a great many

* That is, from freezing to boiling.

bodies, and forms compounds which have received the name of *hydrates*. In uniting itself with acids, bases, and salts, water, in general, does not modify their distinctive characters. Thus, ordinarily we can study the properties of these bodies in their hydrates.

However, in some cases, the water which unites itself to acids, bases, and salts, causes important modifications in their properties.

The state of Water in nature.—The water which we meet with on the surface of the earth, and in its interior, is never pure.

Rain-water holds in solution all the substances which are found in the air, such as oxygen, nitrogen, carbonic acid, and sometimes traces of nitric acid, of carbonate of ammonia, or of nitrate of ammonia. These last salts are especially found in the rain-water of storms. The rain-water that first falls contains, besides these foreign bodies, the dust which is in suspension in the atmosphere. Occasionally rain-water, collected with care, is very pure, and will do to replace distilled water in many chemical operations.

The water of floods, rivers, springs, and wells, is less pure than rain-water. It contains chlorides, sulphates and carbonates of lime, magnesia, and sometimes of soda, potassa, and alumina. The composition of these waters varies with the nature of the ground they have passed over. They are mostly fit to drink, and to cook vegetables, and are without perceptible taste. In this case they have the name of *soft waters*, or *drinkable waters*.

Sometimes waters are not fit for cooking vegetables, or for washing with soap; they are then said to be *hard*.

Soft waters leave but a small residue on evaporation, preserve their transparency when boiled, are limpid, and without taste. They dissolve soap, or at least in its solutions give rise to but trifling precipitates. The bad quality of hard waters is owing to the presence of calcareous salts. They

4 *

curdle solutions of soap, and cannot be applied to all domestic uses. Hard waters are divided into two principal kinds.

Waters called Selenitic.—The greater part of their lime is in the state of sulphate. Such are the waters of the wells of Paris, which are sometimes saturated with sulphate of lime (plaster). They are not rendered turbid by ebullition, and form abundant precipitates with the oxalate of ammonia, and chloride of barium. The hard waters of the second kind contain carbonate of lime, dissolved by an excess of carbonic acid. They render blue the decoction of Campeachy wood, are rendered turbid by boiling, and by exposure to the air, or under the influence of lime-water. They may be rendered fit for drinking, and for domestic use—

1st, by boiling them for some instants, and then letting them stand (the excess of carbonic acid which dissolved the calcareous salt, is disengaged, and the carbonate of lime is precipitated).

2d, by agitating them in contact with the air, which also causes the disengagement of the carbonic acid which is in excess, and the deposit of the carbonate of lime.

3rd, by treating them with lime-water until they are no longer precipitated by this reagent. In this case, the bicarbonate is transformed into a neutral insoluble carbonate of lime.

The selenitic waters are rendered, if not drinkable, at least fit for cooking vegetables, or for washing, when a solution of carbonate of soda is added. Thus is produced an insoluble carbonate of lime, and sulphate of soda. This last salt, though soluble in water, is of no inconvenience in most industrial operations. We may, by the aid of soap, render selenitic water fit for washing. A small quantity of soap is sufficient to precipitate all the lime in the state of insoluble margarate, stearate, and oleate of lime. These precipitates being once formed, the soap will dissolve without further decomposition.

The waters which are considered the most pure, are those of torrents which descend from granitic mountains. We ought, however, to prefer for drinking, the less pure waters, which contain a small quantity of calcareous salts. The experiments of M. Boussingault have clearly established that the lime of drinking water concurs with that which the aliments contain, in the development of the osseous system.

The drinkable waters leave, on evaporation, a residue of which the weight does not in general go above from 1 to 3 decigrammes to the litre. This residue consists chiefly of carbonate and sulphate of lime, and of chloride of calcium.

A litre of sea-water contains about 42 grammes of salts, of which marine salt comprises 26 to 27 grammes.

Stalactites, incrustations, and deposits in boilers.—When water, charged with the carbonate or phosphate of lime, is left in contact with the air, or submitted to the action of heat, these two salts are deposited, because the excess of carbonic acid, which held them in solution, is disengaged. Most stalactites, and many deposits of calcareous carbonates and phosphates, are formed by this sort of slow precipitation. The deposits which certain waters leave in water-conduits, have the same origin.

The great quantities of water evaporated in steam-boilers, deposit on their sides calcareous salts, the hardness of which is one cause of their deterioration, because the hammer must frequently be employed to detach them. This inconvenience may be obviated by introducing into the water raspings of potatoes or clay, chlorhydrate of ammonia, or carbonate of soda.

Air dissolved in Water.—Water which has been in contact with atmospheric air, contains carbonic acid, besides a mixture of oxygen and nitrogen. The presence of these gases is shown by filling with water a glass matrass of about two litres capacity. This matrass is attached, by means of a cork fitted to its neck, to a tube also filled with water, which passes under a bell-glass filled with water or mercury; the water in the matrass is gradually made to boil, and in a short time a considerable quantity of gas is seen to disengage itself, and pass into the bell-glass. 100 volumes of water give about 3·2 volumes of gas.

In analyzing air extracted from water, it is seen to be much richer in oxygen than atmospheric air, containing 32 or 33 volumes of oxygen in 100, instead of 21 volumes which are found in the atmosphere. The presence of this excess of oxygen is easily explained, for oxygen is much more soluble in water than nitrogen, and the small quantity of foreign matters held in solution in ordinary water, does not sensibly modify the solubility of these two gases.

Air dissolved in water serves for the respiration of fish. If water is boiled to deprive it of air, and it is then allowed to cool in vessels hermetically sealed, a fish plunged in it dies in a short time.

It is known, besides, that certain species of fish scarcely ever come to the surface of the water, and that all are furnished with gills fit for absorbing the oxygen in solution. When the proportion of oxygen contained in the water of a pond diminishes, the fish which people it soon perish.

The air which water holds in solution gives to spring waters their fresh and agreeable taste. Those waters deprived of air become heavy, and of a slow and difficult digestion.

Distilled water is insipid and disagreeable, but if it is agitated in contact with the air and saturated with it, it becomes fit to drink.

It is thus that in ships they can use distilled sea-water, having first exposed it to the air.

Distillation of Water.—The distillation of water has for its end the purification of it, by freeing it from foreign bodies which it holds in solution.

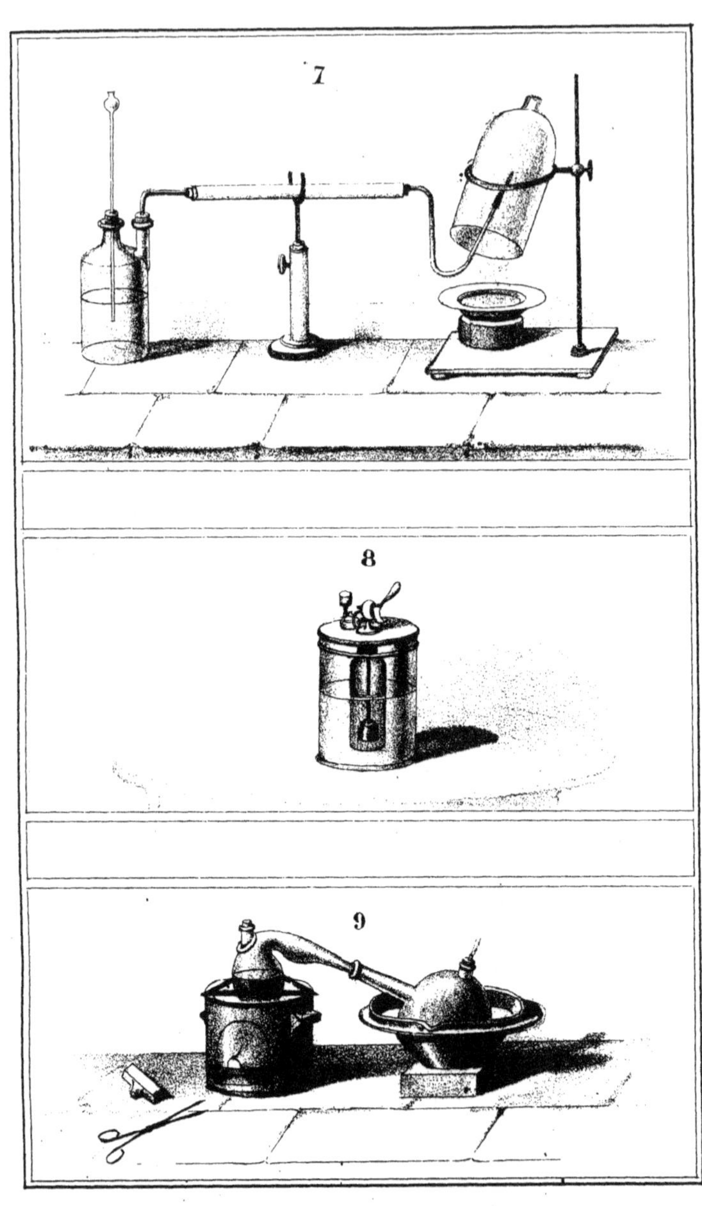
7
8
9

These bodies are of two kinds: one kind gaseous or volatile, as oxygen, nitrogen, ammonia, nitrate and carbonate of ammonia, the other fixed, as salts with potash, soda, lime, magnesia, and alumine for bases.

The first portions of distilled waters carry over with them the gaseous or volatile bodies, and are rejected as impure; the fixed compounds remain behind in the distilling apparatus.

The distillation should be stopped, as soon as the salts held in solution in the water begin to be deposited. If the operation should be prolonged, the distilled water would contain small quantities of these same salts carried over mechanically, or even decomposed.

Distilling apparatus.—There are many kinds of distilling apparatus.

The most simple is composed of a glass retort and a receiver. (Fig. 9.)

The water introduced into the retort, which it fills about three-fourths, is made to boil over an ordinary furnace. The steam condenses in the receiver, which is surrounded with water, carefully kept cool.

The first portions of the distilled water are rejected; those afterwards collected are pure. The distillation ought not to be stopped till about $\frac{4}{5}$ of the water has passed into the recipient.

Water distilled in glass apparatus is sometimes slightly alkaline, because boiling water attacks glass of bad quality and dissolves traces of soda.

Sometimes, also, distilled water contains a little muriatic acid, which proceeds from the chloride of magnesium, decomposed by the concentration, into magnesia and muriatic acid. This alteration of the water is prevented by adding a certain quantity of lime to the water to be distilled; this forms, with the chloride of magnesium, magnesia and chloride of calcium, which is not decomposed by ebullition.

The lime having the ability to absorb the carbonic acid contained in the water, ought to be used in excess: in most cases water is distilled without adding lime.

Generally, distilled water is prepared in an apparatus called an *alembic*, Fig. 10. A copper boiler contains the liquid to be distilled; this is covered with a hood, or movable piece, which together form a sort of retort. The neck is fitted to a curved tube called a *worm*, which is plunged into a cooler, where is kept up a current of cold water admitted *below*, while the warm water passes out above, and may be used to feed the boiler. We owe to Gay-Lussac a small apparatus, which may be applied not only to the distillation of water, but to that of all kinds of liquids.

This apparatus is composed of a matrass of glass (into which the liquid to be distilled is placed,) communicating with a condensing tube, which connects with a receiver. This tube passes through a cooler slightly inclined, which receives cold water from a cock, and loses its warm water by a lateral tube, Fig. 11.

This excellent system of condensation may be employed usefully in the arts.

NITROGEN.

Nitrogen was discovered in 1772 by Dr. Rutherford. In 1773, Lavoisier recognised its existence in a free state in the atmosphere, of which it forms about four-fifths.

Nitrogen is a permanent gas, without color or smell, and is unfit for respiration. It is this property which has given it the name of azote (α, privative, and ζωή, life); but it is not poisonous, and animals die in it for want of oxygen. This gas is not fit for combustion. A lighted taper, plunged into a receiver filled with nitrogen, is instantly extinguished.

The density of this gas is 0·97137.

Nitrogen combines directly with but a small number of

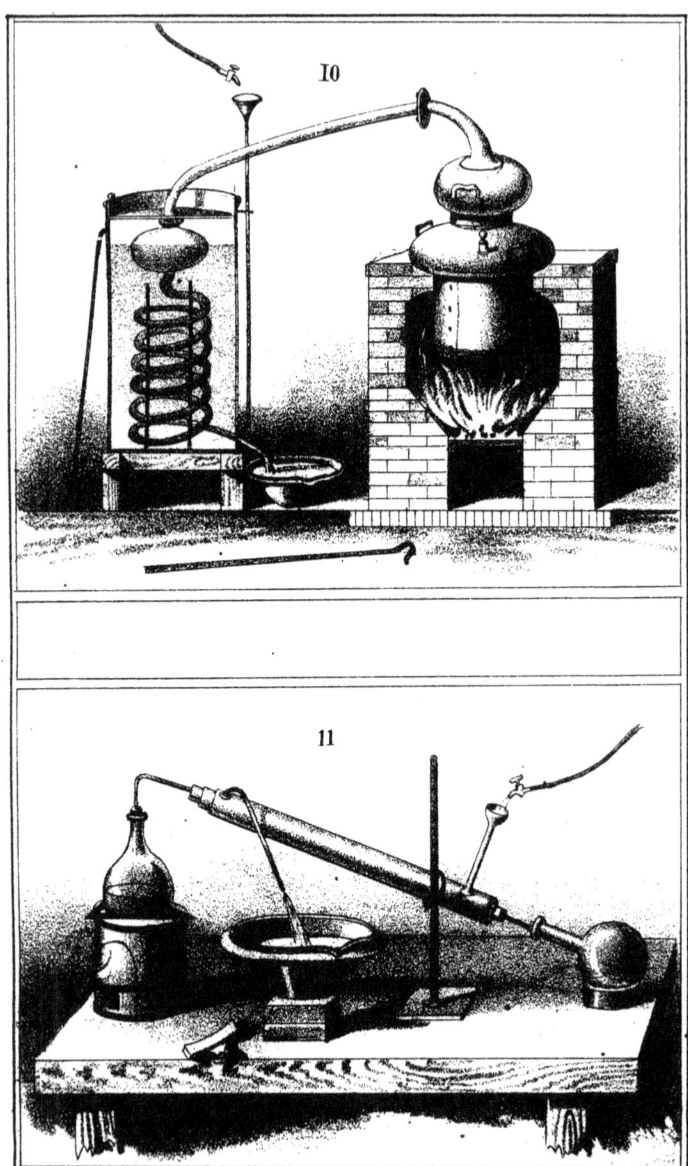
I0
11

bodies; however, when a great number of electric sparks are made to pass through a mixture of oxygen and moist nitrogen, we get a combination known as *azotic* or *nitric acid.*

It is to this reaction that we are to attribute the presence of nitric acid in the rains of thunder-storms. Nitrogen is less soluble in water than oxygen.

Nitrogen exists in a great number of organic substances. The experiments of M. Boussingault have proved that the nitrogen contained in vegetables often comes from the air. This chemist has observed that certain leguminous plants, growing in a soil free from nitrogenous bodies, contain, after their development, a considerable quantity of nitrogen, which has been obtained from the air.

The nitrogen which enters into the composition of animal substances, proceeds from that which their aliments contain; it results, in fact, from the experiments made on warm-blooded animals, that, in the act of respiration, the atmospheric nitrogen does not appear to be sensibly absorbed.

Preparation of nitrogen.—In general, nitrogen is obtained from the atmospheric air. Place on a pneumatic trough a cork to float a small cupel of plaster containing a piece of ignited phosphorus, which must be covered with a bell glass, Fig. 4. The phosphorus, in burning, absorbs the oxygen which is in the bell glass, and we thus obtain the nitrogen.

ATMOSPHERIC AIR.

The ancients considered atmospheric air as an element. Its composition has been known only since the works of Lavoisier and Scheele.

Brun and Jean Rey had shown, a century and a half before Lavoisier, that tin, heated in contact with air, increased in weight.

Bayen, one of the contemporaries of Lavoisier, replaced

tin by mercury, and came to the same conclusion with Brun and Jean Rey.

The increase in weight observed during the calcination of metals, did not indicate whether the air was absorbed entire or in part.

Lavoisier first showed that atmospheric air was composed of two gases, oxygen and nitrogen, of which one only, oxygen, is absorbed by the metals.

We will now describe the memorable experiment which led Lavoisier to the discovery of the composition of atmospheric air.

He introduced mercury into a matrass, the neck of which, very long and curved, came out under a graduated bell-glass placed over a mercurial trough or basin. (Fig. 1.)

The arrangement of this apparatus permitted him to determine with precision,

1st. The volume of air on which he operated.

2d. The volume of the gas absorbed during the operation.

3d. That of the gas remaining.

He kept the mercury heated just to ebullition during five consecutive days; and, although it was evident that, after the five days, the volume of air contained in the bell-glass did not suffer any further diminution, he still continued the experiment some days longer, after which he allowed the apparatus to cool, and showed that 100 volumes of air had been reduced to 73 volumes.

A red crystalline substance was formed on the surface of the mercury; this substance was the peroxide of mercury.

Lavoisier ascertained that the gas which remained in the bell, had properties entirely different from those of atmospheric air—that it was unfit for combustion, and respiration. This gas was nitrogen.

He then introduced into a small retort the peroxide of mercury, which had formed on the surface of the mercury, heated it to nascent redness, and saw that it was decomposed into metallic mercury and a gas, which was, as he has expressed it, *much better fitted than atmospheric air to support combustion and the respiration of animals.* This gas was oxygen.

Lavoisier had then drawn from the atmosphere two different gases; one oxygen, supporting with energy combustion and respiration, the other nitrogen, unfit for combustion and respiration.

After having decomposed atmospheric air, he desired to recompose it, by mixing the two gases which he had extracted from it.

He found that the nitrogen, which remained in the graduated bell-glass, mixed with the oxygen resulting from the calcination of the oxide of mercury formed during the operation, produced a gas absolutely the same as atmospheric air.

While Lavoisier was making these experiments on the composition of the atmosphere, Scheele arrived at the same results.

The Swedish chemist showed that the alkaline sulphurets ab-absorbed one of the elements of the air (oxygen), and left behind a gas unfit for respiration and combustion (nitrogen).

Scheele's experiments have attracted less attention than those of Lavoisier, because they do not present the same evidence; inasmuch as the sulphurets will not, like the oxide of mercury, restore the oxygen which they have absorbed.

We will observe, that the processes of Lavoisier and of Scheele, so remarkable in other respects, are not sufficiently precise.

Thus, in their analysis of the air, these two chemists found more than 27 per cent. of oxygen, while the air really does not contain more than 21 per cent.

Their processes have been perfected in this century, and brought almost to a rigorous exactitude.

Composition of the Air. When the air is analyzed by the most exact methods, it is found to contain in volume

20.80 volumes of oxygen,
79.20 volumes of nitrogen.

And in weight,

23.10 parts of oxygen,
76.90 parts of nitrogen.

These numbers are the results of experiments made by Messrs. Gay-Lussac, Brunner, Dumas, and Boussingault, which agree entirely with each other.

Under ordinary circumstances, the atmosphere contains from 3 to 6 ten thousandths of carbonic acid, and from 6 to 9 thousandths of vapour of water.

The analysis of air, taken at great heights by Gay-Lussac, and those which have been recently made at Paris by Messrs. Dumas and Boussingault, and repeated at Berne, Geneva, Brussels and Copenhagen, appear to establish a uniformity in the chemical constitution of the air, as to the proportion of oxygen and nitrogen which it contains.

M. Lewy has however shown recently, that the air collected over the North Sea contains in weight 22.6 per cent. of oxygen, while the air over the land contains 23 per cent. of it.

M. Lewy ascribes this difference to the fact that oxygen is more soluble in water than nitrogen, and that the animals which people these seas need the oxygen for their respiration. In proportion as these animals use the oxygen dissolved, the surface of the sea, which is in contact with the atmosphere, takes away a new quantity of oxygen from it.

The oxygen and nitrogen in the atmosphere are found in the state of a simple mixture, and not of a combination.

The proportion of vapour of water which the air contains is subject to great variations. It depends in general on the temperature of the air, and of the masses of water which evaporate in certain localities. The proportion of carbonic acid is also very variable.

According to M. Th. de Saussure, a rain diminishes the quantity of carbonic acid contained in the air. In passing through the atmosphere, the water becomes charged with carbonic acid, and carries it into the soil; this gas becomes again disengaged in proportion as the earth dries.

A cold winter, accompanied with frosts which dry the earth, increases the quantity of carbonic acid in the air; a thaw diminishes it.

Over great lakes the proportion of carbonic acid is less than over the surface of the land. The difference is 0.5 for 10,000 parts of air. The quantity of carbonic acid increases in inhabited places.

On elevated mountains the proportion of carbonic acid is more considerable than on the plains, and does not seem to vary day or night.

On the plains the variations are more marked. The proportion of carbonic acid is greater at night than in the day by 0.34 in 10,000 parts of air.

These changes, which generally take place during the first hours after sunrise, proceed from the decomposition of the carbonic acid under the influence of the solar rays by the green parts of the plants.

Messrs. Boussingault and Lewy have confirmed these results, and have assured themselves that the air of a city contains a little more carbonic acid than the air of the country. 10,000 volumes of air collected at Paris contained 3.190 of carbonic acid, and the air taken at Andilly, near Montmorency, only 2.989.

The respiration of men and animals may be particularly mentioned among different causes which produce variations in the composition of a confined atmosphere. According to M. Dumas, a man consumes, in respiration, as well of carbon as of hydrogen, a quantity equivalent to 10 grammes of carbon an hour. Air from the lungs contains, on an average, 4 per cent. of carbonic acid.

Combustion is also a cause of variation. One kilogramme of stearic acid gives off, in burning in a vessel of 50 cubic metres, nearly 4 per cent of carbonic acid.

Thus, many lamps will also cause the composition of confined air to vary.

Organic substances, exposed to the air, decompose and transform the oxygen of the air into carbonic acid.

We see, then, that many causes constantly tend to vary the composition of the air, and diminish the quantity of oxygen which it contains, by transforming it into carbonic acid; these are combustion, the respiration of animals, the spontaneous decomposition of organic matters, &c.

But as the mass of the air is very considerable, the phenomena which take place at the surface of the earth, modify but feebly the composition of the air. The causes of alteration, however, being permanent, we could foresee an epoch when the atmosphere would be found sensibly changed, if the vegetation did not each year decompose the carbonic acid which is produced at the expense of the oxygen of the air. The beautiful experiments of Priestley, of Aimè, and of Th. de Saussure, have proved, in fact, that the green parts of vegetables have the property of decomposing the carbonic acid, under the influence of the solar light, by appropriating to themselves the carbon, and restoring to the air the oxygen engaged in combination with it. Thus the relation which exists in the atmosphere between the oxygen and nitrogen, is kept up.

In comparing the analyses of atmospheric air made by Gay-Lussac some years since, with those which have been lately made, it is established that the proportions of oxygen and nitrogen, contained in the air, have not varied. Analytical methods, however in other respects very complete, not being absolutely exact, it may be that the composition of the air has undergone slight variations, which could not be appreciable unless in a great number of years.

Properties of the Air—Phenomena of combustion in the Air.—It is known that the properties of the atmospheric air are composed of those of the two gases which constitute it. The action the air exercises on a simple or a compound body, is but the *ensemble* of the actions of oxygen and nitrogen on this body.

As to its general properties, the atmospheric air should be considered an elastic fluid, permanent, inodorous, tasteless, and colorless, the density of which, represented by unity, serves as a term of comparison for the density of other gases.

A litre of dry air, under the pressure of 0·760m., and at the temperature of 0°C., weighs 1·2937 grs. Combustion in the air results from the combination of a combustible body, or of its elements, with the atmospheric oxygen. In every combustion, the oxygen is absorbed, and nitrogen undergoes no alteration. The products of combustion are, by themselves, unfit for combustion, and would soon put a stop to it if they were not replaced by a new quantity of air, the oxygen of which goes constantly to support the combustion which had commenced.

Hence the necessity of establishing in fire-places what is called a *draught*, to carry on the combustion.

It is known that wood burns imperfectly when the products of combustion are imperfectly carried off. On the contrary, combustion is energetic in a rapid current of air. In blowing on a body which is burning, we can increase the rapidity of the combustion so that it will burn as in pure oxygen; thus, a bar of iron heated to redness, and presented to the *tuyère* of a forge-bellows, burns, throwing out brilliant scintillations.

The construction of ordinary bellows is based on this principle, as well as that of blowers used in workshops.

Combustion in the air being the result of the combination of different bodies with its oxygen, it may be conceived that it ought to be arrested if the access of the air is suppressed. Thus we put out a coal by covering it with a bell-glass or enclosing it in an extinguisher.

The state of division of bodies exercises a great influence on their combustibility; iron, carbon, the sulphurets, &c., which do not in general burn but at a high temperature, inflame at the ordinary temperature when they are exposed to the air in a state of very minute division. The bodies which exhibit this phenomenon are called *pyrophoric* bodies, their inflammation is caused by the disengagement of heat, which results from the condensation of the air absorbed by their pores.

The combustion of an ignited body continues only when the heat developed by the combustion of one part of its mass brings the parts adjacent to those which are burning to the temperature necessary to make them burn themselves. On the contrary, the combustion ceases whenever the ignited body suffers such a reduction of temperature that it can no longer combine with oxygen. Thus a piece of iron, brought to a red heat, burns in pure oxygen, and is extinguished in atmospheric air, because the nitrogen of the air, by cooling it, stops its combustion. In the same way also, too rapid a current of air, directed on a candle, extinguishes it by reducing its temperature.

A lighted coal is soon put out, when it is placed on a sheet of iron, which cools it.

The gases, like solids, stop burning when they are in contact with bodies which cool them. Thus a fine wire gauze placed in a flame cools it, so that the flame cannot pass through it. This principle gave Sir Humphrey Davy the ingenious idea of his *safety-lamp*. This instrument is composed of an oil-lamp surrounded by a fine wire gauze. When this lamp is placed in an explosive mixture, a detonation is produced in its interior, but the flame does not pass out, being cooled by the metallic gauze.

The miner who works in the coal-pit, and often finds himself in explosive mixtures, is protected from danger when using Davy's lamp: and further, a fine platina wire placed within the lamp, becomes luminous in the explosive mixture which enters the lamp after the explosion, and guides the workman in the darkness.

A flame is always produced by the combustion of a gas, or of a body which becomes volatilized by heat. The illuminating power of a flame varies with the products which are formed during the combustion. When these products remain in the flame in a gaseous form there is but little brilliancy, as is the case with hydrogen, carbonic oxide, and alcohol. But if there should separate during the combustion, a solid body which can become incandescent, the flame is brilliant. Thus the flames produced by the combustion of phosphorus and zinc are very brilliant, because they contain solid bodies, as phosphoric acid and oxide of zinc.

The flames of illuminating gas, and those of candles and lamps, are brilliant, because they are chiefly formed of carburetted hydrogen, which undergoes an incomplete combustion, and gives out the carbon in a very divided state, which becomes incandescent. The presence of carbon in the flame of a lamp or candle may be shown by placing in it a strip of metal, which, by cooling it, is immediately covered with lamp-black.

The presence of hydrogen renders the flame more brilliant. This gas, in burning, produces in fact a great heat, and brings to a white heat the molecules of carbon which give brightness to the flame.

The light produced by a flame may be considerably augmented by placing in it solid bodies, such as platina wire or asbestos. Pieces of quicklime give to the flame of an explosive mixture a brilliancy which the eye can scarcely endure.

The quantity of air which supplies the flame influences its illuminating power. If the air comes in excess, it dims the flame by cooling it; if it is in small quantity, the combustion is incomplete, and the flame becomes smoky.

It may be said that the flame attains its maximum of brilliancy at the moment when it is about giving off smoke. The current of air which nourishes a lamp is ordinarily produced by a chimney, the length and position of which is made to vary according to the appearance of the flame.

The temperature of a flame is independent of its illuminating power; thus the flame of hydrogen, which gives a great deal of heat, is hardly visible.

NITRIC OR AZOTIC ACID.

Nitrogen combines with oxygen in several proportions. We shall only speak here of the most important combination, which has received the name of *Azotic acid* or *Nitric acid.*

The name *nitric acid* or *aquafortis* is given to azotic acid.

Azotic acid is liquid, without color, fuming in the air, and very corrosive; it is considered a violent poison.

It acts on all organic bodies, and destroys them rapidly.

A small quantity applied to the skin disorganizes it, and colours it yellow. Azotic acid produces a similar colour when it acts on most organic matters. This property is often used in the arts, for coloring yellow, feathers, silk, &c.: it often serves in analysis to recognise small quantities of azotic acid.

Azotic acid acts on the tincture of litmus like an energetic acid, and reddens it promptly; it destroys all colouring matters, even indigo. The solution of indigo in sulphuric acid is generally used to recognise the presence of azotic acid in a liquid. Indigo, which resists the action of all acids, even that of concentrated sulphuric acid, is immediately destroyed and coloured yellow under the influence of a small quantity of azotic acid.

Azotic acid* is decomposed, in a great number of cases, into water, into azote and oxygen, or even into oxygen, and into a compound less oxygenated than azotic acid; it is considered one of the most energetic oxidizers.

Light, as well as heat, decomposes azotic acid. The action which the metals exercise on azotic acid is of the highest importance to the arts; it is that, in fact, which enables us to obtain most of the metallic salts.

It may be said, in a general way, that azotic acid is decomposed by nearly all the metals; it oxidizes them by giving up to them part of its oxygen; these oxides once formed, unite to a part of the azotic acid not decomposed, so as to form the azotates. All the azotates being soluble in water, it is seen why azotic acid is generally used for *attacking* the metals; it is to transform them into soluble azotates.

This action of azotic acid on the metals is always accompanied by the production of red vapours, called nitrous fumes, which are caused by the disengagement of a body less oxygenated than azotic acid, and which must necessarily arise, when azotic acid has given up to the metals a part of its oxygen.

The principal metals attacked by azotic acid are iron, zinc, tin, lead, copper, mercury and silver.

Among the metals which azotic acid does not attack, we will cite gold and platinum.

To dissolve metals, which, like gold and platinum, are not attacked by azotic acid, a mixture of azotic and chlorhydric acids is employed in the arts, which bears the name of *aqua regia*.

Aqua regia is obtained by mixing 1 part of azotic acid

---

* As concentrated as possible, azotic acid contains 14.5 per cent. of water.

with 3 or 4 parts of chlorhydric acid. These two acids act upon each other, evolving chlorine: therefore, as chlorine attacks all the metals, even gold and platinum, to form chlorides of gold and platinum, which are soluble in water, we can understand how aqua regia may be employed to dissolve the metals which azotic acid by itself cannot attack.

Preparation.—In the laboratory, azotic acid is obtained (Fig. 14.) by heating in a glass retort a mixture of azotate of potassa (nitre or saltpetre) and sulphuric acid. The azotate of potassa is a salt formed by the combination of azotic acid with potassa. Sulphuric acid has the property of displacing from its combination with potash, the azotic acid, which passes over in the distillation, and becomes condensed in the receiver.

The process used in the large way for preparing azotic acid, is the same as that of the laboratories; only the azotate of potassa is often replaced by the azotate of soda, which is cheaper, and yields by its decomposition a greater quantity of azotic acid.

This operation is always performed in cast-iron cylinders, having a capacity which permits their receiving a charge of from 100 to 150 kilogrammes of azotate of soda. (Fig. 15.)

The cylinder communicates by tubes of earthen or glass with twelve or fifteen tubulated receivers with three necks, containing a small portion of water; the first receivers are often placed in basins and cooled by water.

The application of heat ought to be gradual; and, towards the close of the operation, the cylinder is heated to redness.

The acid which condenses in the receivers is of an orange-yellow color; to render it colorless and fit for commerce, it is boiled for some time in glass or earthen vessels.

The acid of commerce ordinarily marks 36 or 40°; when it is wanted for the manufacture of sulphuric acid, it is employed at 32°.

100 kilogrammes of azotate of soda produce about 130

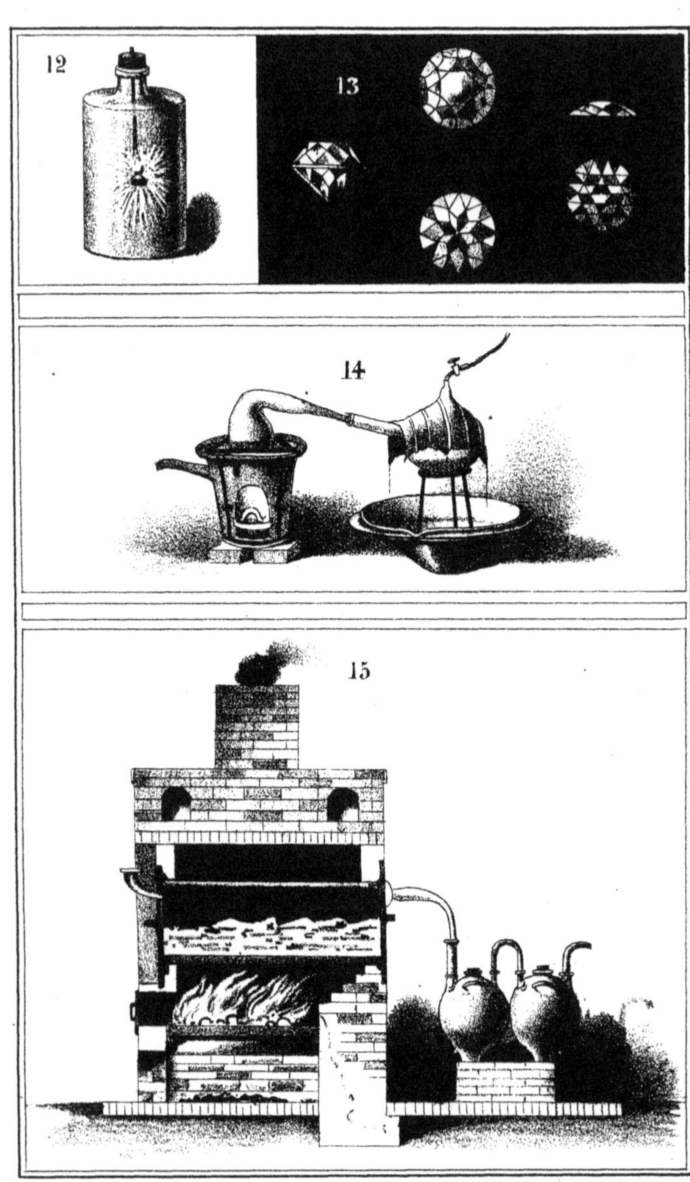
12
13
14
15

kilogrammes of azotic acid at 36°, and 84 of sulphate of soda. This last salt is used in the manufacture of common glass or in the preparation of artificial soda. It contains marked quantities of iron, taken from the material of the cylinders, so that these apparatus undergo a gradual change, especially in that part exposed to the vapours; their wear is made to be more uniform by turning them occasionally.

To preserve the cylinders, sulphuric acid is used which is never below 60°; a weaker acid would destroy them very rapidly, because the water, in decomposing, would oxidize the iron.

Uses.—Azotic acid being an energetic oxidizer, is one of the acids most frequently employed in the arts or in the laboratory.

It serves, in the manufacture of sulphuric acid, to transform sulphurous into sulphuric acid; mixed with chlorhydric acid, it forms aqua regia, which will dissolve gold, platinum, &c.

It is employed to convert starch and sugar into oxalic acid. It is used in dyeing; in engraving on copper and steel; in the assay of money; the polishing, and cleaning off the rust of metals and alloys. It unites with ligneous matters, which it transforms into pyroxyline (gun cotton).

Azotic acid is a valuable reagent; all the salts which it forms with bases being soluble, it is used, in analyses, to dissolve nearly all the metals, and most oxides, carbonates, &c.

It is employed in chemical researches to produce the phenomena of oxidation.

AMMONIA (*Volatile alkali*).

We place here a combination of azote and hydrogen named *ammonia*, which presents the properties of a base; that is to say, which unites with acids to form true salts.

We will speak first of the circumstances under which ammonia can be formed.

Organic substances often contain azote in the number of their elements. They may give rise to ammonia—

1st, When they decompose spontaneously.

2d, When they are subjected to the influence of heat.

3d, When they are heated with a hydrated alkali, potash, for example, all the azote of the organic matter is then disengaged in the form of ammonia.

Azote and hydrogen may unite in their nascent state, to produce ammonia. Thus, when an excess of hydrogen, and an oxygenated compound of nitrogen, are simultaneously passed over spongy platinum, moderately heated, ammonia is produced. In this case, platinum acts only by its presence. It has been shown that in this experiment, this metal may be replaced by the sesqui-oxide of iron. Many metals, and particularly tin, zinc, and iron, treated by azotic acid, produce ammonia. To explain the production of ammonia in the reaction of azotic acid on the metals, it is supposed that, under the influence of this acid, water is decomposed, and hydrogen becomes disengaged, which in the nascent state reacts on the azotic acid, and transforms it into ammonia.

When azotic acid is introduced into a liquid which contains sulphuric acid and zinc, and which produces in consequence hydrogen, the disengagement of gas ceases sometimes completely, and the hydrogen combines with the azote of the azotic acid to form ammonia, which remains in the liquor in the state of an ammoniacal salt.

Ammonia is also developed when iron is exposed to the action of moist air. Water is decomposed, its oxygen unites with the iron to form a sesqui-oxide, and its hydrogen produces, with the azote of the air, ammonia, which, combining with the carbonic acid of the air, gives carbonate of ammonia. Also, rust always disengages ammonia when it is mixed with potassa.

Certain natural oxides of iron and of manganese, some clays, and many earthy matters, contain ammonia. It is found in small quantities in the waters of storms, in the state of carbonate or azotate of ammonia. Traces of ammonia constantly exist in the air: it can thus be seen that porous bodies, in which ammonia is found, have only condensed this gas.

The secretions of animals, their urine, their excrements, contain ammonia, or bodies which easily transform themselves into salts of ammonia.

It may be said, in fine, that hydrogen and azote have a great tendency to unite together to form ammonia, and that this combination is produced with facility when these two bodies, on leaving a combination, meet in their nascent state.

Properties.—Ammonia is a colourless gas, of a sharp and penetrating odour, entirely characteristic.

This gas is unfit for combustion; a candle placed in it is immediately extinguished. It is not combustible in atmospheric air, but when it is introduced by a small opening into a bell-glass filled with oxygen, it will inflame; it then burns with a yellow flame.

Ammonia is unfit for respiration: it produces opthalmia, which is frequently seen among workmen who are exposed to ammoniacal emanations.

The aqueous solution of ammonia, placed on the skin, produces a redness which is soon followed by a blister and a true cauterization; this property leads to the employment of ammonia to cauterize wounds made by the bites of mad animals.

It is sometimes used to restore to consciousness persons who have fainted.

Ammoniacal gas is not permanent. M. Bussy has shown that, in exposing it to the cold produced by the evaporation of liquid sulphurous acid, it can be liquified.

M. Faraday has liquified ammonia, by exposing it to the cold produced by the evaporation, in a vacuum, of a mixture of solid carbonic acid and ether. Solid ammonia is white, crystalline, transparent, heavier than liquid ammonia; its odour is weak, because, at this low temperature, its tension is not great. Ammoniacal gas reacts as an alkali on the red paper of turnsol and the syrup of violets. This property, which does not belong to any other elastic fluid, serves to characterise it; hence its name of volatile alkali. Ammonia is in general known by three characteristics: 1st, by its smell; 2d, by its alkalinity; 3d, by the white fumes of chlorhydrate of ammonia which it produces when it is approached with a glass tube which has been dipped in chlorohydric acid.

If ammonia is made to pass through a tube filled with fragments of porcelain and heated to redness, it is partly decomposed, and gives rise to azote and hydrogen, which are found to be in the proportion of one volume of azote and three volumes of hydrogen; this decomposition takes place more readily if a small quantity of platinum is introduced into the porcelain tube.

Electricity will also decompose ammonia. Ammonia is one of the most soluble gases in water. Water dissolves 670 times its own volume of it. If a receiver filled with pure ammonia is placed in contact with water, the gas is instantly absorbed, and water strikes the top of the receiver with such force as to break it; so that, in making this experiment, it is necessary to envelope the receiver in a cloth, for the force of the glass might wound the operator. (Fig. 21.)

A piece of ice introduced into a receiver filled with ammoniacal gas, absorbs the gas rapidly, and is soon melted. Notwithstanding its great solubility, ammonia does not give white fumes in the air, because it does not combine with water in definite proportions.

Water saturated with ammoniacal gas is used in chemical reactions, in place of gaseous ammonia, which it would be difficult to manage.

This solution, which is often called *liquid ammonia,*

gives up all the gas which it contains when it is heated to 60° Cent., or when it is exposed for a long time to the air: it loses it also in vacuum.

Oxygen acts on ammonia under the influence of electricity, and forms water and azote. M. Kuhlman has shown that, under the influence of spongy platinum, slightly heated, a mixture of ammoniacal gas and oxygen is transformed into azotic acid.

Chlorine decomposes ammonia. If a few bubbles only of chlorine are made to pass through this gas, white fumes are soon formed, which are accompanied with the disengagement of heat and light. Chlorhydrate of ammonia, and nitrogen are produced. Chlorine also decomposes liquid ammonia, but then the action is less energetic. It is not accompanied with light.

Preparation.—The preparation of ammonia is founded upon the property which the fixed alkalies possess of displacing it from its saline combinations. All the ammoniacal salts could be used without distinction in this preparation; but the chlorohydrate of ammonia, which is abundant in commerce, is preferred.

The chlorhydrate of ammonia is a combination of ammonia and chlorhydric acid; when this salt is treated with a base, lime, for example, it is deprived of its chlorhydric acid, the ammonia is set free, and disengaged.

Equal weights of quicklime and *sal ammoniac*,* mixed together, are introduced into a matrass; the mixture ought only to occupy a third, or a half part of the capacity of the matrass; the empty part is filled with small fragments of quicklime, for the purpose of drying the gas. (Fig. 22.) The matrass is fitted with a curved tube, which passes under a receiver. The chloride of calcium cannot be used for drying the ammonia, as it has the property of absorbing large quantities of this gas. The action of lime on the ammoni-

* Chlorhydrate or muriate of ammonia.

acal salt, commences at once; but it soon ceases, unless the matrass is heated by a chaffing-dish of coals. The gas ought to be collected over mercury, and it is known to be pure when it is completely absorbable by water.

Instead of producing ammoniacal gas with lime and an ammoniacal salt, it may also be obtained by heating the liquid ammonia of commerce. A slightly elevated temperature is sufficient for the disengagement of all the gas it contains.

The solution of ammonia (ordinarily called liquid ammonia) is prepared by receiving the ammoniacal gas, first in a bottle for washing it, in which is placed a small quantity of milk of lime to absorb the carbonic acid and foreign bodies which may come over with it, and afterwards in a series of bottles containing distilled water.

As the ammoniacal water is lighter than pure water, the tube which brings over the gas ought to plunge to the very bottom of the bottle: to facilitate the solution of the ammonia, the condensing bottles are surrounded with cold water. The saturation may be considered complete, when numerous bubbles of the gas are seen to disengage themselves from the cool ammoniacal solution.

Uses of Ammonia.—Ammonia is used in laboratories as a reagent. It is employed in medicine, and enters into the preparation of many irritating ointments.

Ammonia is used for dissolving carmine, and many other coloring matters, and for modifying the tints of some colors, such as crimson, and Prussian blue. Scourers consume large quantities of ammonia for taking out grease-spots, and restoring colors changed by acids. The manufacturers of artificial pearls use ammonia for preparing the *Essence d'Orient.* This liquor is obtained by holding in suspension, in liquid ammonia, the small scales of a river fish called *blayfish.* This ammoniacal liquor is blown into the globules of glass; the scales fix themselves around the parietes of the glass, and produce the effect of natural oriental pearls.

M. Thenard has proved that ammonia is serviceable in the treatment of animals affected with a disease known among farmers as *hoven*, especially sheep and cows which have been eating wet grass.

The hoven is owing to the production of a considerable quantity of carbonic acid in the stomach and intestines of animals, which causes death in a short time. As the gas which produces hoven is carbonic acid, and as this gas is absorbable by ammonia, it is sufficient to administer to the animal some liquid ammonia, mixed with water, to cause the disappearance of hoven.

Ammonia combines with all the acids to form *ammoniacal salts*. The most important of these salts is that which results from the combination of ammonia with chlorhydric acid. This is called *chlorhydrate of ammonia*, or *sal ammoniac*. Sal ammoniac is used to prepare ammonia for the requirements of the arts, and for chemical laboratories. It is employed in the manufacture of medicinal sesqui-carbonate of ammonia; and is used for taking off the rust from metals, particularly copper. In this case, the ammonia of the chlorhydrate reduces by its hydrogen a part of the oxide of copper to the metallic state, and the chlorine transforms the other part into protochloride of copper, which volatilizes.

Sal ammoniac is also employed in some processes of dyeing.

The chloride of silver being soluble in an aqueous solution of sal ammoniac, a mixture of these two salts is sometimes used for silvering, without heat, copper and brass. Sal ammoniac is used in the extraction of platinum to precipitate this metal from its solution in aqua regia. Finally, sal ammoniac enters into the composition of a lute used for cementing iron into stone. This lute is prepared by sprinkling iron filings, first mixed with one or two hundreths of sulphur, with a solution of sal ammoniac.

CHLORINE.

Scheele discovered chlorine in 1774. This chemist considered chlorine to be muriatic or marine acid deprived of phlogiston, and named it *dephlogisticated marine acid.*

Then came Lavoisier, who regarded chlorine as being formed of muriatic acid and oxygen, and called it *oxygenated muriatic acid.*

Finally, in 1809, Gay-Lussac and M. Thenard, in France, and Davy, in England, discovered that all the reactions of chlorine could be explained by considering oxygenated muriatic acid as a simple body; and, in fact, this body has never been decomposed.

M. Ampère gave it the name of chlorine, which all chemists have adopted.

Chlorine is a gas of a yellowish green color, as is indicated by its name, derived from the Greek χλωρός; of a strong and suffocating smell, of a caustic taste, and of a density of 2·44. It is unfit for combustion. A lighted bougie, plunged into a bell-glass filled with this gas, burns for awhile, and goes out after having changed color. Chlorine is unfit for respiration, and besides is deleterious. A few bubbles of chlorine, introduced into the lungs, produce a violent suffocation, and may even cause lesions, followed by bloody expectoration.*

Chlorine is not a permanent gas.

Faraday liquefied it by heating to 33°C., in a tube closed at both ends, crystals formed by the combination of chlorine with water. Under the influence of a slight rise of temperature, this hydrate of chlorine is decomposed, and in the bottom of the tube two liquid strata are found; the inferior one is liquefied chlorine, the superior is water saturated with chlorine.

Chlorine is soluble in water. One volume of this liquid

* When *much diluted* may be used with benefit for inhalation in phthisis. (Gregory).

dissolves 3·04 of it, at 8°C. It is at this point that the solubility of this gas is at its maximum. This solubility diminishes rapidly when the temperature increases; at 50°C., it is not more than 1·09. When the solution of chlorine is made to boil, it loses all the chlorine which it contains. This solution is of a greenish yellow, deeper than chlorine, and presents all the properties of gaseous chlorine. It is employed in the laboratories in preference to chlorine, because it can be more easily managed. It must not be prepared at a temperature too low, because the dissolving power of water diminishes rapidly by cooling below + 8°C.; and at 0°, water will not dissolve more than one and a half times its volume of chlorine.

The solution of chlorine ought to be protected from the light, because it is decomposed under the influence of solar radiation. The chlorine reacting on the elements of water, combines with hydrogen to form chlorhydric acid, and oxygen is disengaged.

In cooling to the temperature of 2° or 3° above zero, a solution of chlorine saturated at +8°, it is soon seen to deposit crystals of hydrate of chlorine of a yellow white, the form of which appears to be that of an elongated octahedron with a rhomboidal base.

Chlorine has a great affinity for hydrogen.

The action of these two gases on each other, does not show itself much, when they are protected from the effects of ordinary light and heat. Subjected to the influence of electricity, heat, or solar light, these two gases immediately combine with a violent detonation. According to Mr. Draper, an artificial light, that of a candle for example, will also cause the combination of chlorine and hydrogen. This combination produces chlorhydric acid, which results from the union of equal volumes of chlorine and hydrogen, without condensation: thus, 2 volumes of hydrogen, and 2 volumes of chlorine, give 4 volumes of chlorhydric acid.

Under a diffuse light, chlorine and hydrogen unite slowly

and without noise; but the mixture keeps indefinitely in a dark place.

All the luminous rays do not cause the union of chlorine with hydrogen, the violet rays only having this property; thus we may expose to the direct light a mixture of chlorine and hydrogen, contained in vessels colored red, yellow, or green, without having the combination.

When dry chlorine has been exposed for some time to solar radiation, it possesses more energetic chemical affinities; it combines with hydrogen at the ordinary temperature, and, sheltered from the light (M. Draper). Chlorine acts not only on free hydrogen, but also on a great number of compounds of hydrogen. Thus, when moist chlorine is made to pass into a porcelain tube heated to redness, water is decomposed, and chlorhydric acid and oxygen are formed.

Under the influence, then, of solar radiation and heat, chlorine may take hydrogen from water to form chlorhydric acid. It acts in the same manner on certain organic matters, and deprives them of their hydrogen. This reaction of chlorine becomes particularly evident when it exercises itself on colouring matter.

No coloring matter of an organic nature can resist the action of chlorine. The tincture of turnsol, indigo, ink, are destroyed by chlorine; this property has been applied by Berthollet to the bleaching of cotton and linen cloth.

When chlorine acts on coloring matter, it is easily understood that hydrogen is taken away; but it also sometimes happens that chlorine causes the oxidation of matters subjected to its action. Chlorine then decomposes the water to form chlorhydric acid, and the oxygen being in the nascent state, is brought to bear on the colouring matter to destroy or modify it.

Chlorine can then sometimes be employed as an agent of oxidation, sometimes as an agent of dishydrogenation.

A hydrogenized body, after having been submitted to the

action of chlorine, often keeps some chlorine, which is substituted for the hydrogen, and chlorhydric acid is formed.

The affinity of chlorine for hydrogen, explains its action on miasms, and organic matters in a state of decomposition.

The odor which certain substances give off in putrefying, is due to the presence of a combination of hydrogen and sulphur, called *sulphydric acid*, which the chlorine decomposes.

Chlorine is then used to disinfect substances which disengage sulphydric acid: further, this acid being highly deleterious, chlorine is used to prevent the asphyxias produced by sulphydric acid. But chlorine, being itself deleterious, ought to be employed with caution.

Chlorine does not act on hydrogenated substances only; it combines directly with many simple bodies, such as arsenic, antimony, and potassium, etc., which inflame when they are reduced to a fine powder, and thrown into a jar filled with chlorine. A copper wire, heated at the end, which is plunged into a bottle containing chlorine gas, burns completely, transforming itself into chloruret of copper.

Preparation.—Chlorine is prepared by heating a mixture of oxide of manganese, and an acid formed of chlorine and hydrogen, called chlorhydric acid. The oxygen of the oxide of manganese unites with the hydrogen of the chlorhydric acid to form water, and chlorine is disengaged. It is received in bottles filled with air, as it attacks mercury.

Water of Chlorine is obtained by directing a current of chlorine into a bottle filled with water. This solution is of a greenish yellow, and possesses all the properties of chlorine.

Uses.—The uses of chlorine are numerous. Since the beautiful experiments of Berthollet, chlorine is used for bleaching cloths, taking out the color of paste for making paper, whitening old engravings, restoring old books, and taking out ink-spots. To whiten an engraving, or take out an ink-spot, it suffices to plunge the paper for some instants into water which holds chlorine in solution; the paper is then passed through ordinary water to take out the smell of chlorine: in these experiments, chlorine destroys ordinary ink, but does not in any way attack printers' ink, which has for a base a fat body and lamp-black.

CHLORHYDRIC ACID.

Chlorhydric acid, which is a combination of chlorine and hydrogen, was for a long time called *marine acid*, *muriatic acid*, *hydrochloric acid*.

This acid is gaseous and without colour; it throws out into the air white fumes; its smell is irritating. It excites cough when it is introduced into the air passages; its density is 1.2474. It is not permanent. At +10° C., under a pressure of 40 atmospheres, it transforms itself into a liquid without colour, of a density of 1.27.

This gas is unfit for combustion, very soluble in water, which, at the temperature of 0°, dissolves about 480 times its volume of it. The solution of chlorhydric gas takes place with such rapidity, that when a bell-glass filled with this gas is placed in contact with water, the column of liquid introduces itself with such rapidity into the bell-glass, as sometimes to cause its rupture.

The presence of a small quantity of air very much retards the rapidity of this absorption.

Chlorhydric acid blackens organic matters, and rapidly destroys them. It does not in general act on the metalloids; many metals, such as potassium, iron, tin, &c., decompose it, combining with chlorine, and disengaging hydrogen. The great ease with which chlorhydric acid is decomposed in the cold by iron and zinc, causes it sometimes to be employed in the preparation of hydrogen.

Preparation of Chlorhydric Acid. — Chlorhydric acid is prepared by decomposing marine salt (chloride of sodium) by hydrated sulphuric acid. The water contained in the sulphuric acid is decomposed, and sulphate of soda and chlorohydric acid are formed.

A few grammes of marine salt are introduced into a glass matrass; a tube for collecting the gas is fitted to the matrass, and concentrated sulphuric acid is poured in. The reaction first commences in the cold, but it is afterwards increased by gentle heat. In this preparation, marine salt, first fused and moulded, and then broken into fragments of some size, is used. If concentrated sulphuric acid is made to react on marine salt crystallized, and very fine, it will produce, at the moment when the acid is poured on, a very brisk effervescence, which will cause the mixture to rise up into the tube.

Chlorohydric acid being very soluble in water, should be collected over mercury, which does not exercise any action upon it.

The solution of chlorohydric acid in water, which is often called *liquid chlorohydric acid*, is obtained in laboratories, by means of Woulf's bottles, which serve for the preparation of nearly all the solutions of gas in water. This apparatus consists of a matrass communicating with a series of condensing bottles. As the solution of chlorohydric acid is heavier than water, the gas ought to be brought into each bottle by a tube which plunges but little into the water; in this way the different strata of the liquid constantly mix. The liquid of the first bottle is never pure; it always contains volatile chlorides, and sulphuric acid, which have been carried over in the reaction; but the liquid chlorohydric acid contained in the other bottles is generally pure.

Six parts of marine salt require about five parts of sulphuric acid to decompose them.

The water which absorbs the chlorohydric acid increases

in volume; so that, in commencing the operation, the bottles should not be entirely filled with water.

In the arts, chlorhydric acid is prepared by decomposing marine salt by sulphuric acid, in kilns or cylinders of iron. The chlorhydric acid which is disengaged is condensed in a series of earthen jars communicating with a chimney, having a strong draught. (Fig. 15.)

The chlorhydric acid of commerce (muriatic acid) is not pure; it contains generally all the salts which are found in common water, used for the solution of the chlorhydric acid gas, and besides, sulphurous and sulphuric acids, perchloride of iron, and sometimes arsenious and arsenic acids.

The uses of Chlorhydric Acid.—The uses of chlorhydric acid are numerous and important. This acid, employed as a reagent, serves as a test for the salts of silver, to decompose the carbonates, the sulphurets, as a test for ammonia, &c. It serves in the arts for the preparation of chlorine and the decolorising chlorides, for the extraction of gelatine from bones, &c. It is used alone, or with azotic acid, to dissolve a great number of metals or alloys, and to prepare the metallic chlorides.

BROMINE.

Bromine was discovered in 1826 by M. Balard, who extracted it from the mother-waters of the salt-pits.

This body exists in the sodas of sea-weed, in the water of the sea, in a great number of salt-springs, &c.

Bromine is a liquid of a brownish red, very poisonous, of a penetrating and strong odor. Its name is derived from the Greek βρῶμος, fetid.

In a chemical point of view, bromine resembles chlorine strongly; it has, like it, affinity for hydrogen, and bleaches colouring matters. It forms in uniting with hydrogen a hydracid, known as bromhydric acid.

Bromine has not been yet applied to important uses.

IODINE.

This body was discovered in 1811 by Courtois. Gay-Lussac has traced a complete history of it in one of the most important of his memoirs.

Iodine does not exist in nature in a free state: like chlorine and bromine, to which it has strong analogy, it is always found united to sodium in marine plants, such as sea-weed, the fuci, &c., in sponges, in sea-water, in some saline springs, and in the mineral kingdom in the state of iodide of silver.

M. Bussy has shown its presence in the coal of Commentry, (Allier); and according to M. Duflos, iodine is also met with, mixed with bromine, in the coal of Silesia.

Properties.—Iodine is solid at the ordinary temperature: its odor recalls that of chlorine and bromine, its color is of a metallic gray, it resembles plumbagine.

Iodine crystallizes in rhomboidal scales, large and brilliant, and often in elongated octahedrons.

Iodine fuses at 107° Cent., and boils at 180° C.

The violet vapors which it produces in volatilizing have given it the name of *iodine*, from the Greek word ἰωδής, violet.

When a well-dried matrass is heated, and a small quantity of iodine is thrown in, the matrass soon fills with violet vapors remarkable for their richness and intensity.

Iodine is but slightly soluble in water, which dissolves about 0.007 of its weight at the ordinary temperature; but it is very soluble in alcohol, and gives it a very deep brown color. It is also very soluble in ether. These two solutions, by evaporation, deposit crystals of iodine. They are precipitated by water, which immediately separates from them the iodine, in the form of a brown powder. Iodine dissolves in the sulphuret of carbon, and gives to this liquid a violet colour.

It exercises on organic substances a destructive action,

and colors yellow the skin, paper, &c. This color disappears under the influence of an elevated temperature, if the contact has not been sufficiently prolonged; if so, the organic matter is completely destroyed: iodine combines in this case with the hydrogen of the organic substance, to form iodohydric acid.

Iodine, in reacting on other bodies, behaves in general like chlorine and bromine; but its affinities are more feeble, and these two metalloids displace it from most of its combinations. It slowly destroys coloring-matters, and does not decompose water under the influence of solar radiation. Among the properties of iodine, there is one which enables us to recognise the smallest quantity of this body, and which serves to characterise it.

Placed in contact with starch, in the presence of water, it produces a blue combination, which bears the name of *iodide of starch.* This iodide loses its color at a temperature of 70° to 80°, and recovers its blue color when the liquor is allowed to cool. This curious experiment is due to M. Lassaign.

Preparation of Iodine.—Iodine may be obtained by decomposing an iodide by chlorine, which substitutes itself for the iodine, causing it to precipitate; but it is necessary to be careful to stop the disengagement of chlorine as soon as all the iodine is displaced, or the chlorine will combine with the iodine.

The process which is ordinarily used to prepare iodine, consists in decomposing an alkaline iodide — of potassium, for example — by sulphuric acid, and peroxide of manganese.

Uses of Iodine.—Iodine, free or combined with potassium, is applied in medicine in the treatment of goitres, and scrofulous diseases. It is used, also, in the preparation of daguerrean plates.

SULPHUR.

Natural State.—Sulphur is widely spread in nature, particularly in combination with the metals. It exists, in the native state, in volcanic districts. It enters into the composition of plaster, sulphates of barytes, strontium, &c. United with hydrogen, it forms part of a great number of mineral waters. It is met with, also, in certain animal substances, some essential oils, &c.

The most beautiful specimens of native sulphur come from Sicily, where they are found in crystals derived from the octahedron, and along with sulphate of strontium.

Properties.—Sulphur, at the ordinary temperature, is a solid body, of a peculiar clear yellow, without taste or smell, acquiring by rubbing a characteristic odor: a bad conductor of heat and electricity. It is insoluble in water, and but slightly soluble in alcohol and ether. Sulphur is very brittle; a stick of sulphur, held in the hands when broken, produces a peculiar cracking, which results from the unequal dilatation of its molecules. This body, by rubbing, excites resinous or negative electricity. Its density is represented by the number 2·087. Sulphur fuses at the temperature of 110° C., and boils at 460° C. Its volatility enables us to free it easily by distillation, from impurities which it may contain.

Melted sulphur, at 110°C., presents the appearance of a yellow liquid; and by cooling returns to the solid state and yellow color, as it was before fusion.

In gradually raising the temperature of sulphur, it is seen, between 140° and 150°C., to take a deep yellow color; at 190°C., an orange color, and that its consistence has become viscous; at 260°, it becomes brown; at this temperature, its viscidity is such, that the matrass in which it is fused may be inverted without the sulphur pouring out.

In continuing to raise the temperature, the sulphur is seen to become again somewhat fluid. If, at this moment, it is suddenly cooled by pouring it into cold water, it remains ropy, transparent, preserves its brown color, and becomes elastic, somewhat like caoutchouc, so that it can be drawn out in long threads. It requires some time before the soft sulphur recovers its yellow color, and its original hardness.

Sulphur crystallizes easily, and presents the singular property of taking two incompatible forms. One of these forms is the right octahedron, and elongated with rhomboidal base, the other is the oblique prism, with rhomboidal base.

We sometimes see simple or compound bodies take two or more different and incompatible forms: it is to this property has been given the name of *dimorphism*, or *polymorphism.*

Sulphur has a great affinity for oxygen. It burns in this gas, or in the air, at the temperature of about 150°, producing a beautiful blue flame, with a sharp, characteristic smell, which is that of matches when they are lighted. The product of this combustion is sulphurous acid, which is always found mixed with a small quantity of sulphuric acid.

Sulphur burns brightly, when, after having lighted it, it is placed in a large bottle filled with oxygen.

It combines directly with hydrogen to form an acid, known under the name of *sulphydric acid;* but this acid is never thus obtained, for sulphur and hydrogen do not easily unite except in the nascent state, that is to say, at the moment they each leave a combination.

Sulphur also unites with chlorine, bromine, iodine, and most of the metals; some metals, as iron, copper, silver, will even inflame in the vapor of sulphur, and burn there with as much energy as in oxygen or chlorine.

Extraction of Sulphur.—Sulphur employed in the arts is ordinarily extracted from the earthy soils called *solfataras,* which contain it in the native state. Sulphur is obtained for the most part in Sicily, which produces annually about 50 millions of kilogrammes. The ores of Sicily are very rich, and contain from 30 to 35 per 100 of sulphur. The sulphur is extracted by distillation, or simply by melting, when the mine is very rich.

Two successive distillations are required to purify sulphur completely.

In general, the first is made at the solfatara, in rude apparatus of earthenware. The second is made with more care, in the places where it is consumed, by means of a cylinder of metal which communicates with a chamber of masonry, where the sulphur condenses.

When the distillation is conducted slowly, and when the sides of the chamber do not become heated above 110° C., it condenses in the chamber in a very fine powder, which takes the name of *flowers of sulphur:* if the distillation is carried on rapidly, the sulphur melts, and may be run into moulds of wood, and we thus obtain *roll brimstone.*

The flowers of sulphur usually contain sulphurous and often sulphuric acid; it reddens the tincture of litmus: to purify it, it must be washed in warm water, and then dried at a low heat.

When we wish to prepare pure sulphur in the laboratories, the sulphur of commerce, or roll brimstone, is exposed to the action of heat in a glass retort connected with a glass receiver. This distillation presents no difficulty, and gives very pure sulphur.

Uses of Sulphur.—Sulphur has numerous uses in the arts. It is often used for making moulds and medals, or for taking impressions.

Mixed with carbon and nitre, it constitutes gunpowder. Transformed by combustion into sulphurous acid, it is used for bleaching wool and silk, for the preparation of sulphuric acid, &c.

Sulphur is used in the manufacture of matches.

It is classed among therapeutic agents. It is applied in medicine to the treatment of diseases of the skin.

COMBINATIONS OF SULPHUR WITH OXYGEN.

Sulphur combines with oxygen in several proportions. But we shall only here speak of sulphurous and sulphuric acids, which are the most important, and the only ones which are employed in the arts.

SULPHUROUS ACID.

Sulphurous acid is composed of sulphur and oxygen, and is the result of the combustion of sulphur in the air.

This acid is gaseous, without color, and unfit for respiration and combustion; its odor, irritating and characteristic, is like that of burning sulphur. It acts upon the lungs, and excites cough. Sulphurous acid is soluble in water; this solution presents all the chemical properties of sulphurous gas.

Uses.—Sulphurous acid has the property of decolorizing most coloring matters; thus, violets which are dipped in a solution of sulphurous acid, become in a short time completely white. The property which this acid presents, of acting on certain coloring-matters, is turned to account in the bleaching of silks and wool. These substances cannot be bleached by chlorine, which gives them a yellow color. The sulphurising of silk gives it a peculiar feeling, which is recognised when the silk is taken in the hand. At Lyons, to sulphurize silk, the sulphur is burnt in an apartment, of which the door and windows are closed with care. The sulphur, in burning in the air, changes into sulphurous acid, which reacts on the silk, and bleaches it; the silk is moistened, and suspended on rails placed at the height of 3 metres. For 100 kilogrammes of silk, about two kilogrammes of sulphur are used.

Sulphurous acid is employed in medicine in the treatment of diseases of the skin.

It is also used for bleaching the isinglass, and the straw which is used in making hats. It is also used for taking out fruit-stains, and purifying infected places and lazarettos.

Sulphurous acid is employed to sulphurize casks intended for preserving wine; its presence prevents the wine from forming vinegar. To sulphurize a cask, it is sufficient to burn in it a small quantity of sulphur, which changes into sulphurous acid.

Sulphurous acid will extinguish a fire in a chimney. In this case, throw a sufficient quantity of sulphur on the coals on the hearth; this absorbs the oxygen of the air, and is transformed into sulphurous acid, which is entirely unfit for combustion. It is necessary to be careful, in this case, to close up as tight as possible all the openings of the fire-place, to prevent the access of air. With this view, damp cloths are generally used.

SULPHURIC ACID.

Sulphuric acid is a compound of sulphur and oxygen, more oxygenated than sulphurous acid; thus the termination *ic* has been given to it. Sulphuric acid is made in quantity by oxygenating sulphurous acid by a body which has the property of readily giving up its oxygen. The agent which is used for this purpose is azotic acid.

Formerly, sulphuric acid was prepared in matrasses of glass; but the applications of sulphuric acid have become so important, that this acid is now made in large chambers of lead, the capacity of which has been carried even above 3,000 cubic metres.

The sulphuric acid produced in the leaden chambers is then concentrated in a platina apparatus. Sulphuric acid is liquid, without color or smell; it is almost twice as heavy as water. When it is poured into a vessel containing water, it falls at once to the bottom, and is then dissolved. Its oleaginous consistence has given it the name of *oil of vitriol.*

This acid has a great affinity for water. This affinity is shown either by directly mixing the acid and water, when the combination causes a disengagement of heat which often exceeds 100°C., or in making sulphuric acid react on organic matters.

When, for example, wood is dipped into sulphuric acid, a part of the oxygen and hydrogen of the organic matters combines to form water, and there is produced on the surface of the wood a black matter which contains less water than the wood. When sulphuric acid is exposed to the air, it absorbs moisture, increases in volume, and may thus take up as much as fifteen times its weight of water. When it is desirable to preserve sulphuric acid in its concentrated state, it is indispensable to keep it in a bottle well stopped. This affinity of sulphuric acid for water, is advantageously used to dry gases; it is sufficient, in fact, to pass them over bottles containing sulphuric acid, to have them quite dry.

Sulphuric acid destroys organic matters by depriving them of their water; it rapidly decomposes animal membranes, and acts as a violent poison; a few drops placed on the skin produce deep burns. In cases of poisoning by sulphuric acid, there should be immediately administered to the patient soap-suds, wood-ashes, and above all, magnesia suspended in water or oil. These principles are applicable to all cases of poisoning by the acids.

A great number of metals, such as iron and zinc, are attacked by sulphuric acid. Under the influence of this acid, water is decomposed, and its hydrogen is disengaged, while its oxygen unites with the metal to form an oxide, which then combining with sulphuric acid, produces a salt.

When sulphuric acid, diluted with water, is poured on iron, it forms sulphate of the protoxide of iron, while hydrogen is disengaged.

Other metals, as copper, mercury, and silver, heated with sulphuric acid, form sulphates of copper, mercury, and silver; but in this case, it is not the water which furnishes the oxygen, but the sulphuric acid itself, a part of which is decomposed into oxygen, which goes to the metal, and into sulphurous acid, which is disengaged. This reaction is applied in the laboratories for preparing sulphurous acid. A mixture of concentrated sulphuric acid and copper is introduced into a matrass, to the neck of which a glass tube is fitted, which passes under a receiver full of mercury. Sulphurous acid is collected over mercury, because this gas is soluble in water. By heating the matrass, the reaction of the sulphuric acid commences, and in a short time several litres of sulphurous acid are obtained.

Uses.—Sulphuric acid has many uses, and is employed in nearly all the chemical arts. Its energy and its fixedness render it useful in isolating most of the acids: it is used in the preparation of chlorhydric and azotic acids.

The greater part of the sulphuric acid produced, is applied to the manufacture of sulphate of soda, which serves to prepare artificial soda, of which the arts consume a large quantity.

It is also used in the preparation of alum, sulphate of iron, chlorine, phosphorus, sugar of starch, ether, stearic candles, and in the purification of oils.

Finally, sulphuric acid is the acid which is most used in chemical laboratories.

We find in commerce a fuming sulphuric acid, which is called *sulphuric acid of Nordhausen*, less hydrated than the above, which is principally used to dissolve indigo. This sulphuric solution of indigo is used to dye wool a *Saxon blue.*

ACID SULPHYDRIC.

Sulphydric acid is a combination of sulphur and hydrogen, and was discovered by Scheele; it is often called *hydro-sulphuric acid* or *sulphuretted hydrogen.*

This acid is gaseous, without color; its fetid odor, which recalls that of rotten eggs, constitutes one of the characteristic properties of sulphydric acid.

This acid liquefies under a pressure of about 17 atmospheres, and forms a very fluid, colorless liquid.

Sulphydric acid is very deleterious; according to the experiments of Messrs. Thénard and Dupuytren, a small bird dies immediately in an atmosphere which contains $\frac{1}{1500}$th of its volume of sulphydric acid; $\frac{1}{1000}$th will kill a dog of medium size; $\frac{1}{250}$th will cause the death of a horse.

It is to the presence of sulphydric acid that we must attribute the accidents which unfortunately are of too frequent occurrence to workmen engaged in cleaning privy-sinks.

Sulphydric acid gas is partly decomposed by heat, so that it cannot be directly obtained pure, by passing a mixture of vapor of sulphur and hydrogen through a red hot tube. It is set fire to by a lighted candle, and is transformed into water and sulphurous acid.

When sulphydric acid is burnt in a narrow receiver, it deposits sulphur on the sides of the glass; but a jet of sulphuretted hydrogen, lighted in the open air, burns up completely, with a blue flame and sharp characteristic smell.

Sulphydric acid is slightly soluble in water. This liquid dissolves only about three times its volume at the temperature of 10° C. The aqueous solution keeps a long time without change, when protected from the air; under the influence of oxygen it becomes rapidly turbid, and the sulphuretted hydrogen is transformed into water and sulphur, which precipitates.

Alcohol dissolves about six times its volume of sulphydric acid. Water, saturated with sea-salt, dissolves but a small quantity; thus we could use a solution of salt over which to collect sulphydric acid gas, instead of mercury.

Chlorine acts at the ordinary temperature on sulphuretted hydrogen gas, and decomposes it, forming chlorhydric acid and a deposit of sulphur. This property might serve to combat the asphyxias caused by the absorption of sulphydric acid; but in this case, chlorine, which is itself deleterious, ought to be used with caution. As an antidote to poisoning by sulphydric acid, sulphurous acid might be used; that is, the gas which is produced by the combustion of sulphur in the air. This acid immediately decomposes sulphydric acid, producing water and sulphur. A great number of metals decompose sulphydric acid; some in the cold, some under the influence of heat, form metallic sulphurets, and disengage the hydrogen of this acid.

Sulphydric acid blackens utensils of silver, copper, tin, and oil paintings, when privy-wells are emptied, because it forms in these different cases sulphurets of silver, copper, tin, &c., which are black; eggs cooked in silver vessels disengage during the cooking a small quantity of sulphydric acid, which produces on the surface of the metal black sulphuret of silver.

The natural state of Sulphydric Acid.—Sulphydric acid is found free, or in part combined with earthy or alkaline bases, in the mineral waters called *sulphur-waters.* Organic substances which contain sulphur, produce on spontaneous decomposition sulphuretted hydrogen; it is found in marsh mud, and in pools where sea-water remains. Intestinal gases always contain some of it.

Uses.—Sulphydric acid is one of the reagents most frequently used in analytical researches, to characterise and separate the different metals from each other: when it is free or combined, it precipitates, in the state of different colored sulphurets, the metals which are held in saline combinations. Sulphur-waters are used in the form of baths, in the treatment of diseases of the skin, and many internal affections.

PHOSPHORUS.

The discovery of phosphorus goes back to the year 1669: it is due to Brandt and Kunckel, who extracted this body from the phosphates contained in the urine. In 1769, Gahn and Scheele showed the existence of a considerable portion of phosphate of lime in the bones, and made known an easy process for extracting the phosphorus from it.

Properties.—Phosphorus is without color, transparent, insipid, of a feeble garlicky odor, and of a horny appearance. It is flexible, and sufficiently soft to be indented by the nail. A small quantity of sulphur will render it brittle.

Phosphorus melts at 44° Cent., thus it can easily be melted in hot water. After having been melted, it remains liquid at the ordinary temperature, and even some degrees below zero (Cent). The experiment is made in a glass, where the phosphorus is melted in warm water; this water is taken away and replaced several times by cold water. Phosphorus thus is seen to remain liquid several minutes, in water of the temperature of 12° to 15° Cent.; it solidifies as soon as it is touched with a foreign body, and the thermometer rapidly rises.

Phosphorus presents when perfectly pure, another phenomenon not less curious, which was observed by Thénard.

When it is exposed to a temperature of 70° Cent. and suddenly cooled, it becomes black. This modification is known to be due to a change of molecules; the color disappears by fusion.

Of all the simple bodies, phosphorus presents, in regard to color, the most numerous modifications. When it is submitted to solar rays, either in vacuo, or in gases which do not alter it chemically, as hydrogen, azote, etc.; it rapidly becomes red. This color is due to an isomeric transformation of the phosphorus; the phosphorus thus modified, bears the name of *red phosphorus*, or *amorphous phosphorus*. It is obtained by keeping it at a temperature of 230° to 250° Cent. in an atmosphere which cannot alter it chemically. Red phosphorus does not become luminous in the air, below a temperature of 200° Cent.; it keeps in the air without change. It does not combine with sulphur in fusion; ordinary phosphorus would produce an explosion in contact with melted sulphur. Red phosphorus melts at 250° Cent. At 260° Cent. it passes back to the state of ordinary phosphorus. When it is kept for several days at a temperature a little below 260° Cent., it collects together in a mass, very hard, of a violet brown, which appears even less alterable than red phosphorus. Red phosphorus is insoluble in the sulphuret of carbon, while ordinary phosphorus is largely soluble in this liquid.

Phosphorus loses its transparency in water—it becomes rapidly yellow and opaque. When it is kept a long time in this liquid, it becomes covered with a thick coating of a yellowish white, which appears to be phosphorus in a peculiar molecular state, and sometimes, also, a combination of phosphorus and water, analogous to the hydrate of chlorine. (Pelouze).

Phosphorus does not crystallize by fusion, but its solution in essential oil, in the sulphuret of phosphorus, or in the sulphuret of carbon, deposits rhomboidal dodecahedrons. The best dissolvent of phosphorus, is the sulphuret of carbon, which takes up considerable quantities of it. This solution ought to be handled with precaution, for in evaporating from a great surface, it leaves the phosphorus very much divided, which takes fire spontaneously. Thus a sheet of paper, impregnated with this solution, takes fire as soon as the sulphuret of carbon is evaporated.

Phosphorus may be reduced to powder, by introducing it into a bottle containing warm water, and agitating it rapidly till the water becomes cold; the phosphorus becomes divided into small drops, which become reduced to powder in solidifying.

The property which phosphorus possesses, of becoming luminous in the dark, serves to distinguish it.

It takes its name from two Greek words; φῶς, *light*, and φέρω, *I carry*. Figures or letters traced on a tablet placed in the dark are luminous, and are said to be *phosphorescent*.

The phosphorescence of phosphorus is generally considered as the result of a slow combination of this body with oxygen.

The water in which phosphorus has been kept also glimmers in the dark. When it is agitated, it throws out a glimmering for a short time. Many bodies, such as chlorine, carburetted hydrogen, alchohol, ether, essence of turpentine, prevent phosphorus from shining in the dark. Phosphorus glimmers more in rarefied air than under the ordinary pressure of the atmosphere.

Phosphorus is one of those bodies which has the greatest affinity for oxygen; a slight elevation of temperature is sufficient to make it burn, so that the distillation of it demands particular care.

In distilling phosphorus we cannot use an ordinary distilling apparatus, which might break from the inflaming of the phosphorus, and expose the operator to the particles of the phosphorus thrown off in the combustion. This operation may be performed without danger in an atmosphere of hydrogen. A current of this gas is kept up in a small tubulated retort, into which is placed some twenty grains of phosphorus; this communicates with a receiver also tubulated, and carrying a recurved tube, the extremity of which plunges some millimters into a vessel filled with water. When the air of the apparatus has been driven out, and replaced by the hydrogen, the retort is heated with some live coals, and the distillation of the phosphorus goes on rapidly.

Phosphorus combines at the ordinary temperature with oxygen. A stick of phosphorus, exposed for some time to the air, gives off white fumes, due to the formation of a peculiar acid, which has received the name of *phosphatic acid.*

The heat which is developed during the formation of this acid, is sufficiently great to cause the inflammation of the phosphorus at the end of a few minutes. To preserve phosphorus, it must then be kept from the contact of the air by covering it with water. The slightest friction will cause the combustion of phosphorus, so that it must always be handled under water. The burns produced by phosphorus are dangerous, and a long time in healing, because it leaves in the wound a very corrosive acid (phosphoric), which is the result of its combustion. Phosphorus inflames, under ordinary pressure, at the temperature of 75°C., and burns with a very brilliant flame.

Phosphorus produces, when burning in the air, and especially in pure oxygen, a very high temperature, and light so bright that the eyes can scarcely endure it. It then changes into phosphoric acid. This combustion ordinarily takes place, on plunging into a bottle of 4 or 5 litres capacity, and filled with oxygen, a cupel containing phosphorus, placed on a cork suspended by an iron wire, and then touching it with a hot wire or match. (See Oxygen, Fig. 12.)

It rarely happens that the phosphorus becomes completely converted into phosphoric acid, even when it burns in pure oxygen, and in excess. It nearly always produces a small quantity of red phosphorus, which is preserved from the action of the oxygen by a coating of phosphoric acid, which covers it.

Phosphorus, notwithstanding its affinity for oxygen, does not act at the ordinary temperature on this gas, when it is pure and dry. If a stick of phosphorus is placed in a jar filled with pure and perfectly dry oxygen, the phosphorus will remain without acting on the oxygen, provided the temperature is not raised above 27°C.; but if the pressure is reduced, or another gas is introduced into the oxygen, the combination is soon brought about, and the oxygen is rapidly absorbed by the phosphorus.

Preparation of Phosphorus.—Phosphorus is generally obtained from the phosphate of lime contained in the bones of animals. The bones are composed of carbonate of lime, basic phosphate of lime, and an animal matter used in the preparation of gelatine.

At first, the bones are subjected to a calcination in contact with the air, to destroy the organic matter which they contain.

After the calcination, the bones are white, and very friable; they contain about 77 parts of phosphate of lime, 20 parts of carbonate of lime, and a small quantity of other salts.

They are reduced to a fine powder, of which are taken about six parts, which are mixed with water so as to form a very liquid paste, to which are added, by degrees, four or five parts of sulphuric acid.

The sulphuric acid, by the aid of boiling, changes the carbonate of lime into sulphate of lime, disengaging carbonic acid; at the same time it takes a part of the lime from the subphosphate, and transforms it into an acid phosphate of lime. This last is very soluble in water, while the sulphate of lime (plaster) is scarcely soluble. In treating the mass, then, with water, the acid phosphate of lime is dissolved, and the sulphate of lime almost completely precipitates.

The waters which hold the acid phosphate of lime in solution, are evaporated in a copper basin, or a porcelain capsule; during this evaporation, the greater part of the sulphate of lime held in solution deposits. This salt is taken away with care, and a syrupy liquid is obtained, containing the acid phosphate of lime nearly pure.

This liquid is intimately mixed with the fourth part of its weight of powdered wood charcoal, and dried at a nascent red heat in an iron basin. This drying process is not arrested till the mass begins to give off vapors of phosphorus.

It is then introduced into an earthen retort covered with a coating of fire-clay, which is filled to about three-fourths its volume with the mixture; this communicates by a copper tube with a large, wide-mouthed bottle, half full of water, which has a tube provided to give issue to the gas. The retort is brought to a bright red heat, and the phosphorus condenses in the bottle.

The phosphorus obtained by the method we have just described, is not yet in a state of purity; it contains carbon, and other bodies which have been carried over during the volatilization.

It is purified by melting it in warm water, and mixing it with bone-black in powder, which decolorizes it. It is then taken out with a spoon, and plunged rapidly into cold water, in order to get it in mass. To get rid of the black which it contains, it is tied up in a chamois'-skin, which is plunged into an earthen vessel nearly full of boiling water. By compressing it with pincers, the melted phosphorus passes through the pores of the skin.

Phosphorus is not found in commerce in mass, but in small sticks, some millimeters in diameter.

These sticks are obtained by melting the phosphorus in

water, and plunging into the melted phosphorus a glass tube slightly conical, into which the phosphorus is drawn by suction.

Some water should be left in the tube to cover the melted phosphorus, and prevent it from getting into the mouth of the operator. The tube is then shut with the finger, and placed in cold water; the phosphorus is then taken out of the tube by a slight jar. These sticks, intended for commerce, are kept in water.

Uses of Phosphorus.—The manufacture of phosphorus has become much extended of late years; and its price, formerly so high, is now not more than seven or eight francs the kilogramme.

Phosphorus is chiefly used in the manufacture of friction matches, and in chemical laboratories, to analyse the air, and to prepare phosphurets, phosphoric acid, &c.

ARSENIC.

Arsenic is solid at the ordinary temperature, of a steel gray color, very brilliant when first sublimed, but altering rapidly in contact with the air.

This body is easily reduced into powder; it is without taste, insoluble in water, and its texture is generally crystalline.

Arsenic has no sensible odor at the ordinary temperature; heated to redness, or thrown on a lighted coal, it gives off a garlicky odor, very strong and characteristic. The density of arsenic is 5·75. This body, subjected to the action of heat, volatilizes without becoming liquid. However, arsenic may be fused by heating it in a metallic tube closed at both ends.

Arsenic volatilizes at about 300°C.; its vapors, in condensing, give rise to tetrahedral crystals.

It combines with oxygen, under the influence of a temperature slightly elevated, burns in this gas with a light blue flame, and produces arsenious acid, improperly called in commerce, *arsenic*.

A great number of simple bodies combine directly with arsenic; arsenic in powder, thrown into a bottle filled with chlorine, takes fire, and produces white vapors of chloride of arsenic.

Arsenic is sometimes found in nature in a state of purity. It is prepared by subjecting metallic arseniurets to roasting, which form volatile arsenious acid; this acid, heated with an excess of carbon, is reduced, and gives arsenic which condenses in the receivers.

Arsenic is used for the destruction of insects; it is reduced to a fine powder, and covered with water.

Arsenic introduced into the stomach of an animal will not induce the symptoms of poisoning at once; it is supposed that in this case it becomes poisonous by being transformed into arsenious acid.

ARSENIOUS ACID

Arsenious acid is solid, white, its taste is acrid, nauseous, and excites the saliva; introduced into the stomach in small doses, it produces gangrenous spots, and causes death with extreme suffering.

The antidotes of arsenious acid are the hydrate of the peroxide of iron, and magnesia. These two oxides saturate the arsenious acid, and form with it insoluble compounds, which have no further action on the animal economy.

Arsenious acid is volatile below redness, its vapours are without smell, this can be readily ascertained by volatilizing it on a brick heated to redness. Arsenious acid thrown on burning coals, gives off a garlicky odor, which is that of metallic arsenic. In this case the arsenious acid is reduced by the carbon.

If, in the distillation of arsenious acid, the sides of the condensing vessel become raised to a high temperature, the vapors of arsenious acid form, in condensing, a vitreous and transparent covering. But if the distillation is made into a receiver where the air circulates, the acid condenses in isolated octahedric crystals.

Bodies having an affinity for oxygen, such as hydrogen and carbon, readily reduce arsenious acid.

Arsenious acid immediately after volatilization is found in colorless plates, which have often the transparency of the crystal; if vitreous arsenious acid is kept for some time, even protected from the air and moisture, it is seen by degrees to lose its transparency, and to become transformed into a completely opaque body.

Arsenious acid is obtained for commerce as an accessory product in roasting the ores of tin and cobalt, and as principal product by the roasting of arsenical iron. These operations are performed in reverberatory furnaces, which communicate with chambers where the arsenious acid condenses.

To purify it, it is sublimed a second time in iron vessels. In laboratories, this sublimation is made in glass or earthen retorts.

Uses.—Arsenious acid is used chiefly in the manufacture of printed calicoes and in glass-works; it transforms the protoxide of iron into the sesquioxide of iron, which gives the glass less color than the protoxide.

It is also employed in the seeding of corn, to preserve the seed from the attacks of insects.

ARSENIURETTED HYDROGEN.

Arseniuretted hydrogen is gaseous: it liquefies at — 30° C., but it has never yet been solidified. Its odor is disagreeable and very garlicky. It exercises no action on the tincture of turnsol. Water dissolves about one-fifth of its volume of it.

Exposed to the influence of moist air, it gives rise to water, and a deposit of black arsenic. Heat decomposes it into hydrogen and metallic arsenic. It is on this property is based the use of the apparatus of Marsh.

Electricity causes a similar decomposition.

It is combustible, and burns with a whitish flame; there are formed in the combustion of this gas, water, arsenious acid, and at the same time, a deposit of arsenic. Chlorine, bromine, iodine, decompose it by depriving it of its hydrogen. The action of chlorine takes place with the disengagement of heat and a bright light. The experiment ought to be made with small quantities of gas, and with a good deal of care, to avoid an explosion.

Arseniuretted hydrogen is very poisonous: a German chemist, Gehlen, died, from having inhaled a small quantity of it.

Preparation.—Arseniuretted hydrogen is prepared by the following processes:—

1st. By treating an alloy of arsenic and tin with chlorhydric acid.

2d. By attacking an alloy of arsenic and zinc with hydrated sulphuric acid.

3d. By placing an arsenious solution in the presence of hydrogen in the nascent state.

The gas prepared by these different methods is not pure, it contains a small quantity of hydrogen.

Arseniuretted hydrogen being one of the most deleterious gases known, too many precautions cannot be taken in preparing it; the slightest leak in the apparatus would be dangerous for the operator.

The Detection of Arsenious Acid in cases of Poisoning.—The detection of arsenious acid in cases of poisoning is one of the greatest questions in legal medicine; the number of cases of poisoning by arsenic alone is greater than that of all others put together.

In the chemical researches relating to these kinds of poisoning, arsenious acid may be found, either in a free state in the alimentary substances which have produced the poisoning, or in the matters vomited, in the stools, in the folds of the stomach or alimentary canal. Often, also, it must be looked for in the different organs of the animal economy, where it has been carried by absorption; this last case presents itself when death has followed and the body has been for some days buried.

When the arsenious acid is mixed with solid or liquid matters, it may in general be separated by mechanical means, by washing, or the employment of simple reagents. The body thus extracted is considered as arsenious acid when it is white, and when mixed with carbon, in a small narrow tube, and heated over an alcoholic lamp, it soon produces arsenic, which sublimes in a shining ring, with a metallic appearance. To characterise with still more certainty the sublimed arsenic, it must be taken from the tube and thrown on burning charcoal; it then gives off a garlicky odor, so fœtid and characteristic, that this experiment enables us to recognise the smallest quantity of arsenic. When the compound of arsenic is mixed with organic matters, or when it has been absorbed by the organs, it is indispensable, in order to isolate the poison, to destroy completely the organic matters, because they would mask the reactions proper for recognising the arsenic; they would even, in some cases, present characters which might be confounded with those of arsenic. To destroy organic matters, concentrated sulphuric acid is usually employed; we thus obtain a black char, which is treated with nitric acid, in order to dissolve the arsenic, which thus is transformed into arsenic acid, soluble in water.

After having carbonised the organic substances, and

treated the carbon with distilled water, the solution which contains the arsenic acid is subjected to the reactions which best characterise the arsenic, it is then that the apparatus invented by Marsh, the English chemist, is used; it is called the *apparatus of Marsh.*

The principle of the apparatus of Marsh is very simple, and rests on the following observation: If an arsenical compound is introduced into a flask containing water, sulphuric acid, and zinc—a mixture which disengages hydrogen—the arsenic combines with the hydrogen to form a gas called *arseniuretted hydrogen.*

If this gas is then lighted, and a cold body, such as a porcelain plate, is placed in the flame, the plate will soon be covered with a deposit of arsenic, easy to recognise by its metallic aspect. Again, the presence of arsenic can be proved by passing the arseniuretted hydrogen gas through a glass tube slightly heated; in a short time, the tube is lined with arsenic, proceeding from the decomposition of the arseniuretted hydrogen. This apparatus, in the hands of experienced chemists who know how to appreciate the results obtained, enables them to prove with certainty the slightest traces of arsenic, which might have escaped other methods of investigation.

CARBON.

Chemists give the name of *carbon* to a simple body which, in the state of purity, constitutes the diamond, but which may be black and opaque, and then form *coal,* which is used as a combustible.

Before particularly describing the properties of the principal kinds of carbon, we will first give the characters which are common to all the varieties of this body.

Properties of Carbon.—Carbon is solid, inodorous, infusible, and fixed.

Many of its physical properties, such as color, brilliancy, hardness, density, sonorousness, the faculty of conducting heat and electricity, are eminently variable; so that one might be brought to consider the diamond, graphite, lamp-black, anthracite, coke, charcoal, as bodies belonging to different species, which are, however, but varieties of carbon, since, like it, they combine directly with oxygen under the influence of heat, and produce carbonic acid, which is the characteristic property of carbon.

Carbon burns in oxygen, and in the air, better in proportion as it is lighter; however, in a current of pure oxygen, and under the influence of a high temperature, the hardest and most dense carbon, which is the diamond, burns freely.

Hydrogen, though it forms numerous combinations with carbon, is without direct action on this body. Sulphur, heated with carbon, distils without combining with it; but when the vapor of sulphur is passed over incandescent coal, these two bodies unite to produce a liquid known as *sulphuret of carbon.*

The properties which we shall treat of may be considered as the characteristic properties of carbon, and are exhibited in the different species. We shall now examine each of these species in particular, commencing with the diamond, which is carbon, pure and crystallized.

The Diamond.

The true nature of the diamond remained for a long time unknown.

In 1694, the experimental academicians of Florence found that the diamond burns in the focus of a burning-glass. This fact was confirmed by Francis Etienne, of Lorraine, who substituted for the action of the lens that of a powerful fire of the forge. From 1766 to 1776, many French chemists, and especially Macquer, pointed out that the diamond, kept from the contact of the air, resists the most intense heat. At the same epoch, Lavoisier and Guyton de Morveau remarked that the diamond, in burning in oxygen, produced constantly carbonic acid; from whence they concluded that the diamond ought to contain carbon.

The nature of the diamond was established by Humphrey Davy, who showed that this body gives, in burning, the same quantity of carbonic acid as pure carbon; that in this combustion it produces nothing but carbonic acid; and that, finally, the diamond, in burning in oxygen, does not make the volume of this gas vary. Davy concluded, from his experiments, that the diamond is pure carbon. The diamond is the hardest body known; it cannot be cut, except by its own dust; it scratches, on the contrary, all other bodies, even tempered steel. The diamond is used, in consequence of its great hardness, to form the pivots of fine watches, to polish fine stones, and to cut glass. The diamond is fixed and infusible, and a bad conductor of electricity. When insulated, it becomes very phosphorescent. It becomes electric by rubbing. Diamonds are generally without color, transparent and vitreous, but sometimes they present blue, yellow, rosy, or black tints.

The diamond is ordinarily found crystallized; its principal crystalline forms are octahedron, tetrahedron, and dodecahedron rhomboid; the faces of the crystals are (Fig. 13) often curvilinear. The fracture of the diamond is generally lamellated, from the facility and neatness of its cleavage. Lapidaries make use of this property to advantage in working it. The diamond possesses, to a high degree, simple refraction.

Newton, relying upon the property which combustible bodies possess of refracting light, was the first led to suspect the combustibility of the diamond.

Its refractive and dispersive power gives to the diamond, when cut, its beautiful effects of light. The numerous attempts made to obtain the diamond by artificial processes have failed. This body being *infusible and fixed*, it is conceived that the ordinary processes of crystallization, by fusion and volatilization, cannot be applied to it.

Liquid cast iron is the only body which dissolves carbon and deposits it on cooling; but the carbon which separates from it is graphite, a black and opaque body.

Geological inductions teach nothing on the mode of formation of the diamond; this body is always found in alluvials; it is disseminated through the ferruginous sands which constitute the ancient alluvials.

The cutting of the diamond, unknown among the ancients, was discovered, in 1476, by Louis de Berquem. It added to its natural brilliancy by multiplying the number of its facets.

The cutting of the diamond is executed by giving a rotatory motion to a horizontal plate of steel, covered with the powder of diamond (diamond dust) mixed with oil. The diamond is strongly pressed against this plate, and is thus freed from the rough particles which cover it. The diamond may be cut as *a rose*, or as *a brilliant*. The *rose diamond* has the under surface flat, and the upper elevated, *en dome*, without table, and presenting twenty-four facets. The *brilliant* differs from *the rose*, in that the under surface is cut like the upper, and is composed of symmetrical facets, which correspond with the superior part of the diamond. The brilliant presents, above, a facet called *table*, which is surrounded with many oblique facets, Fig. 13. Diamonds which have a greenish crust are those which possess the most beautiful *water* after cutting. M. Jacquelain has recently arrived at interesting results, by submitting the diamond to the action of strong heat, produced by the pile of Bunsen. Under this influence, the diamond softens, separates into many fragments, loses its transparency, increases in volume, becomes black, and changes into a carbon entirely comparable to coke, and lighter than the diamond. Thus modified, the diamond still scratches glass, but becomes sufficiently friable to be broken up between the fingers. M. Jacquelain thinks that certain black diamonds, called *Savoyard diamonds*, may have been produced under circumstances comparable to those which he has described.

Diamonds are principally found in the Indies, in the Isle of Borneo, and in Brazil. They are extracted by submitting the earth containing diamonds to the action of a current of water, on an inclined plane, composed of a table divided into compartments, in which are detained only the gravel and the diamonds, which are then separated by hand.

There are diamonds called *diamonds of nature*. These bodies are found in a coarse state, in a spheroidal form, and have no cleavage. It is impossible to cut them; they serve to make diamond powder.

The largest diamond known, is that of the Rajah of Matan, at Borneo. It weighs 300 carats, or more than 65 grammes (the carat is equivalent to 0·202 grammes.)

The *regent* diamond of the crown of France weighs 136 carats. It was bought for two and a half millions francs by the Duke of Orleans, then Regent, from an Englishman named Pitt. On account of the beauty of its form, and its perfect clearness, it is estimated to be worth double its cost.

Diamonds which can be cut sell for forty-eight francs the carat (230 francs the gramme), when they do not exceed this weight; when they are larger, their value increases considerably.

Graphite or Plumbagine—Black Lead.

Graphite is sometimes called *plumbagine*, or *black lead*. It contains 95 to 96 per cent. of pure carbon; it is crystalline, soft and unctuous to the touch, and soils the fingers and paper. It burns with as much difficulty as the diamond. Graphite is commonly crystallized in small tables, or in hexagonal spangles, marked with sufficient clearness. This body is found in the oldest transition formations, and chiefly in Bavaria, Piedmont, in the Pyrenees, and in England. Graphite was for a long time considered as a carburet of iron; but analysis has shown that pure graphite contains but a small quantity of iron, which often does not exceed $\frac{1}{200}$th; and the proportion of it varies with different species of graphite. It is now agreed to consider it as a variety of crystallized carbon. Artificial graphite may be obtained by allowing some kinds of melted iron, which are supersaturated with carbon, to cool slowly, and dissolving them in a mixture of chlorohydric and azotic acid. There remains in suspension in the liquor a crystalline body, of a metallic grey, identical with natural graphite.

Plumbagine is used in the arts, reduced to a fine powder, and mixed with oil; it is applied to the surface of castings of iron, which it colors grey, and preserves from rust. It is often used to relieve the friction of carriage axles, of wheels of machinery, and even of clocks. Plumbagine is also used for polishing gun-bullets.

In fine, plumbagine is used, either alone or with clay, to make black-lead pencils.

Anthracite.

Anthracite is a variety of carbon nearly pure, more brilliant than ordinary mineral coal, and blacker than graphite. It is found in Savoy, Saxony, Bohemia, England, and the United States. By its properties and composition, anthracite is intermediate between graphite and coal.

Anthracite burns with difficulty on account of its compactness, and does not kindle unless in large masses, and subjected to an elevated temperature. Isolated pieces go out almost immediately, and do not run together like the fragments of pit-coal. Anthracite decrepitates when heated; this circumstance has prevented the use of it alone in blast-furnaces, because the small fragments it produces in expanding, clog the furnace.

There are two varieties of anthracite: *vitreous anthracite*, and *common anthracite*. The first is purer than the second. Although the combustion of anthracite presents some difficulties, this body ought to be considered as a valuable combustible, which is of great service in the arts.

Lamp-Black.

Lamp-black is produced by the incomplete combustion of certain organic substances rich in carbon. When a piece of porcelain, or a metallic plate, is placed in the flame of a candle, a deposit of black from the smoke soon takes place on the body which cools the flame. Lamp-black is far from being pure carbon. It contains but about 80 per cent. of carbon, and is mixed with resinous matters, and different salts. It is much less impure when it has been strongly heated. Lamp-black is obtained by condensing, in brick chambers, or in large bags, the fumes which come from the incomplete combustion of resinous, bituminous, or fatty matters.

The black thus procured is used in painting. Intimately mixed with dry linseed oil, it constitutes printers' ink.

Lamp-black, mixed with two-thirds its weight of clay, forms the black crayons which are used in drawing.

Metallic Carbon.

The name *metallic carbon* is given to a carbonaceous residue which certain volatile substances deposit in passing through tubes of porcelain or iron heated to redness. This carbon is also produced in blast-furnaces, and in the manufacture of gas for illumination.

In this latter case, the gaseous combinations of carbon and hydrogen, which result from the distillation of coal, suffer a partial decomposition in passing through cylinders strongly heated, and produce metallic carbon. This carbon has often the brilliancy and the ringing sound of a metal. It is very hard, a good conductor of heat, and burns with difficulty.

Coke.

Coke is carbon produced in the distillation of coal. It is often called *refined coal.*

Coal, submitted to the action of heat, gives rise to volatile products, formed principally of water, tar, and gas, and leaves for a residue coke, which has the porous aspect of pumice-stone.

Its color is of an iron gray, with a half-metallic sound. It can be touched without leaving the marks of black on the hands.

Coke draws moisture from the air. In dry times it gives up a part of this moisture.

Coke does not easily burn, except in large masses; and, under the influence of a strong current of air, the incandescent fragments which are taken from the fire soon go out.

Of combustibles, this produces in burning, the greatest heat. In blast-furnaces, it gives results which cannot be obtained with charcoal. Coke weighs less than coal, but more than charcoal. The hectolitre of coke, in pieces, weighs from 40 to 50 kilogrammes.

Coke is used for domestic fires, but it is chiefly used for locomotive fires, and in the melting of metals. In the manufacture of iron, it is used in place of pit-coal, which cannot be employed in the working of blast-furnaces, on account of its easy fusion, and the large quantity of sulphur which it contains.

Coke being much less combustible than charcoal, the carbonization of coal is effected much more easily than that of wood. Coal may be distilled in cylinders, as in the manufacture of illuminating gas; but in this case, the manufacture of coke is only accessory. Coal is often carbonized by a method analogous to that followed in the forests for the preparation of charcoal. In some localities, and principally in the environs of St. Etienne, coal is carbonized in heaps, of a prismatic form, which are 15 to 20 metres long, 1 metre high, 2·50 m. at their base, and 1·75 m. at their upper part.

Finally, coke may be made in kilns of brick, of various forms, which permit the black which forms in the incomplete combustion of the coal to be collected. The smoke is made to pass into a series of arched chambers of brick, were the black is deposited.

As a mean, 100 parts of coal furnish 50 to 60 parts of coke.

*Charcoal.**

Charcoal is the fixed residue left from the distillation of wood, or from its incomplete combustion. Wood, dried in the air, shows about the following composition:—

Carbon, ...	38·5
Water in combination, . .	35·5
Ashes, ...	1·0
Free water, ...	25·0
	100·0

It is thus seen that if, by distillation, the wood could be decomposed into water and carbon, there would be 38·5 per cent. of carbon. But, during the distillation, we cannot avoid producing carburetted hydrogen gas, carbonic oxide, tar, and acetic acid, all which bodies contain carbon; so that

* The object of coking, or making charcoal, is because the natural moisture, and the water formed by the combustion of the wood, absorb much of the heat, and thus prevent the attainment of the high temperature required in furnaces.

the most perfect methods give but 27 to 28 per cent. of carbon. The processes ordinarily used in forests give but 17 to 18 per cent.

Charcoal is made by two different processes. The first, which is the most usual, is done in the open air, and is called the *process of the forests*. In the second process, distilling-vessels are used,by which not only the charcoal is collected, but also the condensed volatile products, rich in acetic acid, and the spirit of wood, which are formed during the distillation of wood. We shall describe here the process of the forests.

Carbonisation in mounds.—Charcoal is ordinarily made in the forests, by a process which is called *carbonisation in mounds*, or *carbonisation of the forests*.

In the process of carbonisation in mounds, three or four uprights, which form a chimney of about 0·30 m. in diameter, are placed in the centre of a plane, of a circular area. Around this, the billets of wood are arranged upright, in a circle. The large pieces are in the centre, the small ones on the exterior. The mound is covered with leaves and dirt. It is lighted by uncovering the chimney, and throwing into the centre of the mound some ignited coals, which are covered with small wood. At the base and exterior of the mound holes are made, which remain open during the carbonisation, for the introduction of air necessary for the operation. The chimney is left open for some hours, to ensure the combustion in the centre, and it is supplied from time to time with light wood, so as to form in the centre of the mound a mass of charcoal. When the combustion is sufficiently active, which varies according to the size of the mound, the chimney is stopped, and then it is left to itself for some hours. Whitish fumes then escape from the surface over the upper part, which begins to settle down. Vent-holes are made in the covering, towards its upper part. A volume of white smoke escapes for some hours; this then becomes bluish and almost transparent, which indicates that the carbonization is done at this place. New vent-holes are made 0·30m. to 0·40m. below the first, and so on, till we get near the holes at the base of the mound, which always remain open. During this operation, the mound settles down considerably, and the wood is converted into charcoal.

General Properties of Charcoal.—Charcoal, made from hard wood, is dense; and very light when made with white wood. Charcoal always keeps the form of the wood from which it is made.

Though friable, charcoal is very hard. It is often employed for polishing copper and bronze.

The combustibility of charcoal varies with its density. Oak coal, which is very dense, kindles with more difficulty than that of black alder, which is very light. The latter is preferred for the preparation of gunpowder.

The method which has been used for charring the wood, also influences the combustibility of the coal; and it may be said that coal prepared by distillation, is always lighter and more combustible than that which has been made in mounds. Charcoal does not begin to burn below the temperature of 240° C.; but at the moment it is taken out of the mound, it is often pyrophoric. Introduced into store while it is warm, it sometimes kindles spontaneously. This inflammability is owing to the power which charcoal possesses of absorbing atmospheric air on disengaging heat.

Charcoal decomposes water at a red heat. The hydrogen of the water is liberated, and its oxygen unites with carbon to form carbonic oxide and carbonic acid. This decomposition is effected in a porcelain tube filled with live coals, which have been first freed, by heat, of gases which they may contain. This tube communicates on one side with a small retort filled with water, and on the other with a glass tube, suitable for collecting the gas, Fig. 6. While the coal is incandescent, the water in the retort is made to boil, and the vapor, in passing slowly over the charcoal, gives rise to gases, which are collected in receivers. Thus water, in its contact with the red hot coal, will give rise to inflammable gases, which are principally hydrogen and carbonic oxide. It is from this cause that the combustion of charcoal is rendered more active by sprinkling it with a small quantity of water; this property is well known to smiths. The proportion of ashes which charcoal leaves, in burning, varies with the quality of the wood from which it is made.

Charcoal is very porous. It absorbs, in cooling, a great quantity of gas and vapor of water. Ordinary charcoal, exposed to the air, contains from 10 to 12 per cent. of water; it is a bad conductor of heat and electricity. By calcination, it loses the gases which it has absorbed, and is transformed into *braise*, or fine coals, which conducts heat and electricity. In this state, it is used to cover the lower extremity of lightning-rods, to facilitate the escape of the electric fluid into the soil.

When two pieces of coal, calcined and cut to a point, are made to communicate with the two poles of a strong pile, if the extremities of the coals are made to approach gradually, there is produced a light whose brilliancy may be compared to that of the sun: this light is no stronger in the air than in vacuo. The experiment is ordinarily made in a glass vessel in which a vacuum has been first made. The point of the carbon, at the negative pole, is seen to be hollowed out; while that at the positive pole is covered with a carbonaceous deposit, which seems to indicate that there is a transfer of the carbon from one pole to the other.

Charcoal has the property of absorbing coloring-matters; it can even cause the separation of a great number of inorganic bodies. It is probable that the bodies absorbed by the carbon adhere to it, and fix themselves to its surface, like the mordants and the coloring-matters to the surface of tissues. The organic matters which have been absorbed by the charcoal may be withdrawn, without having undergone any modification.

The decolorizing properties of charcoal are developed in proportion to its porosity; these are particularly seen in *animal charcoal*, of which we shall speak further on.

Charcoal has also the property of absorbing gases, without, however, combining with them.

This is easily shown by extinguishing, in mercury, pieces of incandescent charcoal, and then letting them ascend into receivers where there are different gases. The gases are absorbed with great rapidity, particularly ammonia and chlorhydric acid.

Theodore de Saussure has shown that the absorption varies with the nature of the gas. It may be said, in a general way, that the gases which are absorbed in the greatest quantity by charcoal, are the most soluble in water.

The absorbed gases are disengaged when the coal is subjected to the action of a vacuum.

The absorbing property of charcoal has been made use of in the arts. On account of this property, charcoal is used as a disinfectant, and to preserve animal matters from putrefaction.

Meats may be preserved for a long time from putrefaction, by covering them with charcoal-dust, which is an excellent antiseptic. Some physicians advise the use of charcoal in the treatment of ulcers.

Charcoal powder has been used with success to disinfect fœcal matters, and convert them into *poudrette*, which may be used at once in agriculture.

The decolorizing and disinfecting properties of charcoal may be used with advantage to purify the most putrid and dirty waters. With this view, it is used in fountains, or charcoal filters. But, as the water, in passing through the charcoal, loses the air which it holds in solution, it is important, before drinking it, to aerate it, by agitating it for some time in contact with the air.

Carbon is entirely unalterable. Thus, Chinese inks, and black paintings, which have charcoal for their basis, are considered indelible. At Herculaneum and Pompeii, manuscripts have been discovered, the black writing of which was perfectly visible. This ink was composed of lamp-black mixed with gum-water.

Charcoal does not alter in moist ground. Acting on this property, wood which is to be placed in moist ground, is charred, for posts and piles.

Animal Charcoal.

The name *animal black*, or *coal*, is given to a mixture of finely-divided carbon and earthy salts, produced by the calcination of bones in close vessels.

This body possesses, in a high degree, the property of decolorizing, which it owes, without doubt, to the great division of the carbon which it contains, for charcoal, compact, and not porous, exercises no action on coloring-matters.

Animal black is chiefly used in the refining of sugar. It is met with, in commerce, in powder and in grains. When it has been used for a certain time for decolorizing syrup, it loses its decolorizing properties. It is revivified by boiling with water acidulated with chlorhydric acid, then with water, and afterwards calcining it, either alone or with bones, in kilns such as are used for making new black. The chlorhydric acid used in the revivification of animal-black, dissolves the lime which has been absorbed by the black in the decolorization of the syrups. Some refiners, in revivifying the black, first leave it to itself for some weeks, when a fermentation is developed in the mass, attributed to the presence of sugar and organic matters. The coal is then calcined. Animal-black may be revivified repeatedly; but, as there is always a quantity of dust formed at each operation, the revivified black is passed over a sieve, which retains the coarse part, and allows the fine to pass through. Often the black is revivified by simply subjecting it to the action of steam, which is passed over it in red-hot cylinders of iron. It may also be revivified by boiling in water slightly alkaline, which dissolves the coloring viscous matters which it has absorbed; it is then washed with acidulated water.

Preparation of Animal-black.—It is ordinarily prepared in large kilns, in which are placed iron vessels, filled with the bones to be burnt.

A small quantity of fuel is sufficient to commence the operation; in fact, as soon as the disengaged gases take fire, the heat is sufficient for the calcination to go on of itself.

When the organic matter is decomposed, the fire is withdrawn from the vessels; and, on cooling, the black is obtained in the shape of the bones used. This is then broken up in a mill, and passed over a sieve. Thus the coarse and fine black is produced.

Combinations of Carbon with Oxygen.

Carbon combines with oxygen in several proportions. We shall here speak of carbonic oxide and carbonic acid.

Carbonic Oxide.

This gas was discovered by Priestly; its true nature was established by Clément Désormes.

Carbonic oxide is a gas without color, taste, or smell, of a density of 0·967; completely neutral, and slightly soluble in water. It is combustible, and burns with a characteristic blue flame, producing carbonic acid. It was for a long time supposed that this gas exercised but little action on the animal economy. But the researches of M. Leblanc show, on the contrary, that this gas is very deleterious, and that an atmosphere containing $\frac{1}{100}$ of it will kill a small bird. The knowledge of this fact is highly important in a hygienic point of view. In fact, carbonic oxide is generated in fire-places whenever there is too much coal; and, if the products of combustion should escape into the room, either from the flues being in an improper condition, or if the damper of a stove should be kept closed, the carbonic oxide produces pains in the head, vertigo, and asphyxia, which were formerly ascribed, improperly, to carbonic acid.

Oxygen, under the influence of heat, transforms carbonic oxide into carbonic acid. Some oxides are reduced by carbonic oxide — oxide of iron, for example. It is chiefly on this property is based the metallurgy of iron.

Preparation.—Oxides difficult to reduce give rise to carbonic oxide when heated with charcoal, while those easy to reduce give rise to carbonic acid. Acting on this, carbonic oxide may be produced by heating to redness, in an earthen retort, charcoal with oxide of zinc. It is thus that Priestly discovered this gas. Carbonic oxide forms in fire-places when there is a deficiency of air. The blue flame which is seen in the upper part of a covered furnace proceeds mostly from the combustion of carbonic oxide. Carbonic oxide may be easily prepared by passing carbonic acid gas over coals heated to redness in a porcelain tube. The process usually employed in laboratories consists in decomposing, in a small matrass, oxalic acid, or the binoxalate of potassa (salt of sorrel), by an excess of monohydrated sulphuric acid. Take one part of oxalic acid, and five parts of concentrated sulphuric acid. Oxalic acid cannot exist in the conditions of the experiment in the anhydrous state; when it is heated with concentrated sulphuric acid, which dehydrates it, it is decomposed into equal volumes of carbonic oxide and carbonic acid. The carbonic acid is absorbed with potassa, and the carbonic oxide remains perfectly pure. A like washing with potassa is necessary, in most cases, when carbonic oxide is prepared, because it is seldom it is formed unmixed with carbonic acid.

Carbonic Acid.

The name *carbonic acid* is given to a gas composed of carbon and oxygen. It exists in the air, and in all waters in contact with the air, in wells, in the galleries of coal-mines, and in a great number of grottos and caves.

Fermentation, combustion, the spontaneous decomposition of organic matters or that which results from the action of heat, and the respiration of all animals, throw into the atmosphere considerable quantities of carbonic acid, which vegetables incessantly decompose, under the influence of light, appropriating to themselves the carbon, and restoring the oxygen to the air. This decomposition of carbonic acid by the green parts of plants, explains the invariableness of the composition of the atmosphere, and at the same time the small quantity of carbonic acid which exists in it. Carbonic acid is found in combination with most metallic oxides, forming marbles, chalk, marls, carbonates of barytes, strontian, iron, copper, &c.

We have to examine the acid in three states, *gaseous, liquid*, and *solid.**

Gaseous Carbonic Acid.

Carbonic acid gas is colorless, of a slightly acid taste, and a pungent smell. It is about one and a half times heavier than the atmosphere. It gives to the blue tincture

* In the soluble carbonates, the alkali is not neutralized; and many carbonates of the bases, especially of ammonia, may be obtained, in all of which the properties of the alkali predominate. It would appear from this that carbonic acid is not a true acid, although it combines with bases.—*Gregory's Hand-Book*, p. 155.

of litmus a slight tinge of red, which disappears on exposure to the air, or by boiling the solution, because, under these influences, the carbonic acid is disengaged. The strongest heat does not change this gas, which, however, is decomposed, by a series of electric sparks, into oxygen and carbonic oxide; a curious phenomenon, inasmuch as under the influence of electricity, oxygen and carbonic oxide unite, and form carbonic acid. The density of this gas being much greater than that of the air, it can be poured from one receiver into another, as readily as a liquid. This fact explains many curious phenomena. Thus, at Pouzzoles, near Naples, in the Grotto du Chien, animals of low stature perish in a few minutes, while men can go into it without danger; the strata of carbonic acid contained in the interior of the grotto not rising higher than about a metre and a half, these animals suffocate, while man escapes.

The phenomenon which takes place in the Grotto of Pouzzoles, may be produced artificially by plunging into a receiver of carbonic acid a close cylinder, which will drive out a certain quantity of the carbonic acid, which, when the cylinder is withdrawn, is replaced by an equal volume of air. In this way two different atmospheres are obtained, which do not mix for a considerable time; one is formed of air, the other of carbonic acid gas. A candle which burns in the first is extinguished in the second.

Carbonic acid may cause asphyxias in cases which, unfortunately, are not sufficiently known. Thus, a vat filled with grape-juice in fermentation, placed at the entrance of a cellar, may disengage sufficient gas to asphyxiate people who might be inside the cellar. In such a case, if we had to rescue a person from such a place, we should first throw in some ammoniacal water, which, uniting with the carbonic acid, would neutralise its action on the economy.

The cellars in the neighbourhood of Paris, certain wells,

or other excavations, often become filled with carbonic acid, the result of the decomposition of organic matters.

Water dissolves about its own volume of carbonic acid at the ordinary pressure; but this solubility increases considerably with pressure. By compressing a mixture of carbonic acid and water, a liquid is obtained which contains five or six times its volume of carbonic acid. This has been applied to the preparation of waters called *gaseous*, and particularly to artificial Seltzer or mineral waters.

These gaseous waters are made by two different processes. The first is a continuous process, in which a suction and force-pump draw from separate reservoirs water and carbonic acid, to force them again into a tight apparatus. The second process is intermittent, the carbonic acid being produced in the apparatus where the saturation is made, and is dissolved in the water by the pressure which it exercises on this liquid. In these two cases, the carbonic acid is produced by the action of sulphuric acid on chalk.

Water, charged with carbonic acid, loses all the gas which it contains, when it is heated, or exposed to the ordinary temperature in vacuo. It loses it, also, but slowly, when it is left to itself in contact with the air.

The property which carbonic acid possesses, of precipitating lime water, is often used to show the presence of this acid in water, or the gases. Carbonic acid, placed in contact with an excess of lime-water, forms with it a white flocculent precipitate, insoluble in water; very soluble, on the contrary, in chlorhydric, nitric, and acetic acids. This precipitate, which is carbonate of lime, being soluble in carbonic acid itself, requires for its formation an excess of lime-water, which is shown with a red paper of litmus or turnsol, which should become blue. Without this precaution, carbonic acid might escape detection, just in proportion as its quantity is considerable.

Carbonate of lime, dissolved in water by means of an

excess of carbonic acid, separates when this gas is disengaged under the influence of heat, by the contact of bodies in minute division, or by the action of the air alone. It is thus are produced calcareous deposits in steam-boilers, and in water-pipes.

Preparation of Carbonic Acid Gas.—Carbonic acid is prepared, 1st, by burning carbon in an excess of air or of oxygen; 2d, by calcining carbonate of lime, which loses its carbonic acid, and becomes caustic lime; 3d, by decomposing carbonate of lime by an acid. In this case, the acid, acting on the carbonate, displaces the carbonic acid gas.

This last process is generally used in laboratories. Carbonic acid gas may be collected over water or mercury.

Liquid Carbonic Acid.

Carbonic acid was first liquefied by Mr. Faraday, by decomposing, in a tube of glass closed at both ends, a carbonate by concentrated sulphuric acid. This chemist found that, at the temperature of 0° C., carbonic acid liquefies under a pressure of 36 atmospheres.

This mode of liquefaction has the double inconvenience of being dangerous for the operator, and of giving but small quantities of liquid carbonic acid.

M. Thilorier, a few years ago, made an apparatus with which several kilogrammes of liquid carbonic acid can be made at a time.

The principle of the apparatus of M. Thilorier is the same as that of Mr. Faraday, except that the carbonate is decomposed in a cylinder of iron, which will bear an enormous pressure. M. Thilorier produced liquid carbonic acid by decomposing the bicarbonate of soda with sulphuric acid.

Properties of Liquid Carbonic Acid.—Liquid carbonic acid is colorless, very soluble in alcohol, ether, and the essential oils; it does not mix with water.

Liquid carbonic acid is very dilatable: in passing suddenly from the liquid to the gaseous state, it produces the extraordinary cold of about 100° C. below zero.

By throwing a jet of liquid carbonic acid into a metallic box pierced with holes, the vessel is seen to fill, almost entirely, with a white flaky matter like snow, which is solid carbonic acid, produced under the influence of the great cold which a part of this acid makes the other part undergo, in passing from the liquid to the gaseous state.

Solid Carbonic Acid.

Carbonic acid solidified as we have explained, keeps for some time in the open air, without being subjected to any pressure.

Solid carbonic acid exists at a temperature of 90° C. below zero, and yet does not produce, on the organized tissues, a frigorific effect as great as one would suppose, no doubt owing to its porosity, and above all to the gaseous atmosphere which surrounds it. The intensity of the cold produced by solid carbonic acid is increased by mixing it with ether. This mixture will congeal, in a few seconds, four times its weight of mercury.

Solidified mercury has the appearance of lead. M. Thilorier was able to make of it pieces of money, medals, &c., and to keep them for some time, in a mixture of ether and solid carbonic acid.

The effect produced on the animal tissues by a mixture of solid carbonic acid and ether is like that of a burn. The fluids become solidified, the blood coagulates and becomes completely hardened; an active inflammation soon shows itself in the organ subjected to the influence of this excessive cold. Mr. Faraday has still further increased the cold produced by this mixture, by putting it under the receiver of an air-pump.

By placing in this mixture tubes of glass or copper, in which gases may be compressed to 40 atmospheres with a force-pump, Mr. Faraday has produced liquefactions and solidifications of gas which could not be produced by other methods.

Combinations of Carbon with Hydrogen.

The combinations of carbon with hydrogen are very numerous. Many essential oils, such as the essence of rose, of citron, of turpentine, &c., and caoutchouc, are formed entirely of carbon and hydrogen.

When organic substances decompose slowly, or are subjected to the action of fire, they disengage gaseous compounds, formed also of carbon and hydrogen; these bodies are known under the names of *proto-carburet* and *bicarburet* of hydrogen.

The proto-carburet of hydrogen, or the *gas of marshes*, arises during the spontaneous decomposition of a great number of organic matters.

Muddy or stagnant waters, when they are stirred up, disengage gaseous compounds, in great part, of proto-carburet of hydrogen, mixed with nitrogen, oxygen, and carbonic acid.

Proto-carburet of hydrogen is found in the galleries of coal-mines, where it is mixed with air and bicarburet of hydrogen, forming an explosive mixture. It is to its presence, especially, that are to be attributed the *fire-damps*, which occasion such serious accidents in mines. Proto-carburet of hydrogen is found in a state of compression, more or less great, in certain specimens of mineral salt, from which it is separated by the action of water, giving rise to a *decrepitation*.

This gas escapes spontaneously from the earth in some places, where it is sometimes employed as a combustible; it also exhales in considerable quantities from the craters of some volcanoes.

The organic matters which, by burning, produce the

greatest quantities of carburetted hydrogen, are coal, fat, and resinous bodies. All these substances are rich in hydrogen and carbon.

The bicarburet of hydrogen, like the proto-carburet, forms in the distillation of most organic bodies; it is this which, mixed with the proto-carburet, constitutes, in great part, gas for illumination.

The bicarburet of hydrogen is obtained pure by heating a mixture of alcohol and concentrated sulphuric acid. The bicarburet is colorless; it burns with a bright white flame. With oxygen or atmospheric air, it forms a mixture which explodes with great force under the influence of ignited bodies.

In this case, the hydrogen contained in this gas unites with the oxygen to form water; while the carbon, also combining with the oxygen, forms carbonic acid.

These detonating mixtures are very dangerous, and are produced whenever illuminating gas becomes mixed with the air of a room.

This gas is also used for inflating balloons.

It unites with chlorine to form an oily substance. This reaction has given it the name of *olefiant gas.*

General Remarks on illuminating by Gas.

An organic matter, subjected to the action of a high temperature, is decomposed into carbon, and volatile or gaseous substances more or less combustible, whose flame is sometimes very brilliant. It is to Lebon, a French engineer, that are due the first experiments on lighting by gas. He published a memoir having for title—"*Thermo-Lamps, or stoves which heat, light with economy, and offer, with several valuable products, a motive power applicable to every kind of machine.*"

Lebon, in 1785, thought to distil wood in close vessels, to extract from it, on one side, carbon and acetic acid, and on the other, gas fit for illumination. He demonstrated that coal is better than wood for making illuminating gas; but notwithstanding this observation, twenty-five years passed before coal-gas was used in the arts.

The first manufactories of coal-gas were established in England by Murdoch.

Attempts were made to replace coal by resins, fat oils, oils of shists,* and by mixtures of tar and steam exposed to a red heat on surfaces of coke, &c. None of these processes could sustain a competition with the distillation of coal. Coal is, in fact, abundant at a low price, and gives two products, of which one is the illuminating gas, and the other coke, one of the best combustibles known.

Nevertheless, even now, in some gas-works, gas is made from oils or resins, or by passing a mixture of tar and steam, over coke heated to redness (*gas of Selligue*).

Gas from oil or resin is brighter than that from coal; it owes this property to the presence of a much larger proportion of bicarburet of hydrogen, and other volatile carburets of hydrogen. Besides, it does not contain sulphuretted hydrogen, sulphuret of carbon, nor ammonia, which are ordinarily found in coal gas badly purified.

The distillation of coal furnishes gas, the composition of which varies with the temperature to which the coal has been exposed.

At the commencement of the distillation, the gas is very rich in bicarburet of hydrogen, and in consequence, very brilliant; the proportion of this gas diminishes as the operation progresses, and at the end the gas contains a considerable quantity of hydrogen and carbonic oxide, which are but slightly illuminating. The gas always contains com-

* The bituminous slate-marl of Autun affords, when distilled, about 10 to 20 per cent. of oily products, two-thirds of which consist of a light oil for the production of gas.—*Knapp*.

bustible vapors, which increase its illuminating power. These vapors, which do not condense entirely in the purifying and the conduit-pipes of the gas, are principally formed of different carburets of hydrogen.

The sulphur, of which the greater part is contained in coal in the state of pyrites, transforms itself, during the distillation, into sulphydric acid. This acid is readily absorbed by the lime in the purifiers. The purity of the gas is tested with a paper impregnated with acetate of lead, which remains colorless when the gas is pure, but becomes black when the gas contains sulphydric acid. The sulphuret of carbon forms only in small proportion, and without doubt condenses with the liquid products; it is rarely met with in coal-gas.

Illumination with coal-gas is at this time an object of great importance. We will briefly point out the details of this manufacture.

This gas is produced when bituminous coal is heated in retorts.

The gas-works use, in general, the coals called *demi-grasses* in preference to those called *grasses*, which are difficult to distil, and besides give tar in great abundance. Poor (*maigre*) coals leave a coke which does not run together well, and give besides but little gas.

The coal-basins of Anzin, Douchy, of Commentry and Mons, supply the gas-works of the north of France, as well as those of the upper and lower Seine. The vein of Mons resembles the *cannel-coal* of Lancashire, which is the most valuable coal in England for the manufacture of gas.

The vein of Mons is inferior to cannel-coal, as to the quality and quantity of gas which it produces; but it leaves a coke less burnt, and more merchantable.

The basin of St. Etienne furnishes coals of excellent quality for the manufacture of gas, which supply the works of the centre and east of France.

The city of Paris has now eight gas-works, which have been successively established since 1818. They had, at their organization, and also for pipes through the streets, a capital of more than 30 million francs. The pipes are laid through a route of more than 450,000 metres. They are made of cast-iron; for some years past, they have also used pipes of sheet-iron, covered with a thick coating of bitumen.

The number of retorts used in the eight gas-works has been put down at a mean of 1200; their capacity varies from one hectolitre to one and a half, and rarely reaches two hectolitres; they are made of cast-iron, or of fire-clay.

The earthen retorts present incontestible advantages over the others. They are less costly, more durable, and the coal distilled in them furnishes a superior gas to that made in metallic retorts.

The coal intended for the manufacture of gas is first broken up, then introduced into the retorts, which are about half filled. This precaution is necessary to allow of the free development of the coke, the volume of which is about one and a third, or one and a half times that of the coal which produces it.

When coal of a good quality is used, the operation generally lasts four hours. It may be estimated that, in a well-conducted distillation, 100 kilogrammes of coal will furnish 25 cubic metres of gas to the gasometer. This quantity of coal is about that ordinarily distilled. A single retort can thus yield, in 24 hours, 150 cubic metres of gas. The retorts are heated and kept to a cherry red, either with coke or tar. The distillation of one hectolitre of coal requires 75 litres of coke.

The gaseous products are conducted through large subterranean pipes, communicating at different distances with cisterns in which are condensed ammonia, ammoniacal salts, water, tar, &c.

When the condensation is complete, the gas is received in the apparatus intended for its purification. The purifiers, the form of which is variable, are ordinarily cases of iron of two and a half to three cubic metres.

The gas runs through three purifiers, and then passes out by a pipe. It is received in the gasometer for distribution to the different parts of the city.

The mean pressure to which the apparatus is subjected is 30 lines. The pressure can be increased or diminished by giving more or less opening to the valves placed in the pipes of exit.

The dimensions of the gasometers vary with the size of the gas-works. They are, however, seldom required of greater capacity than from 70,000 to 80,000 hectolitres. The quantity of gas consumed in Paris, in 1846, was estimated at 25 millions of cubic metres, which were produced by about 100,000 tons of coal. These 100,000 tons furnish 60,000 to 65,000 tons of coke. About one-third of this quantity was consumed for the distillation of the coal. 40,000 tons of coke were put in the market, and mostly applied to domestic uses.

85,000 gas-burners supply Paris for public and private use. The price per burner is six centimes per hour. Each one consumes, on an average, 120 litres of gas per hour, and makes a light equal to about one and a half times that of a Carcel lamp. The price of the cubic metre of gas sold is 43 centimes (1852). The ordinary burners are pierced with 20 holes of the size of a third of a millimeter; the height of the flame is 8 centimetres; that of the glass chimney ought not to exceed 20 centimeters.

The quantity of ammoniacal salts resulting from the condensed waters, may be estimated at more than 100,000 kilogrammes. A more complete purification would increase this quantity still more. It was thought that the tar from gas-works could be applied to the same uses as asphalte and bitumen; but it cannot be prevented softening at a moderate temperature, so that it is used almost exclusively for heating the retorts.

The proportion of tar produced by the distillation of coals, varies with their quality; on an average, it amounts to 4 or 5 per cent. of the weight of the coal.

Endeavors are made to keep down as much as possible the formation of tar, because it is formed at the expense of the quantity of gas, and its power of illumination.

CYANOGEN, PRUSSIC ACID, AND CYANURETS.

The discovery of cyanogen, which is due to Gay-Lussac, is justly considered as one of those which has exercised the most influence on the progress of chemistry.

Gay-Lussac proved that nitrogen and carbon could combine together to form a new body which he named *cyanogen*, and that this compound, in all its relations, acts as a simple body. Cyanogen, like chlorine, forms a hydracid which is cyanhydric, or prussic acid; it combines, also, with oxygen to form acids. It unites with the metals to form cyanides, which may be compared with chlorides.

Cyanogen is formed when azotized organic substances are burnt with potash; thus fibrine, gelatine, skins, &c., heated with potash, produce cyanide of potassium.

Cyanogen was obtained in a state of purity, by Gay-Lussac, by heating cyanide of mercury, which, under the influence of heat, separates into mercury and cyanogen, which last is disengaged.

Cyanogen is gaseous; its odor is penetrating, and irritates the eyes; it is combustible, and burns with a blue, characteristic flame. Combining with hydrogen, it forms cyanhydric acid, often called *Prussic acid.* Cyanhydric acid is one of the most poisonous bodies known; two or three drops being sufficient to produce death.

Cyanogen unites with the metals to form *cyanides*, many

16

of which are used in the arts. We name the cyanide of potassium, which is used in electric gilding, to dissolve the cyanides of gold and silver. The ferro-cyanide of potassium, which may be considered a double cyanide of potassium and iron, is used in dying. It precipitates most metals from their solutions, and is used as a reagent in chemical laboratories. Prussian blue, which is cyanide of iron, is prepared by precipitating a salt of the peroxide of iron by the ferro-cyanide of potassium, and is much used as a blue coloring-matter.

COMBINATION OF BORON WITH OXYGEN.

Boracic Acid.

Boracic acid is a compound of boron and oxygen, which exists in nature. This body presents itself in lamellated, colorless crystals, without smell, of a slightly acid taste; it feebly reddens litmus.

Boracic acid is found in Tuscany, in the small lakes, called *lagunes.* Small craters (soffioni), in the bottom of these lagunes, discharge vapor with boracic acid. This boracic acid dissolves in the waters of the lagunes, and is reduced to a proper degree of concentration by heat; after this, the solution is allowed to cool, and boracic acid crystallizes. The solutions of boracic acid are evaporated by using the heat resulting from the condensation of the vapors of the soffioni; this vapor is conducted through passages of stone, under the evaporating pans. The boracic acid crystallized, and still moist, is first placed in willow baskets, where it drains, and then in brick dryers, likewise heated by the vapors of the soffioni (Fig. 16).

Boracic acid is also extracted from borate of soda dissolved in water; this salt is decomposed by a slight excess of sulphuric acid, and the boracic acid which precipitates is then purified by crystallization. Boracic acid, extracted from borate of soda by sulphuric acid, generally retains a small quantity of this last acid. To purify it, it is washed with cold water, until the water no longer forms a precipitate with a salt of barytes, mixed with weak nitric acid.

Uses.—Boracic acid is used in medicine under the name of *sedative salt of Homberg;* it is used for preparing borax (borate of soda). It enters into the composition of some glasses, that of paste, and of the glazing of common potteries.

COMBINATIONS OF SILICIUM WITH OXYGEN.

Silicic Acid, or Silex.

This is one of the most wide-spread bodies in nature. It forms part of all primitive rocks, of clay, of soils of different formations, of the gangue of a great number of minerals, and of nearly all precious stones. It is met with in small quantities in the ashes of nearly all vegetables.

Some waters contain silex in solution.

Anhydrous Silicic Acid.

Anhydrous silicic acid is white, tasteless, without smell, infusible in the fire of the forge, but, as shown by M. Gaudin, capable of being melted by the oxy-hydrogen blowpipe, and drawn out in very fine threads.

Anhydrous silicic acid, after having been heated to redness, is completely insoluble in water, and the acids. Fluorhydric acid alone attacks it, and transforms it into water and fluoride of silicium. This last property is one of the most characteristic of silex and the silicates.

Hydrogen, carbon, phosphorus, chlorine, and the metals, are without action on silex.

Certain metals, particularly iron, reduce silex in presence of carbon, and form carbonic oxide, and a metallic siliciuret. Thus, when a mixture of oxide of iron and silex are melted in a clay (brasqué) crucible, a residue of iron is obtained, in which the proportion of silicium amounts to 5 or 6 per cent. of the weight of the iron.

Silicic acid is a very feeble acid, but, on account of its fixedness, it can expel from their combinations the most energetic acids; it is thus that, by heat, it decomposes the sulphates.

When silex in powder is thrown on carbonate of soda, kept in a state of fusion, a brisk effervescence takes place in the mass, and carbonic acid is set free.

Potash, soda, and barytes react, when hot, on silex free or combined. The two first bases form with silex, silicates, which are attacked by the acids; this property is made use of to render a great number of mineral substances soluble in the acids.

There are found in nature numerous varieties of silex, bearing the names of *opal*, *rock-crystal*, *agate*, *flint*, *mill-stones*, *tripoli*, *sand-stone*, &c.

Opal is silex, containing 10 per cent. of water. Rock-crystal, so called on account of its limpidity and transparency, is pure anhydrous silex; it crystallizes in hexagonal prisms, terminated by hexagonal pyramids. Rock-crystal, or *hyalin quartz*, becomes electric by friction, it strikes fire with steel, and is sufficiently hard to scratch glass, and even steel.

Quartz is of different colors. When it is colored a clear yellow by the peroxide of iron, it takes the name of *false topaz*, and that of *amethyst*, when it is colored violet by the oxide of manganese.

Agate quartz is concreted, and often presents strata of different colors. The white agates, or those of a pearly, translucid gray, are called *chalcedony; cornelians* are of a blood red, and waved.

Gun-flints are met with in irregular tuberculous masses, of a conchoidal fracture, and great hardness. They are silex, containing two per cent. of water, and one per cent. of alumine. They are almost always covered with a white layer, composed of disaggregated silex and carbonate of lime. Silex enters into the composition of most potteries.

Tripoli is earthy silex, in very fine grains, united together by the force of adhesion, assisted by compression.

Sandstone is a quartz sand, agglutinated by a calcareous or silicious cement.

METALS.

GENERALITIES ON THE METALS.

The metals are solid at the ordinary temperature, with the exception of mercury, which is liquid.

Most metals possess a characteristic brilliancy, which they lose when brought to a state of minute division. Their powders, which are ordinarily black or gray, become brilliant again when rubbed with a hard body.

Metals, taken in mass, are all opaque; but the light passes through them, if they are reduced into leaves of an extreme thinness. It is thus that gold-leaf appears green, when placed between the eye and the light.

The ordinary color of metals is a grayish white; gold, however, is yellow, and copper of a peculiar red.

They are, in general, without smell; however, tin, copper, iron, and lead, exhale a disagreeable odor, particularly when they are rubbed with the hand. Some metals have a peculiar and disagreeable taste, as iron and tin. The metals, with the exception of sodium and potassium, are heavier than water; hammering ordinarily increases their density.

The hardness of metals is variable: some, as lead and tin, are very soft; others, as iron and antimony, are very hard. The presence of small quantities of carbon, arsenic, and phosphorus, increase in general their hardness. Metals possess the property of ductility, that is, of being elongated into wires by being passed through the drawing-plate.

Malleability is the property they possess of being hammered or rolled out into thin sheets or leaves. Metals or their alloys, which have undergone the action of the hammer, the drawing-plate, or the roller, become nearly always hard and brittle; to continue to reduce them into wires and sheets, they must be annealed from time to time, and allowed to cool slowly. Metals have different degrees of ductility and malleability. We will here class the principal metals in the order of their ductility and malleability.

ORDER OF DUCTILITY.	ORDER OF MALLEABILITY.
Gold,	Gold,
Silver,	Silver,
Platinum,	Copper,
Iron,	Tin,
Copper,	Platinum,
Zinc,	Lead,
Tin,	Zinc,
Lead.	Iron.

Malleability and ductility are in general increased by heat.

Tenacity is the power which prevents their rupture. This property differs in different metals.

The tenacity of metals is compared by finding out what weight will break wires of the same diameter, but of different metals.

Metallic wires of two millimeters diameter break with the following weights:

Iron................	249·159	kilogrammes.
Copper	137·399	"
Platinum	124·000	"
Silver	85·062	"
Gold	68·216	"
Tin	24·200	"
Zinc................	12·710	"

When metals are elastic and sonorous, they have these properties developed in proportion as they are harder. This remark appears to apply to the alloys; thus, bronze formed of copper and tin, is harder and more sonorous than either of these metals.

Their fracture is to be considered in several different aspects, for it is often a means of distinguishing one from the other. Thus, the fracture is *lamellated* in bismuth and antimony, *granulated* in tin, &c. Metals can assume regular crystalline forms, which are in general octahedral, the cube or forms derived from them.

Metals conduct heat and electricity the best of all simple bodies.

Their fusibility is very variable. Some, like lead and tin, fuse much below a red heat; others, as platinum, rhodium, and iridium, do not melt except with the aid of powerful lenses, or the oxyhydrogen blow-pipe.

The following table shows the order of fusibility of the principal metals:—

Metal	Fusing point	Metal	Fusing point
Mercury	— 39° C.	Manganese, between iron and cast iron	
Potassium	+ 58° C.	Nickel, do.	
Sodium	90° C.	Forged iron	2118° C.
Tin	230° C.	Palladium,	Almost infusible, running together only in the fire of a powerful forge.
Bismuth	246° C.	Molybdenum,	
Lead	312° C.	Uranium,	
Cadmium	360° C.	Tungsten,	
Zinc	370° C.	Chromium,	
Antimony	432° C.	Titanium,	
Silver	1022° C.	Cerium,	Infusible in the fire of the most powerful forge; fusible with the oxyhydrogen blow-pipe.
Copper	1092° C.	Osmium,	
Gold	1102° C.	Iridium,	
Cast iron	1587° C.	Rhodium,	
Steel, between iron and cast iron		Platinum,	

The action of Oxygen, of Atmospheric Air, and of Water, on the Metals.

Some metals, as potassium and sodium, absorb oxygen at the ordinary temperature; but most metals are not oxidized except at a higher temperature. Others, as gold, platinum, palladium, rhodium, and iridium, do not absorb oxygen at any temperature.

Dry air acts on the metals like oxygen, but with less energy; moist air oxidizes them more rapidly than dry air; it then forms oxides which are ordinarily hydrated and carbonated.

Several metals decompose water at the ordinary temperature, as potassium and sodium; others, as iron, zinc, tin, antimony, &c., do not act on water, except at a temperature approaching redness. Some metals, as gold and platinum, do not exercise any action on water, even under the influence of a red heat.

Acids sometimes bring about the decomposition of water by the metals; the oxygen of the water, in this case, unites with the metal, to form an oxide, which combines with the acid, while hydrogen is disengaged. Some acids, as nitric and concentrated sulphuric, will even give up a part of their oxygen to the metals.

CLASSIFICATION OF METALS.

The best classification of metals is that proposed by M. Thenard. We shall adopt it with the modifications introduced by M. Regnault, which, however, leave undisturbed the basis of the classification of M. Thenard.

The metals are classed in six sections, according to their degree of affinity for oxygen.

This affinity is shown,

1st. By the action which oxygen exercises on the metals.

2d. By the action of heat on the oxides, and by the more or less easy reduction of these oxides.

3d. By the decomposition of water which the metals give rise to, either directly or in the presence of acids.

First Section. — The metals of the first section absorb oxygen at a low temperature; their oxides resist the most elevated temperatures, and are decomposed with difficulty by bodies having a strong affinity for oxygen. They decompose water in the cold, disengaging hydrogen. The metals of this section are: *potassium*, *sodium*, *lithium*, *barium*, *strontium*, and *calcium*.

Second Section.—The metals of the second section absorb oxygen at a higher temperature; their oxides are, in general, as difficult to reduce as the preceding. But these metals do not decompose water till between 100° and 200°C., and sometimes only at a low red-heat. The metals of this section are: *glucinium*, *aluminium*, *magnesium*, *zirconium*, *thorium*, *yttrium*, *cerium*, *lanthanium*, *didymium*, *manganese*, *uranium*, *pelopium*, *niobium*, *erbium*, and *terbium*.

Third Section.—The metals of this section do not absorb oxygen, but at a moderately-elevated temperature; their oxides, indecomposable by heat, are easily reduced by hydrogen, carbon, and carbonic oxide. These metals do not decompose water, except at a red heat, or at ordinary temperature in the presence of acids. The metals of this section are: *iron*, *nickel*, *cobalt*, *zinc*, *cadmium*, *chromium*, and *vanadium*.

Fourth Section. — The metals of this section are distinguished from those of the preceding, by not decomposing water in the presence of acids, though they do so at a red heat. But as they have a great tendency to acidify, they decompose water in the presence of energetic bases like potash. The metals of this section are: *tungsten*, *molybdenum*, *osmium*, *tantalum*, *titanium*, *tin*, and *antimony*.

Fifth Section. — The metals of the fifth section do not

decompose watery vapor but slowly, and at an elevated temperature; their oxides are not reduced by heat. These metals are *bismuth*, *lead*, and *copper*.

Sixth Section. — This section comprises the metals called *noble*, which do not decompose water; their oxides are reduced by heat. These metals are: *mercury*, *silver*, *rhodium*, *palladium*, *rhuthenium*, *platinum*, and *gold*.

It may be remarked that metals of the first section form the most energetic bases. Those of the second give less energetic bases, and often acids. In the third are found, among oxides of the same metal, both bases and acids. The fourth section gives principally acids.

The metals are sometimes divided into —

1st. *Alkaline Metals*, which are: potassium, sodium, and lithium.

2d. *Alkaline-earthy Metals*, which are: calcium, barium, and strontium.

3d. *Earthy Metals*, which are: aluminium, magnesium, glucinium, zirconium, yttrium, erbium, terbium, thorium, pelopium, niobium, cerium, lanthanium, and didymium.

4th. *Into Metals properly called*, which are: manganese, iron, chromium, zinc, cadmium, cobalt, nickel, tin, titanium, antimony, bismuth, lead, copper, uranium, molybdenum, vanadium, tungsten, tantalum, mercury, silver, gold, platinum, osmium, iridium, rhodium, palladium, and rhuthenium.

METALLIC OXIDES.

This name is given to the binary compounds formed by the combination of a metal with oxygen. The oxides are divided into four classes, viz:

1st. Basic oxides;

2d. Acid oxides (oxacids, metallic acids);

3d. Indifferent oxides;

4th. Saline oxides.

Basic Oxides belonging to the metals of the first section, have the property of neutralizing acids, of turning green the syrup of violets, of restoring the blue color to litmus reddened by acids, and of changing to a reddish brown the yellow color of turmeric.

The Oxacids possess acid properties, neutralize bases, form salts with them, and often redden litmus.

The Indifferent Oxides are those which combine neither with acids nor bases. The bioxides of *barium*, of *calcium*, *manganese*, *strontium*, &c., are indifferent oxides.

The Saline Oxides are those which result from the combination of two oxides of one metal, one acting as an acid, the other as a base.

Action of Heat on the Oxides.—The oxides of the metals of the sixth section lose their oxygen, and are reduced to the metallic state, by the action of heat.

None of the other oxides are completely reduced by heat; but certain metallic acids, as chromic acid, ferric acid, manganic and permanganic acids, plumbic acid, some peroxides, as those of manganese and copper, lose a part of their oxygen when they are heated. Moreover, the metallic oxides are nearly all fixed. Most of these melt only at a very high temperature.

Action of the Pile.—All the oxides, with the exception of the earthy oxides, may be decomposed by the pile. When an oxide is placed in contact with the two poles of a pile, the metal reduced is soon seen to appear at the negative pole.

When the metal will form an amalgam, the decomposition of the oxide is facilitated by using mercury. The oxide, slightly moistened, is formed into a cup, which is filled with mercury; this is placed on a metallic plate, which communicates with the positive pole of the pile, while the negative pole is plunged into the mercury; at the end of a short time an amalgam is obtained, which by distillation gives the metal which formed the oxide.

Action of Oxygen.—Many oxides absorb oxygen when they are in contact with this gas or with air, either at the common temperature, or at a high temperature. Such are the protoxides of potassium, of sodium, of barium, of iron, of manganese, of tin, of copper, of lead, &c.

Action of Hydrogen.—Hydrogen, under the influence of heat, reduces the oxides of the four last sections; excepting, however, the oxide of manganese, and the oxide of chromium.

Hydrogen brings to the state of protoxide, the peroxides of the two first sections, as well as the peroxide of manganese.

Certain oxides, particularly those of the last section, are reduced by hydrogen at a somewhat elevated temperature.

The oxides reduced by hydrogen always leave the metal pure; it is thus metals are often prepared in laboratories.

Action of Carbon.—Carbon reduces the metallic oxides at a more or less elevated temperature, excepting the earthy oxides, or those of the second section, and the alkaline-earthy oxides.

Carbon, in its action on the oxides, produces either carbonic acid, or carbonic oxide, according to the proportion of carbon used, and the affinity of the metal for oxygen. If the oxide is easily reduced, like the oxides of copper and silver, carbonic acid is always obtained. If the reduction does not take place except at an elevated temperature, and if the carbon is in excess, carbonic oxide is produced. When the reduction is made at a temperature approaching redness, carbonic acid and carbonic oxide are both produced.

Carbon is used in metallurgic operations, to extract metals from their oxides. The carbon produces, in burning, the heat necessary for the reduction and at the same time takes away the oxygen of the oxide, which it transforms into carbonic oxide, or carbonic acid. Metals extracted from their oxides by carbon, ordinarily retain a small quantity of carbon. Iron, for example, obtained in blast-furnaces, may contain from two to six per cent. of carbon.

Manganese and chromium, reduced by carbon in a clay crucible, also retain carbon.

SALTS.

Before Lavoisier, the name *salt* was applied indiscriminately to a certain number of bodies, whose composition and properties often did not present any analogy. It was enough that a body was solid, crystallizable, transparent, and soluble in water, to give it the name of salt. Lavoisier was the first to fix the true nature of salts, and gave the following definition of them:

A salt is a body formed by the combination of an acid with a base, in which the properties both of the acid and the base are found to be more or less neutralized.

At the time Lavoisier proposed this definition of salt, the hydracids were not known. It was thought that a salt resulted necessarily from the combination of a base with an oxacid, and ought to contain the elements of the acid and the base.

Later, the existence of a new class of acids was demonstrated, the *hydracids*, which in uniting with the bases, form water and binary compounds. Chemists then found themselves placed in the alternative of abandoning the definition given by Lavoisier, or of rejecting from the class of salts, bodies which, like marine salt, have all their general properties, but differ in composition. Berzelius proposed the name of *haloid salts*, to binary compounds resulting from the reaction of hydracids on bases. The chlorides, bromides, iodides, fluorides, cyanides, sulphurets, were considered by Berzelius as *haloid salts*. The salts formed by the oxacids, and which are called *oxysalts*, may unite together to form double salts; it is thus the sulphate of potassa combines with the sulphate of alumine, and constitutes alum.

Phenomena of Saturation.

When a base is made to act gradually on an acid, the properties of the acid and the base are seen gradually to disappear. There arrives a time when these two bodies have lost their characteristic taste, their action on turnsol (litmus), &c. The acid is then said to be saturated by the base.

At first, the term *neutral salt* was given to the saline compounds in which the respective properties of acid and base were *neutralized*. But the expression *neutral salt* afterwards had another signification.

The moment when the neutrality is perfect is recognised by the aid of coloring substances, which readily become modified under the influence of acids or bases. Thus the tincture of turnsol, syrup of violets, the solution of the coloring matter of Campeachy wood, turmeric, &c., may be used when *neutralizing a solution*, by seizing upon the moment when the acid and the base shall have ceased to act on these colored reagents. As the tincture of turnsol is the colored reagent most frequently used to test the presence of acids and bases, it is necessary to know its composition. We borrow from M. Chevreul the following observations on this reagent.

Blue turnsol ought to be regarded as a true salt, resulting from the combination of a base with an acid which is red.

An acid reddens turnsol, because it isolates the red acid which exists in the tincture of turnsol.

The sulphate of potash does not react upon turnsol, because sulphuric acid and potash have so strong a mutual affinity, that the coloring principles of the tincture of turnsol cannot unite to the base or the acid, so as to form combinations of another color than that of the coloring principles in a state of purity. If there existed a coloring matter sufficiently energetic to remove the potash from the sulphuric acid, the sulphate of potash, in presence of this coloring matter, would have an acid reaction.

General Properties of Salts.

Salts are nearly all solid. Their color is variable, and depends, in general, on the nature of the base they contain. The alkalies, the earthy oxides, and some metallic oxides, form colorless salts, when the acids with which they are united are themselves colorless. Most of the metallic oxides, as those of copper, iron, cobalt, nickel, chromium, gold, platinum, &c., give colored salts. When the acid which enters into the composition of the salt is colored, as chromic and permanganic acid, the salt has a color, in general, like that of the acid.

The taste of salts is often characteristic, and depends nearly always on the base. Thus the salts of magnesia are bitter; those of alumine, sweet and astringent; those of lead, sweet and styptic. It may happen, however, that the taste of the salts partakes of the nature of the acid. The sulphites, sulphurets, and cyanides, have a taste and properties affecting the senses and organs generally, which depend especially on the nature of the acids with which they were formed. Certain acids may modify, or even change completely, the taste of a base; thus, the citrate of magnesia has not the ordinary taste of the salts of magnesia.

Action of Heat.

Heat produces effects on salts which vary with the nature of the acid and that of the base. When a salt contains much water of crystallization, it readily fuses without losing this water, and thus presents the phenomenon of *aqueous fusion.* In continuing to heat it, the water of crystallization volatilizes, the salt returns to the solid state, and may enter into fusion a second time; it then undergoes *igneous fusion.*

Some salts, submitted to the action of heat, give rise to a peculiar noise, which is called *decrepitation.* When sea-salt is thrown on live coals, it is thrown about on all sides, producing a series of slight detonations. For a long time, the decrepitation was attributed to the sudden expansion of the water contained in the crystals; but it is now demonstrated that the volatilization of the water is not the only cause of the decrepitation; for some salts decrepitate by heat, even after they have been dried a long time in vacuo, and thus lost the small quantity of water contained between their molecules. The decrepitation ought then to be attributed to an unequal distribution of heat between the molecules of the salt, which causes the rupture of the crystals.

Some salts are rendered phosphorescent by heat, as the fluoride of calcium, some sulphurets, &c.

Some salts and certain oxides, throw out a bright light when their temperature is gradually raised; they then display new properties, and are in general more difficult of solution.

Action of Electricity on Salts.

All salts are decomposed by the pile, when they are moist or dissolved. The acid goes to the positive pole, the base to the negative.

It often happens that the base is decomposed, and the reduced metal goes to the negative pole, while the acid and the oxygen of the base go to the positive pole.

Action of Metals on Saline Solutions.

When we place in a saline solution, containing one of the metals of the four last sections, another metal, belonging also to one of these sections and having a greater affinity for oxygen than that which is in the solution, this metal, in general, substitutes itself for that of the salt, and precipitates it.

Generally the precipitated metal attaches itself to the precipitating metal, with which it forms an element of the pile which causes the complete decomposition of the salt. The metal, in slowly depositing, sometimes assumes beautiful crystalline forms.

The most remarkable crystallization, is that called *arbor Saturni*, which is made by placing a plate of zinc in a solution of the acetate of lead.

This is done by pouring into a wide-necked bottle, water containing the 30th part of its weight of acetate of lead, first rendered acid by acetic acid. A piece of zinc, attached to a stopper by brass or copper wire, is introduced into the bottle. Shortly the zinc and the wires are coated with brilliant and long crystals of lead. The name of *arbor Dianæ* is given to the crystallization which is obtained by the precipitation of nitrate of silver by mercury; the body which crystallizes is an amalgam of silver.

Hygrometric action of the Air on Salts.—Those salts which attract moisture from the air, when in contact with it, and become liquid from being dissolved in the water which they absorb, are called *deliquescent salts.*

All very soluble salts are deliquescent in an atmosphere saturated with moisture.

There are, on the contrary, salts which give up to the air, either in whole or in part, their water of crystallization. These are called *efflorescent salts.*

It may happen, however, that certain anhydrous and fused salts become efflorescent by absorbing water from the air. Thus the sulphate of soda, fused, absorbs water in contact with the air, and crumbles into powder; it remains in this form, because the hydrated sulphate of soda is not deliquescent. Some salts, as the sulphate of soda, lose in dry air all their water of crystallization, while others, as the carbonate of soda, always retain a certain quantity, no matter how dry the air may be.

Action of Water on Salts. — The solubility of salts in water is very variable. Some salts, as the sulphate of barytes, the phosphate of lime, &c., are insoluble; other salts often require less than their weight of water to dissolve them.

Anhydrous salts, in forming solid hydrates with water, in general develop heat, when placed in contact with this liquid.

Salts which do not combine with water, or those which contain all their water of crystallization, produce, on the contrary, cold, when passing from the solid to the liquid state in contact with water; such are the chloride of potassium, nitrate of ammonia, and sulphate of magnesia.

An absorption of heat takes place from the same cause in the first case; the heat developed by the combination is, nevertheless, often sufficiently intense to raise the temperature.

The cold produced is great in proportion to the rapidity of the solution; so that dilute acids often are used instead of water, as they dissolve hydrated salts more quickly.

A most intense cold is obtained by mixing hydrated salts with pulverized ice, or, better, with snow. This easily explains why ice or snow, in melting, absorb a considerable quantity of heat.

All these mixtures are called *freezing mixtures.* Water is said to be *saturated* with a salt at a given temperature, when it is no longer able to dissolve the smallest quantity of the salt at this same temperature. A mother-water, which has deposited crystals on cooling, or even a solution agitated for a long time with an excess of the salt in powder, ought to be considered as *saturated.*

A saturated solution of a salt may dissolve a new salt.

Water saturated with nitrate of potash can still dissolve a considerable quantity of sea-salt, and even a certain quantity of a third, or even a fourth salt, provided the mutual action of these different salts does not produce other compounds which would precipitate.

A saturated solution of a salt sometimes deposits some of this salt, when it dissolves a new salt.

It is thus that water, charged with nitre, precipitates a part of this salt, when it is agitated with chloride of potassium. Many industrial operations, and some analytical processes, are founded on the property which water possesses, when saturated with a salt, to dissolve several other salts.

Variations of temperature modify the solvent power of water. This generally increases with the temperature. There are, however, certain salts which are more soluble in cold than warm water.

The air in general does not exercise any influence on the crystallization of salts. The sulphate and seleniate of soda do not crystallize when they are preserved from contact with the air; a bubble of air is sufficient to cause the sudden crystallization of these two salts.

Certain saline solutions have also the singular property of remaining surcharged with an excess of salt for some time; such are the solutions of the nitrate of silver, and acetate of lead. When they are agitated, or a solid body is introduced, the solution sometimes thickens into a mass. The cause of this supersaturation is yet unexplained.

Whatever process is used to crystallize saline solutions, the crystals which form retain some water.

When the water is combined with the salt in definite proportions, it is called *water of crystallization*, or of *combination*. If the quantity of water retained by the crystals is small, and is not found in simple relation with that which enters into the composition of the salt, the name of *water of interposition* is given to it.

Exposure to the air, or a few instants in vacuo, or even pressure between the folds of unsized paper, will remove the water of interposition, which is not an integral part of the salt, and exists in it only in variable and very feeble proportions.

Phenomena which determine the Composition of Salts.

We are indebted to Berthollet for valuable observations on the laws which determine the decomposition of salts. This chemist has proved that, when a salt is placed in presence of an acid, a base, or even another salt, a decomposition always takes place if, by the reaction either of the acid or the base, or the salt, on the elements of the first salt, new compounds would be formed, more volatile, less soluble, or more fusible, than those which were placed together.

Thus, by consulting the physical properties of bodies which ought to form in the preceding mixtures, it will always be easy to foretell, in advance, the nature of the bodies which will be formed.

When sulphuric acid is poured on marble (carbonate of lime), we might, relying on the laws of Berthollet, say in advance that there will be effervescence; that the sulphuric acid will expel the carbonic acid from its combination with lime, because carbonic acid is less fixed than sulphuric acid.

What we have said of the influence of volatility is applicable to the influence of insolubility, or to that of fusibility.

We shall now examine the principal metals and their compounds.

POTASSIUM.

Potassium was first isolated in 1807, by Humphrey Davy. This discovery, one of the most important which has been made in chemistry, has fixed the true nature of the alkalies and earths.

The properties of potassium were then studied with the

greatest care by Gay-Lussac and Thenard, who were the first to explain a practical process for preparing this metal.

Properties.—Potassium is solid at the ordinary temperature, and possesses a metallic brilliancy. Melted in the oil of naptha, it is as white as silver; but when exposed to the air, it tarnishes rapidly, and becomes of a bluish-gray color. It is softer than wax, and may be moulded between the fingers. This experiment ought to be made under naptha, because potassium takes fire in the air, even at ordinary temperatures. The density of potassium, at 15° C., is 0·865; it is then lower than that of water.

Potassium is, next to mercury, the most fusible of all metals; it fuses at 58° C., and is volatile at a red heat. It may be volatilized in a glass tube by means of a spirit-lamp; its vapor is green. This experiment ought to be made in an atmosphere of nitrogen, in order to prevent the oxidation of the metal.

Potassium has a great affinity for oxygen. When its temperature is raised, when it is touched, for example, with a rod of red-hot iron, it burns with vivacity, and is transformed into oxide of potassium (potash).

Potassium may be kept in perfectly dry oxygen, or atmospheric air, without change. It decomposes water at the common temperature, and deprives it of its oxygen. Thus, when a globule of potassium is thrown into a jar filled with water, it is seen to dance about rapidly, and become incandescent; it then combines with the oxygen of the water to form potash, which remains in solution while the hydrogen of the water is set free. The reaction of the potassium on the water developing a very high temperature, the hydrogen takes fire in contact with the air, and reproduces water.

To show the production of hydrogen in the preceding experiment, a small quantity of water is introduced into a tube filled with mercury, and a globule of potassium is made to pass up to it. As soon as the metal comes in contact

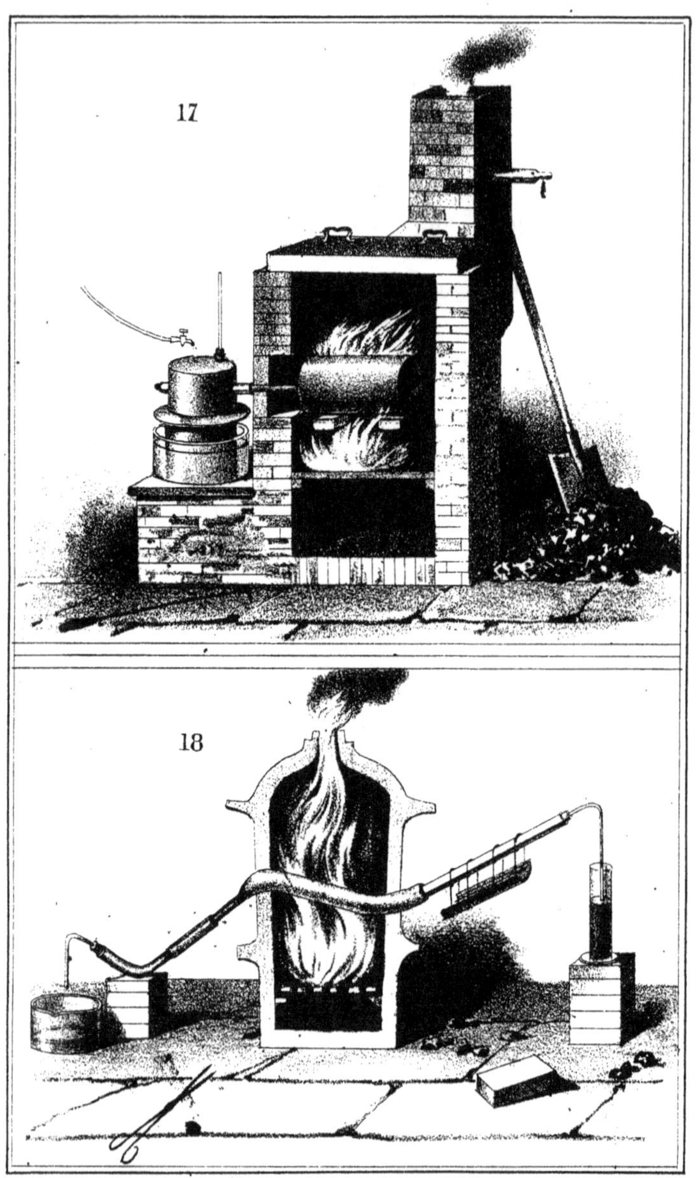
17
18

with the water, the reaction takes place; the hydrogen, in disengaging, depresses the column of mercury contained in the tube, and in a few moments the tube is filled with hydrogen.

Preparation.—Davy isolated potassium by subjecting the hydrate of potassa to the action of a strong pile. He made a cavity in a piece of the hydrate, and filled it with mercury; he placed this upon a metallic plate which he made to communicate with the positive pole of a pile of 150 pairs, while the negative pole was plunged into the mercury.

The hydrated potassa was decomposed under the influence of the electric current; the oxygen of the oxide, and the oxygen of the water, went to the positive pole, while the potassium and the hydrogen went to the negative pole.

The potassium, finding the mercury at the negative pole, forms with it an amalgam; submitting this amalgam to distillation in a small glass retort, the mercury volatilizes, and the potassium remains in the retort in a state of purity.

As this experiment only gives small quantities of potassium, this metal is always prepared by reducing the hydrate of potassa by iron, or by decomposing the carbonate of potassa by carbon.

We shall first speak of the mode of preparing potassium by means of the hydrate of potassa and iron, which we owe to Gay-Lussac and Thenard.

In this method, the hydrate of potassa, in fusion, is brought to react on the turnings of iron heated to redness, placed in a curved gun-barrel (Fig. 18).

The hydrogen, proceeding from the decomposition of the water of the hydrate of potassa, is disengaged; the iron absorbs the oxygen of the water, and of the oxide of potassium, while the potassium set free volatilizes, and condenses in the receiver.

The potassium should be taken from the receiver with an iron rod, and placed under liquid carburet of hydrogen, which preserves it from oxidation. It is generally preserved in the rectified oil of petroleum.

The process of M. Brunner consists in decomposing, in an iron vessel, the carbonate of potassa, by carbon, which completely reduces the potassa at a very high temperature, and transforms the carbonic acid of the carbonate into carbonic oxide. The potassium distils and condenses in a cool receiver containing the oil of naptha, Fig. 17.

The retort is covered with a refractory lute; it is placed over two horizontal bars of iron, in a blast-furnace, furnished with a chimney having a strong draught; it is filled from above, first with charcoal, afterward with a mixture of charcoal and coke.

The operation is commenced by heating strongly the iron retort, and the receiver should not be adapted until the vapors of potassium begin to disengage. Often the oil of naphtha, in the receiver, is fired, to absorb the oxygen of the air which it contains, and avoid the oxidation of the potassium.

The potassium thus obtained is not pure, it always contains carbon. To purify it, it is first filtered in a cloth, under oil of naphtha heated; then it is distilled in an iron vessel, or a refractory glass retort, covered with an earthy lute. The vapors are condensed in the oil of naphtha.

This operation lasts about three hours, and gives 30 to 40 grammes of potassium; it is more easy of execution than the preceding, but it gives a metal less pure. We shall speak now of the principal compounds of potassium, commencing with the protoxide of potassium, which is called *potash*.

Potash.

Hydrated potash is solid, white, caustic, very alkaline, and unctuous to the touch; it instantly alters the skin; in its contact with organic substances, it develops a peculiar odor, which is that of *lye*.

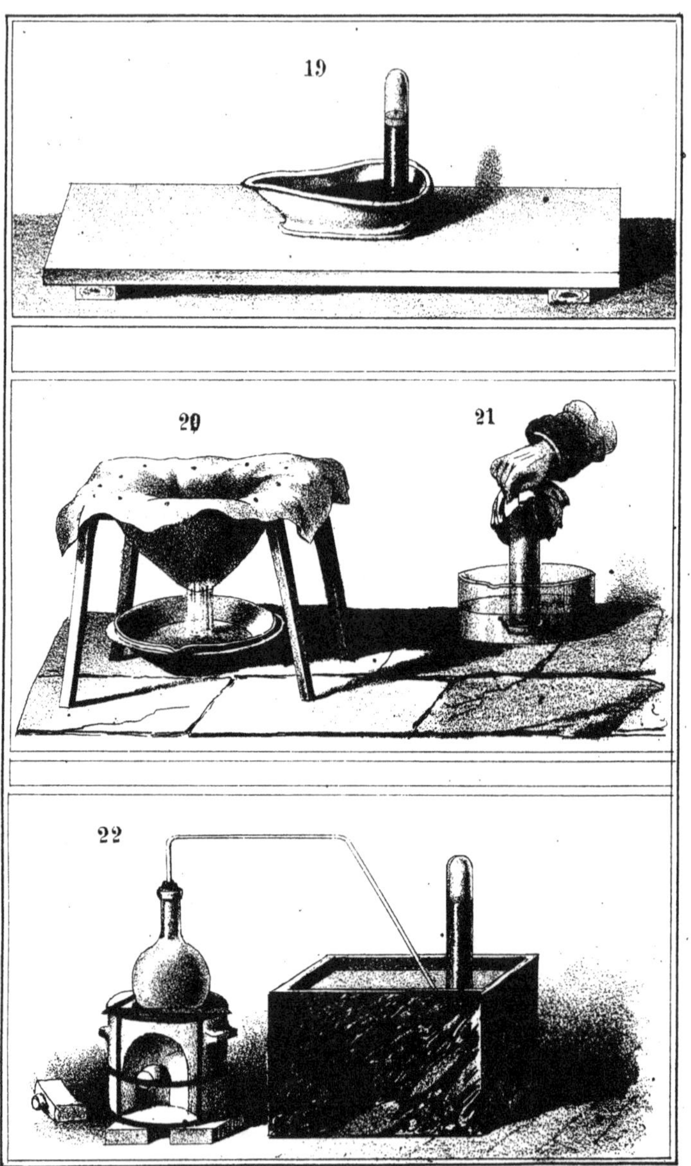
19
20
21
22

Potash acts on nearly all organic substances, and destroys them rapidly. It decomposes or dissolves a great number of animal substances, such as the skin, silk, &c.; it saponifies fat bodies. Potash fuses below a red heat, and then volatilizes, producing white vapors. It has a great affinity for water; exposed to the air, it attracts moisture, and falls into deliquescence. Potash dissolves alumine and silex, and attacks glass and porcelain; so that it can never be concentrated in vessels either of glass or porcelain. This operation ought to be done in capsules of silver.

Preparation.—Potash is extracted from its carbonate: this salt may be obtained by different methods. The ashes of vegetables contain different salts of potash, and chiefly the carbonate. By treating the ashes with water, a great part of the potash is taken up, combined with carbonic, sulphuric, chlorohydric, and silicic acids, and the insoluble residue is thrown upon a cloth. It is principally composed of silex, and phosphate and carbonate of lime. (Fig. 20.)

The liquors evaporated to dryness give a residue called *calcined potash*, or *saline.*

To extract from this the carbonate of potash, the mass is dissolved in boiling water, and then evaporated to crystallization. The foreign salts are deposited, while the carbonate of potash remains in the mother-waters.

The carbonate of potash having been once obtained, the potash is extracted from it by submitting it to the action of the hydrate of lime, which, according to the rules established by Berthollet, decompose the carbonate of potash, because the carbonate of lime is insoluble. The carbonate of lime is then formed, and the potash remains free.

In order to decompose the carbonate of potash by lime, dissolve one part of carbonate of potash in ten or twelve parts of water; put this solution in an iron vessel, carry it to ebullition, and add to it a boiling liquor of hydrate of lime. The carbonate of potash requires for its decomposition about its own weight of lime.

The lime ought to be added so gradually as not to stop the ebullition, to facilitate the deposit of the carbonate of lime.

To be sure that the carbonate of potash has been completely decomposed, a small quantity of the liquid is taken out and allowed to stand; to this nitric acid is added. If all the carbonate of potash has been decomposed, there will be no effervescence, and it ought no longer to precipitate lime-water.

As soon as the decomposition of the carbonate is complete, the liquor is allowed to cool protected from the air; and when clear, it is decanted, and evaporated in a capsule of iron, or, better, of silver. When the hydrate of potash is melted, it is poured on plates of iron, or into cylindrical moulds, made in two parts, so that they can be separated one from the other, and the potash, when it has become solid, can be taken out.

If the carbonate used in this preparation is pure, the potash which is extracted is equally so; at least it will only contain a small quantity of its carbonate, produced during the evaporation, at the expense of the carbonic acid of the air. If, however, the carbonate of potash of commerce is used, the potash always contains chlorides, sulphates, and carbonates.

To purify it, alcohol is used; which only dissolves the hydrated potash, and precipitates the foreign salts. With this end, the impure potash is evaporated to the consistence of honey; to this is added a quantity of alcohol at 33°*, which represents about one-third of the original weight of the potash; the mixture is stirred and boiled for some minutes, and put into a bottle with a ground-stopper.

The liquor, left to repose, divides into three layers: the lower of which is formed of sulphate of potash and anhy-

---

* See article Alcohol, in the Organic Chemistry.

drous lime; above is a solution of sulphate, carbonate and chloride of potash; the superior is an alcoholic solution of potash. This last is decanted and distilled, so as to drive off about two-thirds of the alcohol it contains; it is then rapidly evaporated in a silver capsule.

The potash thus prepared is called *potash à l'alcohol;* it is nearly pure, and ordinarily contains only traces of chloride.

Uses of Potash.—Hydrated potash is a valuable reagent; it is used in the preparation of a great many oxides, and also to attack, by the dry method, the silicates, and to render them soluble in the acids.

It is used in medicine as a caustic, from which it has received the name of *caustic potash.* It is also used in the manufacture of soaps, glass, &c.

Natural State of Potash.—Potash exists largely in nature; it is always combined with acids; it is met with in nearly all rocks, and principally in feldspar. It is often found in considerable quantity in arable land and in clay; it saturates, in part, vegetable acids, and forms different organic salts, which, by calcination, produce carbonate of potash, which is found in their ashes.

Nitrate of Potash.

Nitrate of potash, also known by the names of *nitre, salt of nitre, saltpetre,* and *nitrate of potash,* is white, without odor, of a taste at first pleasant, but soon sharp and bitter; its crystals are very friable; when they are kept some time in the hand, they break, giving rise to a slight noise. Nitre is unalterable in the atmosphere in ordinary circumstances; it does not deliquesce, except in an atmosphere almost saturated with moisture. It fuses at about 300°, and in cooling leaves a white compact mass, known by the name of *crystal mineral.* This mass pulverizes more readily than the crystals of nitre, which always have a certain degree of elasticity.

The nitrate of potash is hardly soluble in alcohol of 90° of the alcoholometer, and is altogether insoluble in absolute alcohol; it dissolves freely in water. The solubility of nitre, which increases considerably with the temperature, permits it to be purified with great facility, and to be freed, by crystallization, from the foreign salts which it may contain.

A mixture of nitre and carbon burns with vivacity, when it is heated, or when it is touched with an incandescent body: the carbon changes into carbonic acid, at the expense of the oxygen of the nitric acid; a part of this carbonic acid is disengaged; another part remains united with the potash, and nitrogen is liberated.

Sulphur reacts on nitrate of potash, under the influence of heat. This reaction always gives rise to a lively disengagement of heat.

If the nitrate of potash is in excess, the sulphur is entirely changed into sulphuric acid, which remains combined with the potash.

The name of *detonating powder* is given to a mixture of 3 parts of nitre, 2 of potash, and 1 of sulphur. When a few grammes of this powder are slowly heated in a spoon, the mass at first fuses, and then detonates with violence; the detonation is due to an instantaneous disengagement of gas.

The *flux of Baumé* is a mixture of 3 parts of nitre, 1 part of sulphur, and 1 part of saw-dust; it has the property of determining the fusion of different metals. It acts thus, not only on account of the elevated temperature which is produced by the reciprocal action of the bodies which it contains, but also because a part of the sulphur unites directly with the metals, and forms with them fusible sulphurets.

Acids more fixed than nitric acid decompose nitre, under the influence of heat. Clay itself may bring about this decomposition. For a long time nitric acid was prepared by decomposing nitre with clay.

Natural State of Nitre.—Nitre abounds in nature; it is principally found in Egypt, in India, in America, and in Spain. In these countries, the nitre effloresces on the surface of the soil; it is collected with long brooms made of twigs, which are called *houssoirs;* hence the name *saltpetre de houssage* (swept saltpetre).

In France there are many places which produce saltpetre; thus in the departments of Seine and Oise, and in la Roche-Guyon, these efflorescences are found, which are very rich in saltpetre.

This salt crystallizes on the walls of old buildings, in stables, and in places inhabited by animals; plaster from old buildings also contains saltpetre.

Though France does not possess deposits rich in saltpetre, it can nevertheless furnish it in great quantities; thus, at the time of the most active wars, the annual production of saltpetre in France has been 1,900,000 kilog. Paris furnished seven-twentieths of it; Tourraine, two-twentieths; all the other provinces, ten-twentieths; the artificial nitreries, one-twentieth.

In France they always extract nitre from the saltpetre materials, which, besides the nitre, contain nitrates of lime and magnesia. According to Gay-Lussac, the plaster saltpetres of Paris contain about 5 per cent. of nitrates.

The soluble part of the saltpetre materials has in general the following composition:—

Nitre, ...	25
Nitrate of lime, . .	33
Nitrate of magnesia, .	5
Sea-salt, ...	5
Other salts, ...	32
	100

M. Kuhlmann has also found a notable quantity of nitrate of ammonia in them.

The operation which transforms the nitrates of lime and magnesia into nitrate of potash, is called *saturation of the liquors.*

The waters in which the saltpetre materials have been washed, containing the nitrates of lime and magnesia, are mixed with carbonate of potash, which forms insoluble carbonates of lime and magnesia, and soluble nitrate of potash.

In this reaction, the nitrate of ammonia is also decomposed into nitrate of potash and carbonate of ammonia, which disengages during the concentration of the liquors.

Washing the Saltpetre Materials.—In order readily to wash the saltpetre materials (old plaster and mortar), they begin by breaking them up, and placing them on hurdles or sieves, of willow; they are then mixed with ashes, or any other substance containing the carbonate of potash intended for the *saturation of the liquors.* The materials mixed and broken up, are placed in casks having one head out, and placed over a trough called a *recipient.*

A quantity of water, equal in volume to one half of the solid matter, is poured in; it remains there ten hours; after this time, it is allowed to run off; the plaster retains the half. Suppose that 500 litres of water are poured over one cubic metre of plaster containing 40 kilogrammes of saltpetre; then 250 litres of water are drawn off, containing 20 kilogrammes of saltpetre. To draw off the remaining 20 kilogrammes, 250 litres of fresh water are poured in; and 250 litres run off, containing 10 kilogrammes of saltpetre. In continuing to add successively 250 litres of fresh water, each operation will give solutions less and less strong, but will reduce to one-half, the quantity of saltpetre retained by the solid materials. When this quantity becomes so small as not to cover the expense of the evaporation of water used for the solution, the operation is stopped.

The waters from the washing are not sufficiently rich in to be subjected to evaporation; they are made to pass over new materials, following the same course as before. The waters are not evaporated unless they contain 14 kilogrammes of saltpetre per hectolitre.

These last waters ought to mark about 12° on the areometer; they are then *evaporated.*

Boiling.—The evaporation of the saltpetre waters, called *boiling*, is made in large boilers of iron or copper.

The liquors, in concentrating, deposit carbonate of lime, sulphate of lime, and animal matters; these deposits are called *boues* (impurities). During this concentration, there is a strong ammoniacal odor.

Boues, or Impurities.—Boiling brings the impurities from the circumference of the boiler to the centre; so that by placing in it, at some distance from the bottom, a vessel into which the insoluble deposits will settle, they can be easily taken away.

When the liquor arrives at a certain point of concentration, it drops crystals of chloride of potassium and sodium, which are not near so soluble as nitre in boiling water. These salts are taken off with skimmers. The deposit of chlorides generally takes place at the time the liquor marks 42° on the test liquor; the evaporation ought to be pushed to about 45°. At this time, the evaporation is stopped; and it is known that it has been carried sufficiently far, if a drop of the solution solidifies when placed on a cold body.

The liquor is then carried to the crystallizer, and as it crystallizes, it is agitated with wooden rakes.

Thus is obtained a crude saltpetre, crystallized in small needles, which bear the name of saltpetre of the *first boiling.* This first operation is performed in France by the licensed manufacturers, who have not the right to refine saltpetre. Crude saltpetre is delivered in this state to the government-shops; it contains about 25 per cent. of foreign matter. It is purified from this by the process of *refining.*

Refining Saltpetre.—To refine saltpetre, the crude salt is dissolved in its own weight of warm water. The liquor is clarified with bullock's blood, which forms a scum that is taken off with care, and then it is crystallized. This operation gives saltpetre of the *second boiling*.

This is, however, not yet sufficiently pure for the manufacture of gunpowder. It ought to be subjected to a third crystallization, which gives saltpetre of the *third boiling*, the only one which can be employed to make powder.

In the crystallization of saltpetre, the solution should be continually agitated till the moment of crystallization, in order that the salt may deposit in small crystals, which are purified more easily than large ones.

To clear the crystals of saltpetre from the mother-water, charged with foreign salts, with which they are impregnated, they are treated with a water saturated with nitrate of potash, which deposites the chlorides and sulphates, &c., and leaves the nitre pure.

Nitre is chiefly used for the preparation of gunpowder. We shall treat with some minuteness of this important fabrication.

GUNPOWDER.

Powder is an intimate mixture of nitre, sulphur, and carbon. It is distinguished into three principal kinds, *gunpowder*, *sporting* or *hunting powder*, and *mining* or *blasting powder*.

Sporting powder, made in the government works, is composed as follows: —

Nitre,	78
Carbon,	12
Sulphur,	10

Mining powder: —

Nitre,	62
Carbon,	18
Sulphur,	20

The composition of gunpowder for war is nearly the same in all countries. It may be conceived that there must be but little difference in the composition of a substance which, like powder, has to fulfil certain invariable conditions. The following table shows the composition of the gunpowder used for war by different countries.

	Nitre.	Charcoal.	Sulphur.
France, Prussia, United States of America,	75·00	12·50	12·50
England	75·00	15·00	10·00
Russia	73·78	13·59	12·63
Austria	76·00	11·50	12·50
Spain	76·47	10·78	12·75
Switzerland (round powder)	76·00	14·00	10·00
Holland	70·00	16·00	14·00
Sweden	75·00	9·00	16·00
China	75·00	14·40	9·60

Properties of Powder.—It is well known that powder easily takes fire under the influence of heat, and that it almost instantaneously develops a considerable volume of gas, which then acts as an energetic power.

It is not only the proportions of bodies used in the composition of gunpowder which influence the volume of gas developed at the moment of combustion; the physical condition of powder also exercises a great influence on its projectile effects; it is necessary then carefully to point out the physical properties of powder.

Powder ought to be so hard that the rubbing together and jarring, which the grains are subjected to during its transportation and in the manufacture of munitions, will not produce dust in such quantities as would be hurtful to the rapidity of the inflammation of the powder. A good powder ought to be so hard that the grain will not break down under the fingers, nor soil the back of the hand.

Before the grains have been glazed, the slightest rubbing

suffices to detach a considerable quantity of dust, which diminishes greatly the projectile power of the powder.

The glazing takes off the asperities from the grains, and gives them a lustre and polish by hardening their surface: this however should not be too prolonged, because the grain would be less easy to inflame, and would not be homogeneous, especially if the glazing is done on a very moist grain.

The graining will cause the qualities of powder to vary. In comparing grained powder with powder in mass, it will be seen that the first inflames almost instantly, because the flame penetrates the interstices between the grains, while the powder in mass burns slowly, and in layers. Powder in mass, fired in a gun, would *hang fire*.

The size of grain ought to be adapted to the nature of the arms; in small arms, powder of fine grain is always used.

The size of the grain varies with the kind of powder, and mode of manufacture.

Density exercises great influence on the quality of powder. Dense powders inflame less easily than when they are light and porous, and give less waste in their transport.

Powders which give off their gas too rapidly are called *brisantes* (bursting). They expend a great part of their effect on the sides of the gun, which they may burst without their projectile force being increased in any considerable proportion. In this respect, they have an analogy with fulminating powders.

The causes which contribute to render powders too explosive, are the employment of too inflammable charcoal, the lightness of the grain, or want of pressure in the intimate mixture of the nitre, sulphur, and charcoal.

A hard charcoal makes powder not sufficiently inflammable. To make a good article, there should be observed a certain relation between the condition of the charcoal, the density of the powder, and the size of the grain.

The same mixture of nitre, sulphur, and charcoal, produces a powder of good quality, or one too explosive, according to the size, form, and density of the grain. It is by varying these different conditions, that powder of such a quality as may be required is obtained.

The best powder is that which is completely on fire before the projectile has left the gun, and of which the combustion takes place successively as the projectile is displaced.

Powder detonates when it is inflamed by an electric spark, by a violent shock, by the contact of an ignited body, or by a heat of about 300° C. The flame of alcohol, or that of hydrogen, is not always sufficient to inflame powder.

Good powder ought to inflame rapidly, without leaving an appreciable residue, on a sheet of paper, which it ought not to set on fire.

It was for a long time thought that iron was the only substance which, when struck against a very hard body, would inflame powder; but experience has shown that blows of copper against copper, and of iron against marble, will also cause it to detonate.

The shock of a leaden ball, projected from a gun, inflames powder placed against lead, or even against wood.

Different causes favor or retard the inflammation of powder. Moist powders always burn more slowly than those that are dry; angular powders inflame more rapidly than those which are round; unglazed powders are more inflammable than glazed powders.

Powder does not inflame unless brought suddenly to a red heat; if it is subjected to the action of heat whose intensity increases progressively, the sulphur which it contains fuses and separates from the mass. The sulphur may even be distilled by heating powder in vacuo.

From the fact of charcoal, on account of its porosity, having the property of absorbing moisture, powder, whatever be its quality, cannot be preserved perfectly dry even in the best magazines. Water, penetrating the grain of the powder, causes its projectile effects to vary, also causes an efflorescence of the saltpetre on the grain, and destroys the intimate mixture and aggregation of its components.

Powders made with pure saltpetre absorb moisture with a rapidity and quantity proportioned to the quantity of charcoal and dust they contain. The absorption of moisture is also greater in powders made with red coal than with black.

In general, fine powder absorbs moisture more rapidly than when coarse.

The solid residue from the combustion of powder forms in the gun a coating, which increases in thickness with each discharge.

This is a very great inconvenience, especially for sporting-powder, or that for musketry, and does not permit of a brisk fire being kept up for a long time.

The quantity of dirt or coating which powder produces, depends on the purity of its ingredients and their proportions; an excess of sulphur, or an incomplete trituration, are causes of dirt; the degree of dryness, and the rapidity of the combustion, also exercise a great influence.

Powders with very large grains, or those which are moist, leave a good deal of crust.

The products of the detonation of powder are solid and gaseous.

The solid products are mostly carried off, or even volatilized, by the high temperature which results from the detonation of the powder.

It has been attempted to ascertain the nature of the gases proceeding from the combustion of powder, by inflaming some in a small cartouch-box of copper, and then emptying it under a bell-glass filled with mercury; a gaseous mixture is thus obtained, formed principally of carbonic acid, nitrogen, and carbonic oxide. We should, however, observe that the gases which are thus produced cannot be compared to those produced in guns, because the conditions of combustion are not the same.

Gay-Lussac analyzed the gases from powder, by dropping it, grain by grain, into an incandescent tube; he found that these gases were formed of 53 parts of carbonic acid, 5 of carbonic oxide, and 42 of nitrogen.

The quantity of gas which powder produces in burning is subject to variations, the cause of which has not yet been explained.

A litre of powder, weighing 900 grammes, gave to Gay-Lussac 450 litres of gas, supposed at a pressure of 0·76m., and at the temperature of zero.

The moment the explosion takes place, these gases are at a very elevated temperature, which dilates them considerably; it may be said that 1 volume of powder gives, in burning, at least 2000 volumes of gas.

The temperature produced at the moment of explosion is very high; it may be estimated at more than 1200°C. It is sufficient to melt gold, pieces of money, and copper; it does not fuse platinum.

In order that this elevation of temperature should have full effect, it is necessary that the combustion of the powder should take place rapidly, so that the heat may act on the gaseous mixture, dilate it, and thus augment its elastic force.

We will now point out the conditions which the different substances entering into the composition of powder ought to present.

Nitre.

This body is always delivered to the French powder-works, by the government refineries, in a state of almost absolute purity; it ought never to contain more than three one-thousandths of foreign bodies.

In France the nitre is used in small crystals; in England, fused nitre is preferred; this is more easily pulverized, and keeps better.

Sulphur.

In France, distilled sulphur is used, collected in masses in barrels, into which it is poured when liquid. Melted sulphur is preferred to the flour of sulphur, which is never pure, containing a quantity of sulphuric and sulphurous acid.

All the sulphur intended for the manufacture of the different species of powder, comes from the refinery of Marseilles.

Charcoal.

The choice and preparation of charcoal for the manufacture of powder are of great importance.

All woods are not equally fitted for furnishing charcoal for powder.

The wood of the black alder is almost exclusively used in France for the fabrication of charcoal intended for gunpowder for war and the chase. The woods of the black alder, the poplar, the white alder, the aspen, the linden, and the willow, are about equally fit for the fabrication of the charcoal to be used in the composition of mining-powder. When there is no choice, the charcoal of any of these different kinds of wood will answer for all sorts of powder.

The shoots of the alder are cut when they have five or six years' growth at the most; their bark is taken off, as in burning it leaves more ashes than the sapwood, and shoots are chosen whose diameter varies from 0·015m. to 0·030m.

The charcoal is prepared either in trenches or tight vessels.

In the carbonization in trenches, a trench is made 3 metres long and 12 decimetres deep, the sides of which are of brick; fagots of wood are introduced, weighing 15 kilogrammes, sustained by a longitudinal rest, leaving in the trench a vacuity, which is filled with branches. This is set fire to, which soon spreads through the mass, and causes the wood to settle down. As this takes place, the trench must be filled with new fagots.

When there is no more flame from the combustion, the carbonization is considered complete; to stop the combustion, the trench is covered with wet woollen cloths, then with a quantity of earth. It should not be emptied for two or three days after the fire is extinguished; without this precaution, the charcoal, still hot, would inflame in the atmosphere. This process produces charcoal which should be sorted with the greatest care, to separate foreign bodies, and parts not well charred, which are called *brulôts*.

The yield of charcoal fit for powder is 20 per cent. The sides of the trenches, being of brick, rapidly burn out. Vessels of iron have been substituted for these, on which are placed sheet-iron covers, to smother the fire. The charcoal used for gunpowder is now prepared in this way.

Carbonization in close vessels. — This is performed in cylinders of iron, like those used for the manufacture of gas for illumination. The volatile parts escape by a tube which leads into a chimney. This distillation is made at a temperature which never reaches redness. In preparing charcoal for choice powder, it ought to be conducted slowly; the distillation of wood, for the government powder, lasts about twelve hours. Thus is obtained *red charcoal.* It is a charcoal which contains much hydrogen and oxygen, and is rather roasted wood, or half-charred wood, than pure charcoal. Red charcoal does not contain above 70 to 72 per cent. of carbon.

100 parts of dry wood give about 40 parts of *red charcoal* (charbon roux).

Manufacture of Powder. — There are two principal

methods of manufacturing powder. 1st, by pestles or stampers; 2d, by millstones.

The Process by Pestles.—This mode of trituration is effected by means of pestles in mortars of wood.

These pestles are made of pieces of beech-wood, of the weight of about 20 kilogrammes, fitted with a metallic shoe at the end, weighing also about 20 kilogrammes. The mortars are hollowed out of a stout piece of oak wood. The pestles are moved by a hydraulic wheel, Fig. 25.

The preparation of the composition commences by breaking up the charcoal; the sulphur is pulverised in the triturating vessels. To each mortar is added 1·25 kilogrammes of charcoal, 7·5 kilogrammes of saltpetre, and 1·25 kilogrammes of sulphur. These three materials are then submitted to a *battage* (stamping), which lasts about eleven hours, and is not interrupted except by the changes necessary to prevent the adherence of any of the material to the bottom of the mortar. After eight changes, the battage continues two hours more without interruption. The material should be moistened to get it into mass. On taking out the pestles, it is in the form of a moist cake, called *galette*. This, before being grained, is subjected to two preliminary operations, *le guillaumage* and *l'essorage*. The ingredients are then passed over a sieve, called *guillaume*, the holes of which are about 8 millimeters in diameter. The workman, by a backward and forward, or rotary motion, given to the sieve, gives to a disc of hard wood placed in the sieve, a circular movement; this disc of wood breaks up the cake.

The *essorage* consists in exposing to the air the ingredients of the mass, which is the commencement of the drying that facilitates the *graining*.

This gives to the powder a uniform grain. It is done by means of a disc of wood in sieves of skin, the holes of

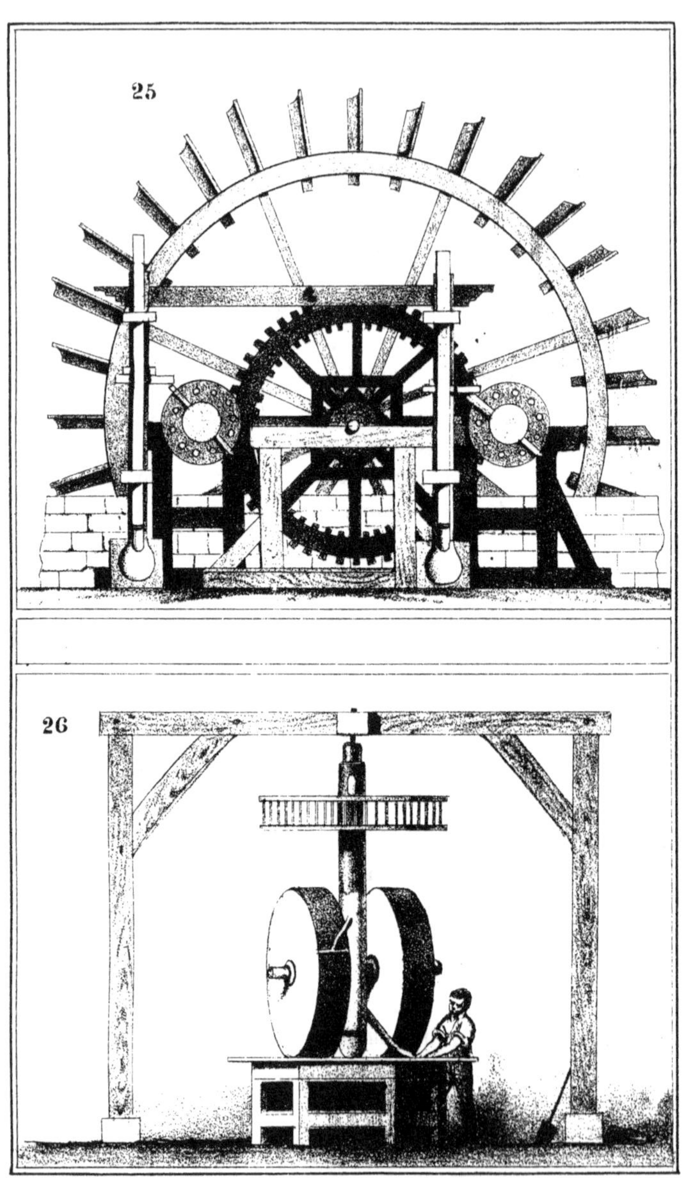
25
26

which are 2·40 millimeters in diameter for cannon-powder, 1·50 millimeters for musket-powder.

The dust, or fine grain, is taken out by passing it over still finer holes.

After these different manipulations, the powder is subjected to desiccation, either in open air, spreading it on cloths, or in the drying-rooms, or large closed apartments, into the interior of which warm air is blown, which passes over the surface of the powder, taking up its moisture.

This last operation producing a great deal of dust, the powder is afterwards cleared of the dust, and has then to be tried and barrelled up.

Process of the Mills.—This process is used for sporting powder:—

1. The carbon and sulphur are pulverised in drums of wood containing 120 kilogrammes of bronze balls of a diameter of 6 to 7 millimeters.

The charcoal is first introduced by charges of 21 kilogrammes, and triturated alone for eight to twelve hours. Fifteen kilogrammes of sulphur are then added, which is turned with the charcoal for four hours.

2. Six kilogrammes of this binary mixture, and 20 kilogrammes of saltpetre are placed with 60 kilogrammes of bronze balls, of about 5 millimeters in diameter, in a drum of leather. The trituration of the ternary mixture is performed in the space of twelve hours, at the speed of about 20 to 25 revolutions per minute.

3. On being taken out of this, the mixture is wet with one to two per cent. of water, and placed under millstones of iron rimmed with bronze, of the weight of 2500 kilogrammes, which are called *light millstones*, to distinguish them from the *heavy millstones* used at the Government works. (Fig. 26.) The basin in which the millstones turn is of wood, and is charged with 50 kilogrammes of mixture, which is triturated during two hours. In checking the progress of

the mills at the end of the trituration, cakes (*galettes*), are formed, which contain two to three per cent. of moisture.

4. These are reduced into grains by means of a machine which gives motion to eight sieves or bolters, mounted on the same stand, and which grains about 80 kilogrammes of cake an hour.

5. The glazing is done in barrels of wood, divided into three or four compartments by transverse divisions, each of the capacity of 100 to 150 kilogrammes of powder. The grains, by rubbing on themselves and on the sides of the vessel, acquire, in the space of twenty-four hours, the hardness and polish required.

6. The drying is done by exposure to the sun, or in an artificial drying-house.

CARBONATE OF POTASH.

This salt is often known in commerce by the name of *vegetable alkali, salt of tartar, dulcified alkali*, or simply *potash*.

It is formed by the combination of carbonic acid with potash. It has a sharp, and slightly caustic taste; it is very soluble in water, and deliquescent; water, at the ordinary temperature, dissolves its own weight of it; it has an alkaline reaction. It is insoluble in alcohol, fusible at a red-heat, and indecomposable by heat.

Carbon, at a red-heat, acts on carbonate of potash, and gives rise to potassium; it is upon this reaction is based the preparation of potassium by the process of M. Brunner.

Lime, in presence of water, transforms the carbonate into hydrate of potash.

Preparation.—Vegetables contain potash united to different organic acids, such as *acetic, malic, oxalic, tartaric*, &c. When these salts are calcined, they are decomposed into carbonate of potash, which is found in the ashes of the vegetable.

The name of *potash of commerce* is given to the soluble part of ashes, evaporated to dryness.

Carbonate of potash from the lye of ashes, is not pure; it is always mixed with different soluble salts, such as the sulphate, chloride, and silicate of potash.

The quantity of real potash contained in the potash of commerce, varies with the constituents of the wood which is used for making the ashes.

The purest potash is that made from the ash of the birch, and the least pure from that of the pine. One hundred kilogrammes of ashes ordinarily yield about ten kilogrammes of soluble residue called *saline*. The salts which accompany the carbonate of potash being much less soluble than this last salt, the saline is often purified by treating it with its own weight of cold water, which dissolves the carbonate of potash, and leaves, for the most part, the other salts. The solution, evaporated to dryness, leaves carbonate of potash purer than saline. This is generally colored brown by organic matters; after it has been calcined in contact with the air, it becomes white, and is called pearlash.

In commerce, potashes bear names which refer to their origin. They are known as potashes of America, Russia, Vosges and Trèves, &c.

Uses.—Neutral carbonate of potash is used in the manu facture of soft soaps, of crystal glass, and Prussian blue. It is also sometimes used to transform into nitrate of potash, the nitrates of lime and magnesia contained in the saltpetre materials.

The neutral carbonate, submitted to the influence of an excess of carbonic acid, is changed into the bicarbonate, which is used in the treatment of gout and gravel.

SODIUM.

Sodium has a great resemblance to potassium. This body was isolated by Sir H. Davy, by decomposing soda with the pile. A short time after, Gay-Lussac and Thenard showed that sodium could be obtained by the action of iron on soda, under the influence of an elevated temperature.

Sodium is actually prepared by the process of M. Brunner, by decomposing the carbonate of soda by carbon, with the apparatus described in treating of the preparation of potassium.

Sodium is of a silver white, of a metallic brilliancy when recently cut; but it tarnishes in contact with the air. Its density is equal to 0·92. It fuses at 90° C., and volatilizes at a red-heat. Sodium is less volatile than potassium. It decomposes water like this last metal, at the ordinary temperature. When a piece of sodium is thrown upon water, hydrogen is disengaged; but the heat produced by the reaction of this metal on water, not being so great as by potassium, does not inflame the gas.

If water is thickened by dissolving gum in it, to impede the movements of the metal, or if the sodium is thrown into a glass containing but a few drops of water, there is less loss of heat, the metal becomes incandescent, and the hydrogen inflames. The other properties of sodium resemble those of potassium.

There is also a great analogy between the protoxide of sodium (soda), and the protoxide of potassium (potash).

CHLORIDE OF SODIUM.

The chloride of sodium, often called *sea salt*, is white, colorless, and of a salty but agreeable taste. It is slightly soluble in alcohol. Sea salt is nearly as soluble at the ordinary temperature, as at the boiling-point of water which is saturated with it; thus, the boiling and saturated solution of chloride of sodium, only deposits small quantities of it on cooling.

This property permits sea-salt to be separated from most other salts, and particularly from nitrate of potash, the solubility of which, in water, increases greatly with the temperature.

In fact, in treating a mixture of sea-salt and nitrate of potash with boiling water, a great part of the nitrate deposits on cooling, while the sea-salt remains in solution in the water.

Sea-salt crystallizes in cubes, or in aggregated forms produced by the symmetric arrangement of a number of small cubes; these crystals are anhydrous, and decrepitate strongly when heated to 200° or 300°. They will keep in the atmosphere in dry weather.

The chloride of sodium is fusible at a red-heat, and volatilizes at a still higher temperature, producing white fumes. This vaporization increases considerably in a current of gas.

Sea-salt fused, crystallizes in cubes on cooling. In this state, it does not decrepitate when heated.

Some oxides, and especially the oxide of lead, decompose sea-salt dissolved in water, producing a metallic chloride and caustic soda. This reaction takes place so readily, that it has been proposed in the arts to prepare soda by treating sea-salt with litharge; but then the soda always contains in solution a considerable quantity of oxide of lead. The process of Leblanc presenting, besides, incontestable advantages over all others, the use of litharge has been given up in the preparation of soda.

When a mixture of sea-salt well dried, and silex, are heated together, no reaction takes place; but if a current of steam is made to pass over the mixture, there is formed a silicate of soda, and chlorhydric acid. It is on this reaction is based the use of sea-salt in glazing some kinds of potteries, such as stone-ware. A quantity of sea-salt is thrown into the oven; this volatilizes, and meeting with the silex which exists in the paste of the pottery, and with steam, produces silicate of soda, which forms a vitreous coating on the surface of the pottery.

Uses.—The uses of sea-salt are numerous. It is used in the preparation of artificial soda, and of sulphate of soda; also in the glazing of potteries. It is used in preparing chlorhydric acid, in the manufacture of the decolorizing chlorides, and to produce chlorine. It is used for domestic purposes, and in agriculture, in large quantities. It is one of the most widely-spread salts in nature. It exists in large quantities in sea-water, in lakes, and salt springs. It is also deposited in abundant strata in the earth, and is then called *mineral* or *rock salt.*

ROCK SALT.

Rock salt is crystallized, and often in transparent masses of a milky white; it presents an easy cubic cleavage. It is sometimes met with in fibrous masses. It is generally colored grey by a small quantity of bitumen, and often has a reddish color, due to the presence of the oxide of iron.

Rock salt is sometimes of great purity, as is the case with that of Wieliczka; but often is mixed with sulphate of lime, clay, &c.

Some specimens of the salt of Wieliczka have a curious peculiarity, which has been examined by MM. Dumas and Rose.

When this salt is placed in water, a series of decrepitations are heard, and there is a disengagement of gas, which sometimes appears to be pure protocarburet of hydrogen, and sometimes a mixture of this gas with hydrogen and carbonic oxide. It is probable that the gas is imprisoned under more or less pressure between the molecules of the salt, the layers of which it breaks as soon as they are reduced in thickness by the action of water, and then produces a decrepitation. In France, attempts have been made to introduce pulverized rock salt into the market; but it has not yet been introduced into consumption, on account, no doubt, of the presence of foreign bodies which it always contains, and the slowness with which it dissolves in water. Rock salt has all the properties of common sea-salt. It is, however, attacked much more slowly than this last salt by sulphuric acid, and does not decrepitate with heat; in both these respects, it resembles melted sea-salt.

Extraction of Chloride of Sodium.—Rock salt is extracted either in the solid state, by shafts and galleries, or from its solution.

Extracted in the solid state, rock salt is exposed for sale in blocks, or after having been first broken. When it is not very pure, it is dissolved in water, and crystallized by evaporating the solution.

To extract the sea-salt held in solution in salt-springs, the waters are first evaporated in the open air in places called *drying-houses*, so arranged as to offer great surface for evaporation.

These drying-houses are composed of large sheds, in which are piled up bundles of thorny sticks. The salt water is brought to the top of the sheds in canals, which communicate with troughs leading to lateral gutters, which tumble the water over the fagots; this, in falling, is divided, and evaporates rapidly. The course of the water is changed to suit the direction of the wind, which has a great influence on the rapidity of the evaporation.

The drying-houses are generally divided into two sections. The first receives the water from the spring; the second, the waters which have already circulated over the fagots. Pumps placed at intervals, and moved by water-wheels, raise the water from the lower reservoirs to the conduits, which pour it out on the fagots.

As the water concentrates, it deposits on the fagots sulphate of lime, generally mixed with carbonate of lime and oxide of iron. These deposits are removed from time to time.

When the water is concentrated to a point corresponding to about 15 to 20 per cent. of salt, the evaporation is finished in boilers.

The salt contained in sea-water is extracted by subjecting this water to a spontaneous evaporation in large reservoirs, called *salt-pans.*

In cold countries, where the method of salt-pans cannot be used, the salt is extracted by exposing the sea-water to a very low temperature. The water divides itself into two parts; one solidifies first, and is nearly pure water; while the other remains liquid, and retains in solution all the soluble salts. In taking away, from time to time, the ice which forms, water is finally obtained highly charged with salt, which is then evaporated in boilers.

BORATE OF SODA.

This is called *Borax* in commerce. It exists in nature, and is found in Persia, India, and China.

Native borax crystallizes in hexahedral prisms. These crystals are impure, and are often found mixed with a fat matter whose composition is unknown. It is sometimes known in commerce by the name of *tincal.*

To purify tincal, it is treated with lime-water, which forms, with the fat matter, an insoluble compound; and the salt is crystallized in vessels of wood or lead. Borax is generally produced by uniting boracic acid directly with soda, which comes from Tuscany. The borate of soda is white, and of an alkaline taste and reaction; when it is heated, it melts, and forms, in cooling, a solid, transparent, and vitreous mass.

Uses.—Borax is used in soldering. When oxydizable metals are to be soldered, they are covered with borax, which, in melting, prevents oxidation, and dissolves traces of oxide which would prevent the soldering.

Borax enters into the composition of many glasses. It is chiefly used in the fabrication of very fusible glasses, and in the glazing of some potteries.

CARBONATE OF SODA.

Carbonate of soda is a white salt, inodorous, of a sharp and slightly caustic taste, and of an alkaline reaction. It is very soluble in boiling water, and crystallizes in large prisms which are hydrated.

Exposed to the air, it loses part of its water of crystallization, and effloresces.

Carbonate of soda is decomposed at a red heat by steam, which disengages all its carbonic acid, and produces soda.

Lime decomposes the carbonate of soda, by depriving it of its carbonic acid, and isolating the soda.

Preparation of Carbonate of Soda.—For a long time, the carbonate of soda used in the arts was extracted either from marine plants, such as fuci and sea-weed, or from certain terrestrial plants, as *salsola soda*, or *barilla*, which grow on the sea-shore. These plants are burnt, and from their ashes is extracted, by lixiviation and evaporation, salts more or less rich in carbonate of soda, which are called *sodas of sea-weed*, *of Alicant*, *of Carthagena*, *of Malaga*, *of Narbonne*, &c. But little has been done at working these natural sodas, since Leblanc discovered a method of obtaining carbonate of soda artificially, by decomposing by chalk and carbon, the sulphate of soda, which is produced by treating sea-salt with sulphuric acid.

This discovery is considered, justly, as one of the most

important which has ever been made in the industrial arts. The process of Leblanc, perfected by D'Arcet and Aufrye, is now exclusively used in the manufacture of carbonate of soda. We will describe it somewhat in detail.

Into a reverberatory furnace of an elliptic form, with a bottom of large surface constructed of fire-brick, is introduced a mixture of 400 kilogrammes of anhydrous sulphate of soda, 400 kilogrammes of dry chalk, and 135 to 140 kilogrammes of coal.

These materials are stirred, from time to time, with an iron poker; they soften down at a red heat, and acquire by degrees a pasty consistence, giving off a great quantity of gas, which burns with a blue flame. After four or five hours of calcination, the semi-fluid mixture is stirred anew, brought to the front of the furnace with an iron rake, and put into a sort of wheel-barrow of thick sheet-iron, where it is left to cool. This product is called *crude artificial soda.* The above mixture usually yields from 550 to 600 kilogrammes of crude soda, of 38 to 40°. Two workmen can manufacture many thousand kilogrammes per day.

Crude soda is of a bluish grey; it is slightly porous; exposed to moist air, it deliquesces, and becomes friable. When first made, it is hard. It is then pulverized or broken up, and subjected to the action of warm water, which dissolves all the soluble parts which it contains. The sulphuret or oxysulphuret of calcium, and the excess of charcoal which it contains, are separated by decantation. The solution is evaporated in iron boilers. The carbonate of soda precipitates to the bottom of the vessel; it is then taken out with ladles, as it deposits, and placed to drain.

The carbonate thus deposited, after its calcination in a reverberatory furnace, is often exposed for sale in this state.

To perfect its purification, it is sometimes subjected to a new solution, and then evaporated to dryness.

The product thus obtained is called, in commerce, *sal soda.*

Uses.—The carbonate of soda is used in the manufacture of glass and soaps. These consume it in enormous quantities.

It is also used in some operations of dyeing, and chiefly in the washing of threads and tissues. Crude soda, mixed with quick-lime and lye, furnishes a liquor used in the manufacture of Marseilles soap.

The neutral carbonate, treated with carbonic acid, is transformed into the bicarbonate of soda.

The bicarbonate is used in medicine in the preparation of pastilles of vichy, and in the treatment of calculous affections.

CALCIUM.

Calcium was isolated by Davy, with the pile. A small cupel of moist lime was filled with mercury; the positive pole was placed in communication with the metallic plate on which the cupel was placed, and the negative pole with the mercury: he thus produced an amalgam of calcium. This amalgam, distilled, gave pure calcium.

Potassium will decompose lime at a high temperature, and free the calcium.

Calcium is white, and of a metallic brilliancy; burnt in the air, it is transformed into *lime.*

LIME.

Lime is a base formed by the combination of oxygen with *calcium*, and is oxide of calcium. Lime was known to the ancients; it entered into the composition of the mortars used by them.

This base is white, caustic, and very alkaline. It greens the syrup of violets. When it is plunged into water and taken out, as soon as the air contained in its pores escapes, it hydrates, giving rise to a disengagement of heat as high even as 300° C., and produces a hissing noise, accompanied with thick vapors of water.

The high temperature developed by lime, with water, is sufficient to inflame gunpowder. Lime which is *fallen*, that is, reduced to powder by absorbing water, is often called *slaked lime*, to distinguish it from *quick-lime*, which is anhydrous lime. Lime suspended in water, constitutes *milk of lime*.

When lime is of good quality, and is placed in water, it increases in volume; such lime called *fat lime*. Lime-water is much used in laboratories as a reagent. To prepare it, some lime is put into a bottle, which is then filled with water. The excess of lime deposits, and the lime-water remains clear; this first solution is never pure, mostly containing some potash which existed in the lime. To prepare a pure lime-water, the lime must be washed three or four times, before leaving it definitively in contact with water.

Lime, in the anhydrous or hydrated state, absorbs carbonic acid from the air, and produces carbonate of lime; in carbonating, it becomes excessively hard, and transforms itself into a substance which has the composition of limestone, and often its hardness. It is from this important property that lime is used in making mortars.

Natural State of Lime.—Lime is never met with in nature in a free state; if this base was for a moment isolated, it would immediately absorb the water and carbonic acid contained in the air. It is found combined with carbonic acid, constituting all the varieties of carbonate of lime, known as chalk, marble, Iceland spar, &c. The shells of mollusca are almost entirely carbonate of lime. Lime, combined with sulphuric acid, constitutes plaster, a body widely spread in nature.

This base enters into the composition of bones, and is there found in the state of phosphate and carbonate of lime.

It is, besides, combined in nature in different proportions

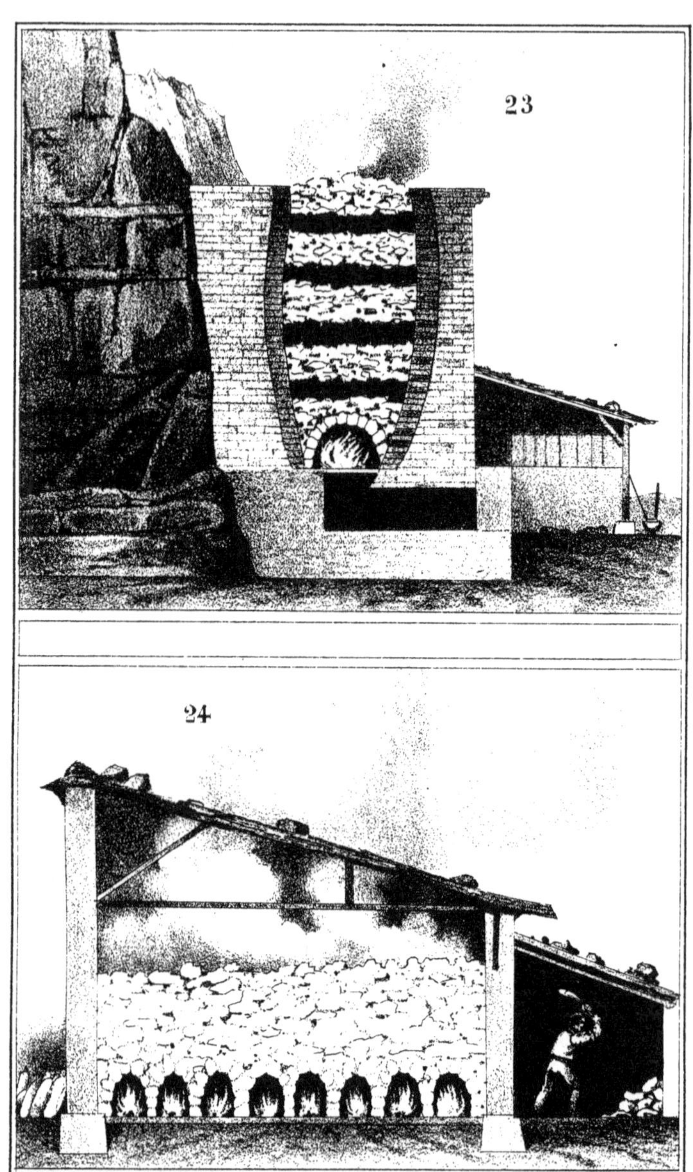
23
24

with silex. Most vegetables contain lime united with organic acids.

Extraction of Lime.—As lime is indecomposable by heat, those salts of lime are used for its preparation, whose acids can be driven out by a high heat. The nitrate of lime would do; but this salt not being abundant, the carbonate is always used.

Pure carbonates furnish, by calcination, *fat lime.* If it is impure, or especially if it is clayey, it leaves a *poor lime,* and hardens by contact with water.

When pure carbonate of lime is calcined, its decomposition is slow, and it requires a high heat. Different gases, such as the carbonic oxide, hydrogen, and especially steam, influence the decomposition of the carbonate of lime. When a porcelain tube is filled with pieces of lime, and exposed to a low red heat, no disengagement of carbonic acid gas is seen to take place; but if, as Gay-Lussac has shown, a stream of gas or steam is made to pass into the tube at this temperature, the decomposition of the salt soon takes place.

The influence of steam on the decomposition of carbonate of lime, has been long known to lime-burners, who know that a stone, while moist, decomposes much more easily than one which has been dried in the air. They often throw into their lime-kilns a small quantity of water, which, in volatilizing, hastens the decomposition of the lime-stone.

Different kilns are used for burning lime. In some places, the kilns are made of the lime-stone itself, resembling in some respects that used in forests for making charcoal. This is covered with turf and earth, leaving within a central chimney, by which the gases of combustion escape.

Sometimes, also, an intermittent kiln is used, the vault of which is made also of lime-stone.

These are now replaced by *continuous, perpetual,* or *draw kilns,* which are formed of truncated and inverted cones, Fig. 23. Lime and coal are alternately introduced into these kilns, in the proportion in bulk of four parts of stone to one part of coal. These kilns are filled up to the level of the upper opening. The fire is lighted below by means of fagots, and communicates gradually; the carbonate of lime loses its carbonic acid, and the calcination is considered finished whenever the smoke ceases to come out. The lime is then taken out below, and fresh quantities of lime and coal are added above.

Uses of Lime.—The uses of lime are numerous. Lime is the basis of mortars; it is used in tanning, and in the purification of illuminating gas, to absorb sulphydric and carbonic acids.

It is also used in the preparation of potash and soda, to carry off the carbonic acid from the alkaline carbonates.

Lime is used in the saponification of fat bodies intended for the manufacture of stearin-candles, and in the manufacture of sugars, in the operation called *defecation.* It is used in agriculture as a manure. When a soil is too clayey, a quantity of lime is mixed with it, which, by absorbing water and carbonic acid, slakes and makes the ground lighter, and assists vegetation.

This addition of lime also restores the calcareous element which vegetation carries off with each crop.

CHLORIDE OF LIME.

The name *chloride of lime, bleaching chloride,* is given to a body prepared by treating the hydrate or milk of lime with chlorine. This body is white, amorphous, pulverulent, and of a strong odor recalling that of chlorine; water dissolves it freely.

When treated with weak acids, it gives off chlorine.

Chloride of lime acts on organic substances and decomposes them: it destroys coloring matters: its action however on these matters is slow when it contains an excess of lime. Thus the blue color of litmus and chloride of lime may be mixed together, without the color being sensibly changed; but the intervention even of a very feeble acid displaces hypochlorous acid, and thus causes the decolorizing of the litmus.

Preparation.—To prepare chloride of lime, a stream of chlorine is made to pass through milk of lime or into slaked lime. The mass mixed through water, and decanted or filtered, gives a concentrated solution of chloride of lime.

Uses.—Chloride of lime is used in large quantities for bleaching cloths, and the paste for making paper, and in the manufacture of painted stuffs.

The miasmas of hospitals and dissecting rooms may be destroyed by chloride of lime; this should always be used in small quantity, because chlorine in excess, mixed in the atmosphere, would exercise an unwholesome influence on the lungs.

The name *eau de Javelle*, is given to a liquor obtained by passing chlorine into potash. Eau de Javelle disengages chlorine under the influence of acids, and may be used as a decolorizer and disinfecter.

Chloride of lime and eau de Javelle then act like Chlorine when treated with acids; in the arts they are preferred to chlorine, because their odor is less active, their action slow, successive, and continued, and finally, they are more easily transported.

SULPHATE OF LIME, GYPSUM, PLASTER.

Gypsum, or plaster, is sulphate of lime; this salt is crystallized in large transparent tables, which are ordinarily grouped in *fer de lance.*

When the crystals of sulphate of lime are opaque, they are called *gypsum alabaster.* This must not be confounded with real alabaster, which is carbonate of lime.

Hydrated sulphate of lime is insipid, or of a slightly bitter taste; it is colorless, and indecomposable by heat.

This salt is equally soluble in hot or cold water, for at +10° to 100° C., 1000 parts of water dissolve 3 parts of plaster. It is more easily dissolved in concentrated sulphuric acid; it is completely insoluble in alcohol mixed with water: it contains 20.9 per cent. of water: it is entirely dehydrated at a temperature below 200°, particularly in a current of gas.

The hydrated sulphate of lime has the hardness of stone; after it is dehydrated, it becomes pulverulent and like flour. When sulphate of lime dehydrated is placed in contact with water, it hydrates anew, combines with 2 equivalents of water which heat had caused it to lose, and resumes its original hardness. On account of this property, sulphate of lime is used in building; when it is calcined, it dehydrates; when water is added, it takes up precisely the quantity necessary to restore it to its original hardness.

The calcining of plaster is done in furnaces, the vault of which is made with the stones of the plaster itself. (Fig. 24.)

The temperature ought not to be very high, for a temperature of 200° C. is sufficient; after the calcining, it is reduced to powder in mills.

Calcination at too high a temperature causes the plaster to undergo a fusion (sorte de fritte) which prevents its hydrating easily.

Plaster once calcined ought to be protected from the moisture of the air, or it will by degrees hydrate, *become dead*, and then lose a part of its good quality.

Plaster well prepared ought to produce heat when mixed with water. Often, indeed, its quality is tested by the temperature it produces in hydrating.

Plaster often gives off sulphuretted hydrogen when mixed with water, owing to the presence of a quantity of sulphuret of calcium produced by the action of carbon or carburetted gases on the sulphate of lime; this sulphuret gives off traces of sulphuretted hydrogen under the influence of water and carbonic acid.

Plaster, in solidifying, increases in volume; this property renders it eminently fit for moulding, for in dilating itself it takes the impression of the finest marks.

The most esteemed plaster is that found in the vicinity of Paris.

The hardness of plaster depends altogether on that of the hydrated sulphate of lime which it produces. This hardness is found to be, after mixing, what it was before calcining, that is, what it was in the stone itself.

The sulphate of lime transforms itself into sulphuret of calcium, under the influence of organic matters in a state of decomposition. This sulphuret, afterwards decomposing by the action of carbonic acid, gives rise to the disengagement of sulphydric acid. In this way can be explained the presence of sulphydric acid in certain waters which originally contained sulphate of lime. M. Chevreul was the first to show that a like decomposition might take place in the soil in some large cities, which contain, like that of Paris, a large quantity of sulphate of lime. This sulphate, in changing into sulphuret, might become in time a cause of insalubrity. M. Chevreul proposed aerating the soil of large cities, so as to change the sulphurets found there into sulphates.

Uses.—Sulphate of lime is used in buildings as a cement, as it presents the advantage that it solidifies in a few minutes.

The name of *stucco* is given to plaster mixed with water containing glue, and sometimes gum in solution. Stucco is easily polished, and has the appearance of marble.

Various colors may be given to it: often pieces of marble are put into the stucco before solidification, which are then polished with the stucco.

Stucco does not resist moisture, but may be used in the interior of houses.

Stucco with lime is a composition obtained by mixing lime

with pulverized marble; in its composition it presents no analogy with that made with sulphate of lime.

For some time, a peculiar cement has been made with plaster, called *aluminated plaster*. This body, which is prepared by calcining plaster with alum, has, like plaster, the property of solidifying rapidly when slaked with water; but it becomes harder, and produces a mass which has both the hardness and semi-transparency of marble. It appears to resist the influence of moisture.

M. Kuhlman has shown that plaster may be made very hard by subjecting it to the action of a solution of silicate of potash, which produces on the surface of the plaster a coating of silicate of lime.

Plaster is used in agriculture: it increases the growth of some plants, particularly leguminous plants.

CARBONATE OF LIME.

Carbonate of lime, from its numerous uses and its abundance in nature, is one of the most important salts.

It presents itself in different states bearing the name of *Iceland spar*, *aragonite*, *limestone*, *marble*, *chalk*, *alabaster*, and *lithographic stone*.

Carbonate of lime assumes variable forms, which, however, may all be reduced to two principal ones, which are the form of calcareous spar and that of aragonite.

These two forms are incompatible.

Properties.—Pure carbonate of lime is perfectly white; small traces of foreign matters are sufficient to color it. The color of marbles is generally ascribed to metallic oxides.

The hardness of carbonate of lime varies greatly with the different varieties of this salt. Thus, it is well known that marble is harder than lime-stone, and the hardness of this latter is greater than that of chalk. This depends, probably, on the mode of formation of the carbonate of lime.

It decomposes, at a red heat, into carbonic acid and lime.

It is on this property is founded the manufacture of lime; nevertheless, this decomposition does not take place when the carbonate is burned in a hermetically closed vessel.

Hall showed that if chalk is introduced into a gun-barrel closed at both ends, the calcareous carbonate, instead of decomposing, fuses, and forms a body which has all the properties of marble.

It was attempted, some years ago, to make artificial marble, by the fusion of an amorphous carbonate of lime. A manufactory was established at Paris, where white or variously-colored marbles were made by melting pure chalk, or chalk mixed with the metallic oxides. This enterprise fell through; but the problem of the manufacture of artificial marbles is nevertheless resolved.

The experiment of Hall, too, has explained the presence of crystallized carbonate of lime in soils of igneous origin. Carbonate of lime is insoluble in water, so that it can easily be produced by treating a soluble carbonate with a salt of lime. When this double decomposition takes place at the common temperature, a crystalline precipitate is obtained, which, under the microscope, presents small crystals having the form of Iceland spar. If the decomposition is made hot, the crystals which are produced belong to the crystalline system of aragonite.

Carbonic acid dissolves the carbonate of lime, and gives rise to a bicarbonate soluble in water.

This property accounts for many natural phenomena. Whenever water, holding carbonic acid in solution, passes over calcareous deposits, it dissolves the carbonate, and gives rise to a bicarbonate of lime. As the bicarbonate is not very stable, and as it readily loses half of its carbonic acid, this salt may, in a great number of circumstances, form deposits of insoluble carbonate of lime.

Neutral carbonate of lime, proceeding from the decomposition of the bicarbonate, produces—

1. The calcareous deposits which often obstruct water-pipes, particularly about the joints.

2. The calcareous deposits in steam-boilers.

3. The crystalline deposits called *stalactites* and *stalagmites.* When water, holding carbonic acid in solution, passes over calcareous rocks, it dissolves, as we have said above, the carbonate of lime; then, on reaching the interior of caves, each drop evaporates, and deposits the neutral carbonate of lime. When these concretions form under the roofs of caves, they are called *stalactites;* if they are produced on the ground where the water falls, they are called *stalagmites* (Fig. 27). When the stalagmites have yellow and red zones, they form the *oriental alabaster*, which will often take a beautiful polish.

4. The calcareous incrustations are also produced by the carbonate of lime; when different objects, such as fruit, birds'-nests, &c., are exposed to the action of certain mineral waters holding bicarbonate of lime in solution, these objects become covered with calcareous incrustations, often known as *petrifactions.* The most celebrated incrusting springs are the Sprudel at Carlsbad, which produce a calcareous deposit, zoned, and of great fineness, which is used for objects of ornament. The waters of San Filippo, in Tuscany, and of St. Allyre, in Auvergne, are also examples.

5. The calcareous tufs or gravel stones, which are very abundant in some countries, and which are used for building stones, have also the same origin. Many towns in Italy are built with calcareous tufs riddled with small holes, and evidently produced by calcareous deposits formed from the decomposition of bicarbonate of lime.

MARBLES.

Marbles belong to two calcareous varieties: the saccharoid and compact varieties.

The saccharoid carbonate of lime is formed of small crys-

27

tals white and brilliant like sugar. It is rarely colored; it often passes into the lamellated texture like the Parian marble. This calcareous variety furnishes the statuary marbles; that of Carrara, the grain of which is very fine, is the most esteemed. It also furnishes ornamental marbles, among which we will mention as the chief, the *blue turquin*, colored yellow by traces of bitumen; *yellow antique marble*, colored yellow by the hydrate of the peroxide of iron; *cipolin marble*, marked with wide undulating belts white and green, resulting from the association of the white calcareous saccharoid, and green talcy schist. The compact calcareous varieties are very numerous, and in general are ornamental marbles. This variety is distinguished into

1st. The *black antique*, which is a uniformly black marble.

2d. The *small granite*, which is blackish with bright tints.

3d. *St. Anne marble*, which has white veins on a black ground, or a dark grey.

4th. The *small antique*, showing a mixture of white and black spots, principally found in Belgium.

5. *Portor marble*, found in the neighborhood of Genoa.

6th. *Griotte* (spotted) *marble*, brown ground with red spots.

7th. The *marble of Sarancolin*, found in the Pyrenees.

8th. The *marble of Languedoc*, or carnation marble.

9th. *Florentine marble* or uniform, which is argillaceous, compact, of a yellowish grey, etc.

MAGNESIA.

Magnesium only combines in a single proportion with oxygen, forming an oxide called *magnesia*.

This base is prepared by precipitating a salt of magnesia by potash in excess, or, better, by subjecting the carbonate of magnesia to calcination.

Magnesia is white, pulverulent, insipid, inodorous, fixed,

infusible in the fire of a forge, and almost insoluble in water. It possesses a feeble alkaline reaction, turns to green the syrup of violets. When placed in contact with water, it hydrates slowly; if exposed to the air, it absorbs both carbonic acid and moisture.

All the salts of magnesia, except the citrate, have a bitter taste; this property sometimes gives to magnesia the name of *bitter earth.*

Uses of Magnesia.—Magnesia is used in medicine to saturate the acids which are developed in the stomach during indigestion: it is useful in cases of poisoning by the acids, even by arsenious acid.

M. Bussy has shown that magnesia combines directly with arsenious acid, forming with this acid an insoluble compound, without action on the animal economy, and that it cannot be replaced as an antidote to arsenious acid, by the carbonate of magnesia, which does not act on this acid.

ALUMINE.

Alumine is a base formed by the combination of oxygen with a metal called *aluminium.*

This oxide exists largely in nature; it is found in clays, marls, feldspar, mica, etc., and in a great number of minerals. When alumine is pure, it is called *corundum.* Corundum, after the diamond, is the hardest substance known.

It is called *corundum hyaline*, or *white sapphire*, when it is transparent. If it is colored red, it is called *oriental ruby;* if blue, it is called *sapphire;* if green, *oriental emerald;* when it is yellow, it is called *oriental topaz;* and if violet, *oriental amethyst.* Under these varieties, corundum constitutes the precious stones whose value is often comparable to that of the diamond. Emery is corundum which contains a quantity of iron. It is employed for cutting agates, polishing glass, metals, &c.

Pure alumine is white, and adheres to the tongue. It is infusible in the highest temperature which can be produced in furnaces; it fuses under the oxyhydrogen blow-pipe, and becomes very fluid; it cannot, like silex, be drawn into threads. M. Gaudin obtained *artificial rubies* by melting it with traces of chromate of potash. Alumine is insoluble in water, and dissolves in acids, if it has not been calcined; but if it has been subjected to a high temperature, it dissolves with difficulty. Alumine is completely soluble in potash and soda.

It is indecomposable by heat. When heated with nitrate of cobalt, it forms a compound of a beautiful blue, which has received the name of *Thenard blue.*

Alumine, exposed to the air, does not absorb carbonic acid. Carbonate of alumine is as yet not known.

Hydrated alumine is obtained by precipitating a salt of alumine by ammonia, or, better, by the carbonate of ammonia. It forms a gelatinous precipitate. Anhydrous alumine does not combine directly with water; but the hydrate obtained by precipitation retains its water with force, and only entirely gives it up at a red heat.

Alumine condenses a considerable quantity of moisture, and its weight increases 15 per cent. Agriculture profits greatly by this property; it is alumine, in fact, which, existing in variable quantities in different soils, preserves the moisture, which is useful in vegetation.

Hydrated alumine can combine with most coloring-matters, and give rise to insoluble compounds called *lakes.* If, for example, alumine is precipitated in a solution of Brazil wood, the coloring-matter forms, with this base, an insoluble compound, and the liquor is completely decolorized. This property extends to the salts of alumine, which are used, in dyeing, to fix the coloring-matter in the stuffs; they are called *mordants.*

ALUM.

Salts may sometimes be combined together to form what are called *double salts*. Of these, one of the most interesting on account of its applications, is *alum*, which results from the combination of sulphate of potash with sulphate of alumine.

This salt is white, and its taste is astringent and acid. It is very soluble in warm water.

Alum, exposed to the air, effloresces slowly. It crystallizes in octahedrons, or in cubes.

Cubic alum appears to have the same composition as octahedral, and when it is dissolved in cold water, it deposits by spontaneous evaporation, octahedric crystals of alum. These two forms often combine.

Alum, subjected to heat, fuses at a temperature of 92° C. Cooled in this state, it preserves its transparency; it is called *rock alum*. On continuing to heat it, it loses water, swells out greatly, and forms a voluminous and opaque body, resembling somewhat a mushroom. It is used in medicine as a caustic, under the name of *burnt alum*.

Preparation.—There exists at Pozzoli, near Naples, a stone which contains alum ready formed. It is pulverized, and submitted to the action of water; the liquors deposit, by evaporation, octahedric crystals of alum.

There is found in Italy, at Tolfa near to Civita-Vecchia, a stone called *alum stone* or *alunite*, which resembles ordinary alum in its composition, combined with an excess of hydrated alumine. When this stone is slightly calcined, almost two-thirds of the alumine which it contains is dehydrated, and this excess of base becomes insoluble. The heat should be watched, for too high a temperature would decompose the alum completely. The residue taken up by water, gives the alum known as *Roman alum*.

The greatest part of the alum used in France, Germany,

and England, is extracted from an aluminous schist containing sulphuret of iron, and bituminous matter; the sulphuret of iron exposed to the air is converted into sulphate of iron and sulphuric acid. This oxidation, produced in presence of schists containing alumine, gives rise to mixtures of sulphate of iron and sulphate of alumine. The mass is taken up by water, and the solution evaporated; the sulphate of iron deposits in crystals, while the sulphate of alumine remains in the mother-waters; a salt of potash is then added, which causes the precipitation of the alum; this operation is called the *brevitage* of the liquors. The alum is then purified by crystallization.

Manufacture of Alum from clay.—Alum is sometimes made by treating clay calcined with sulphuric acid. This calcination is intended to peroxidize the iron found in the clay, to render it less soluble in sulphuric acid. The sulphate of alumine thus obtained is precipitated by sulphate of potash; crude alum is formed, which is purified in the same way as alum produced from the aluminous schists.

Uses.—Alum is used in the manufacture of prints, as a *mordant;* its presence determines the combination of the coloring matter with the tissue; it is also used in the preparation of sheep-skins, the sizing of paper, and the clarification of liquids.

FELDSPAR.

Mineralogists comprehend, under the general denomination of *feldspar*, those minerals formed by the combination of silicate of alumine with different silicates.

Orthose, (ordinarily called *feldspar*, *petuntze*, *adularia*, *ortho-clase*,) scratches glass; it fuses in the fire of a kiln for porcelain, and produces a glass which is always milky. Orthose is used in preparing porcelain, of which it forms the coating: it is rarely pure, and is ordinarily found mixed with quartz.

KAOLIN.

The *Kaolins*, in the crude state, are friable minerals, often very white, giving a short paste with water.

They are in general formed of grains of quartz or sand, of small fragments of silicates with different bases, and of a kaolinic clay which forms the chief part of them.

The experiments of M. A. Brongniart have demonstrated that the kaolins proceed from the decomposition of feldspar rocks, which are slowly transformed into silicates of potash soluble in water, and into basic silicate of alumine, which constitutes kaolin.

There exists in France, in the neighborhood of St. Yrieix, near Limoges, a bed of kaolin, which supplies a large number of porcelain factories.

CLAY (ARGIL).

The name of *clay* is given to aluminous substances which are eminently plastic when saturated with water.

Clay, being freed by washing from the foreign substances which it contains, may be represented by a silicate of alumine of very variable composition, containing 18 to 39 per cent. of alumine, 46 to 67 per cent. of silex, and 6 to 19 per cent. of water.

Clay ordinarily contains foreign matters, such as the debris of feldspar rocks, of quartz, of pyrites, of carbonate of lime, of traces of organic substances and free silex, etc.

The proportion of potash contained in clay may reach, according to M. Mitscherlich, four per cent.

Clay at first does not mix readily with water, but it then tempers, and forms a pliant and ductile paste. This property causes it to be used in the manufacture of pottery. Subjected to calcination, it loses its water, cracks and shrinks considerably, and becomes hard enough to strike fire.

If clay were entirely pure, it would be infusible in the

highest temperature of the furnace; but the lime, potash, and oxide of iron which it contains, always render it fusible.

MARL.

Marls are earthy matters chiefly composed of clay, carbonate of lime, and silex, in variable proportions. They are used in the manufacture of delft-ware and earthen-ware. They effervesce with acids, make a short paste with water, and are more or less fusible.

They are known as *argillaceous, calcareous*, and *muddy* marls.

Marls which have the property of crumbling in the air, are used in agriculture to render clayey lands friable; they besides furnish them with the calcareous element useful to vegetation.

OCHRE.

The name of *ochre* is given to clay colored by the hydrate of the peroxide of iron. Yellow ochre calcined forms red ochre; sometimes, natural red ochres are found.

Umber earth is a hydrate of the peroxide of iron, mixed in variable proportions with clay, and the hydrate of the peroxide or sesquioxide of manganese.

FULLERS' EARTH.

A clay used in the scouring of woollens and cloths, is called *fullers' earth.* Before being used for this purpose, it is washed, to free it from the pebbles which it generally contains. By placing the fullers' earth on a greased cloth, which is then passed over a cylinder, the clay absorbs, by its capillarity, all the greasy matter from the cloth.

GLASS.

The name of *glass* is given to a substance fusible at a high temperature, brittle, hard, transparent, and insoluble in water, formed by the combination of silicate of potash, or silicate of soda, with one or more of the following silicates:

silicate of lime, silicate of magnesia, silicate of barytes, silicate of alumine, silicate of iron.

When the silicate of lime is replaced by the silicate of lead, the glass is called *crystal*. Potash is always its base.

General Properties of Glass.—All glass undergoes a complete fusion with heat. The nature and proportion of the bases which it contains, exercise a great influence on its fusibility; it may be said, in a general way, that potash, soda, and the oxide of lead, increase the fusibility of glass, while alumine and lime diminish it. Glass, with soda for a base, is more fusible than that with potash. Glass is elastic, and very sonorous.

Glass having several bases, when heated under some circumstances, undergoes an alteration called *devitrification.*

The devitrification of glass takes place when it is melted, and left to cool very slowly, or when it is heated to the point of softening, and kept in this state of demi-fusion for a long time, and then left to cool gradually. Devitrified glass is very hard, fibrous, opaque, less fusible than transparent glass, and a better conductor of heat and electricity than common glass.

This alteration in glass was first noticed by Réaumur, and afterwards studied by Messrs. Dartigues, D'Arcet, and Dumas. It is due to a crystallization of silicates in definite proportions, which is infusible at the degree of heat which suffices to fuse or soften glass.

The production of silicates difficult of fusion, and crystallizable, proceeds from the volatilization of a part of the alkaline base which is in the glass, or a simple separation between the silicates which constitute the glass.

Devitrified glass possesses the hardness of stoneware, and often the whiteness of porcelain. It strikes fire with steel, and bears changes of temperature better than glass. Reaumur noticed that devitrified glass presented, to a certain degree, the aspect and infusibility of porcelain; so that it is often called *porcelain of Réaumur.*

The devitrification of glass, and especially of very calcareous glass, is easily effected by heating it strongly in sand, which absorbs the portion of the alkali which volatilizes.

The glasses most fit for devitrification are those which contain the most alumine; then those charged with lime. Glass, with potash and oxide of lead for a base, devitrify with difficulty.

Glass heated to the point of softening, and suddenly cooled, becomes very brittle; when it has been subjected to a very slow cooling, on the contrary, it will resist, without breaking, rather sudden changes of temperature. Glass, suddenly cooled, undergoes a sort of hardening, and is found to be in a peculiar physical condition. By dropping some melted glass into cold water, and letting it harden, small ovoid masses, terminating in a point, are obtained, which are called *Prince Rupert's drops.* The vitreous mass is then in a forced equilibrium, which is preserved by the solidity of the parts on the surface, and which is destroyed so soon as a solution of continuity in the enveloping surface is produced, or if a part is broken off; so that, when the point is broken off, the *Prince Rupert drops* are immediately reduced to powder with a slight explosion.

The same effect is produced with what is called the *philosophic phial.*

This is a sort of short and thick tube, closed at one end. This tube is made by suddenly cooling some of the glass which the workman takes up on the end of his blow-pipe, to judge of the quality of the matter contained in the melting-pot. The interior layers of the tube being more slowly cooled than the outside, the glass is found to be in about the same condition as that of *Prince Rupert's drops;* so that the slightest jar within the tube, a small marble, for example, dropped in, is sufficient to cause its rupture.

The same phenomena are produced in thick glass exposed for sale, without having been slowly cooled.

In order to avoid these effects, glass should be subjected to a very slow cooling process, called *annealing.*

The annealing is done either in an oven, the temperature of which is gradually reduced, or in long galleries heated from one point, in which the glass is placed in sheet-iron cases, moving slowly along in an endless chain. The glass to be annealed is put into the oven at one side, and taken out at the other.

Glass which has not been annealed, cracks very easily with sudden changes of temperature. The workmen apply this property for the purpose of detaching from their blow-pipes articles of glass which they are forming. After the glass has been annealed, it is cut with a diamond with curvilinear edges.

Glass, in passing from the liquid to the solid state, has the property of remaining for a long time in a soft condition. The workman profits by this malleable property of glass, shaping it after various patterns; it may even be reduced into threads which have the fineness of silk, and may be made into cloth. The density of glass varies with the nature of the bases which enter into its composition. The calcareous alkaline glasses are the lightest, and those with lead for a base are the heaviest. The great density of these last glasses, renders it difficult to obtain them in a homogeneous state; heavy strata are always formed in their mass, which are with difficulty distributed uniformly through them. The deoxygenating bodies, under the influence of heat, act on glass containing oxides of iron, manganese, copper, and particularly lead. In this case, the oxide is reduced, and the glass becomes of a black color. This phenomenon manifests itself especially when glass is heated in a current of hydrogen, or simply in an enameller's lamp.

Well-made glasses are considered to be insoluble in water; nevertheless water after a time acts on them, and tends to decompose them into soluble alkaline silicate, and into insoluble earthy silicate.

The glass of old houses has on its exterior a rough surface, which is produced by the action of water on the glass. Some glass, such as crown glass, glass for tumblers, is often sufficiently hygrometric to be covered with a deposit of water, when it is exposed to moist air. Alkalies in excess, heated with glass, render it attackable by acids.

Acids after a time act on all glasses; they tend to seize upon the bases, and separate the silex. This action is shown by letting sulphuric acid remain in a glass bottle: this acid forms sulphates with the bases of the glass, and sometimes after a time pierces the bottle.

Manufacture of Glass.—Glass is ordinarily composed of silex, carbonate of soda or potash, sulphate of soda, carbonate of lime, and red lead. It has also been proposed to employ sulphate of barytes, volcanic lavas, and feldspar.

Silex, under the influence of heat, decomposes the alkaline and calcareous carbonates, and forms with their bases fusible double silicates.

The carbonate of soda may be replaced in the manufacture of glass by sulphate of soda, which is cheaper. The sulphate of soda not being decomposable by silex, except at a temperature which would cause a rapid alteration of the melting-pot, a small quantity of carbon is mixed with it; this takes away a part of the oxygen of the sulphuric acid, and facilitates the formation of the silicate of soda.

In general, 1 part of carbon and 13 parts of sulphate of soda are used. It must, however, be understood, that though the manufacture of glass with the sulphate of soda is more economical, the products made are not so beautiful as those made with the carbonate of soda.

The matters to be vitrified are nearly always fritted, and then subjected to a bright red heat in pots of fire-clay.

As far as possible to drive out the bubbles, which are constantly produced in the manufacture of glass, the melted glass is kept at a very high temperature for a long time before using it; the disengagement of the bubbles of gas is also increased by introducing into the glass arsenious acid, which, while acting as an oxidizing body, volatilizes and carries off the bubbles which would remain suspended in the vitreous mass.

Besides these bubbles, the glass may also contain *white knots, spots, streaks* and *lines:* the white knots are pellets of unmelted sand, the spots, streaks and lines, proceed from a want of homogeneity in the vitreous mass, and are due to a cold fusion (fusion froide), or to an incomplete mixture of the components before being put into the furnace.

Certain foreign bodies contained in the alkaline salts used in the manufacture of glass may also injure its transparency. Such are the sulphates which remain in the vitreous mass in consequence of a cold fusion, and form what are called *fiel de verre*, or *salt of glass*, (*sandevir*).

When the sand used in the manufacture of glass is highly ferruginous, the oxide of iron is reduced by the carbon which is added to decompose the sulphate of soda, and forms a silicate of the protoxide of iron, which gives to the glass a deep green tint. This glass is bleached by peroxidizing the protoxide of iron, and transforming it into sesquioxide, which forms a silicate almost colorless. With this end the peroxide of manganese is used, which goes by the name of *soap of glass-houses.*

The peroxide of manganese ought not to be used in excess; it would give to the glass a violet color.

To conclude, in order to avoid all the defects in glass, the substances with which it is made ought to be first purified with care; and above all, the temperature in the furnaces should be raised as high as possible. This is the only way to obtain handsome products in glass-works.

We shall now examine the principal kinds of glass.

Bohemian Glass.

This glass, remarkable for its lightness and its whiteness, is a double silicate of potash, of lime and alumine.

The silex used in Bohemia for the manufacture of fine glass, is hyaline quartz, generally obtained from the round pebbles of torrents. The quartz is broken (*etonné*) by being heated in furnaces, and then thrown into a large vessel full of water.

Bohemian glass is made in open pots; the fusion is always done with wood, which is abundant in this country. By its use, the coloration which is produced by the smoke of coals is avoided.

Bohemian glass is very hard to fuse; it owes this quality to the large proportion of silex which it contains. Tubes for organic analysis are made of this glass in preference to all others, on account of the difficulty of fusing it. Common glass may be melted in Bohemian glass, without deforming it.

Crown Glass.

This glass, like the Bohemian, is a silicate of potash and lime. It is used chiefly for making optical instruments; united with flint glass, it forms achromatic object-glasses.

Crown glass ought to be of a perfect limpidity, free from streaks or bubbles, and sufficiently colorless not to show coloration, even in very thick masses.

The preparation of crown glass offers very great difficulties. It is hard to find large blocks of glass which have not differences of density from one point to another in the mass.

Crown glass, fit for the fabrication of lenses, is made by stirring the melted glass, with cylindrical agitators of pure fire-clay, until it becomes entirely solidified. The mass is then cut so as to choose the parts of the melting-pot which contain the most homogeneous glass. This ingenious method of stirring glass by means of an agitator which does not color the mass, was used for the first time by M. Guinand. It has been perfected by Messrs. Guinand, Jr., and Bontemps.

Window Glass.

This glass, as it is made in Paris, is always a double silicate of soda and lime. The soda in window-glass proceeds from a mixture of sulphate of soda and carbon. We will describe the process followed in the French glass-works, to manufacture window-glass.

When the glass is purified and skimmed off, the blower plunges repeatedly an iron blow-pipe into the melted glass, and takes up a mass of glass of considerable size. By blowing into the pipe he forms a spheroid, and at the same time gives a motion to the blow-pipe like that of a bell-clapper. The globe elongates considerably, both by its own weight, and by the blowing. The blower then places the closed extremity of the elongated globe in the furnace, keeping his finger on the other end of the pipe. That part of the glass which is heated, fuses; the air contained in the glass globe dilates, and bursts it. He then turns it briskly, so as to enlarge the opening, detaches it by throwing some cold water on the end of the pipe, and then proceeds to spread it out.

He traces with water a straight line along the length of the cylinder, and then passes a red-hot iron over the wet mark; the cylinder splits in two, and is carried to the oven to be spread out.

As it heats, it opens, and soon forms a plate of glass, which is spread out by means of a plane of wet wood. The plate is then placed in the furnace for annealing.

Looking-Glass.

This glass, like the preceding, is a silicate of soda and lime, and differs only in its proportions.

Looking-glass ought to have great transparency, and should have neither bubbles, striæ, nor knots.

Great care is taken in the choice and purification of substances used for the preparation of looking-glass; this glass always has a green tint, which characterizes all glass with soda for base.

Two sorts of melting-pots are used in the manufacture of this glass, called *pots* and *basins* (*cuvettes*).

The substances to be melted are placed in the first, which remain there sixteen hours; they are then poured into the *cuvettes*, where they refine for sixteen hours; at the end of thirty-two hours, the glass is fit to be poured out.

The glass flows out on iron tables which are first heated. It is spread out on the table by means of a cylinder or roller.

The glass is introduced into the annealing furnace, and there divided by a diamond into pieces; the defective parts are put to one side, and then follow the processes of *smoothing* and *polishing*.

The glass reduced to the desired dimensions, is fixed with plaster to a stone table. It is rubbed with a smaller glass, placing between, at first coarse sand, and then finer sand and emery mixed with a large quantity of water. This operation is called *grinding*.

The last operation, called *polishing*, is done by rubbing the glass with a heavy polisher, covered with felt: there is interposed between the glass and polisher colcothar (oxide of iron) of different degrees of fineness.

GLASS FOR BOTTLES.

Glass for bottles contains a little potash and soda, a large quantity of lime and alumine, oxide of iron, and a little oxide of manganese.

Yellow sands are used in this manufacture, new ashes, and the soda of sea-weed, spent ashes, which are called *charrée*, the residuum from the lye of the soda of commerce and common clay

The green color of glass for bottles does not injure the sale; the essential point is to produce economically glass of great toughness. The brown color is due to the protoxide of iron or to the intermediate oxide of iron. The glass for bottles made on the borders of the Rhine is colored yellow by a mixture of peroxide of manganese and peroxide of iron.

It is common to introduce into the composition of glass for bottles a large quantity of *calcin* (broken bottles.) But it now seems to be understood that the *calcin* makes the glass more dry (*sec*) and brittle. Good glass-works do not permit any but new materials to enter into the fabrication of glass bottles.

The blower who forms the bottle, collects a quantity of the melted glass by means of his pipe; he blows while turning the glass in a mould which has the form of the bottle, then takes it up, *thrusts in the bottom*, detaches it, and makes the rim for strengthening the neck by drawing around it a drop of melted glass. The bottle is then taken to the furnace for annealing.

CRYSTAL.

The name of *crystal* is given to glass which has potash and oxide of lead for a base. The high price of the materials which enter into the composition of crystal, and the great care necessary in its fabrication, make crystal glass an expensive luxury.

The substances used for making crystal ought to be very pure; the silex ought to be as free as possible from iron and organic matters. The sands of Aumont, Etampes and Fontainbleau, are those which are preferred in glass-works; the great object is to procure fine sand. The fineness of the grain is an essential condition for a good mixture with the solvents. The best way to determine the purity of the sand to be used in the manufacture of glass, is to heat it in contact with air at a high temperature. During the calcination, the iron contained in the sand passes to the state of peroxide, and produces a reddish color. The sand which becomes the least colored is the purest; it is however almost impossible to find sand entirely free from iron.

The carbonate of potash requires a preparatory purification, which consists in dissolving this salt in water, and freeing it by crystallization from the sulphates and chlorides which it may contain. The carbonate of potash being more soluble than the preceding salts, remains in the mother-waters, from which it is extracted by evaporation to dryness.

Vessels and instruments of iron only should be used in the manufacture of crystal; vessels of copper would be attacked by the potash, and would color the crystal green. Attempts have been made, in vain, to make white crystal glass by replacing the carbonate of potash by the carbonate of soda. This last salt always gives a greenish tint to the mass.

The litharge of commerce cannot be used, because it contains oxides of copper and iron, which would color the vitreous mass. It is replaced by minium, which is purer.

The proportion of substances used in making crystal glass, varies with the nature of the fuel used, and with the temperature of the furnace. The higher this temperature is raised, the smaller the quantity of solvents which will be required; which not only produces a better quality of glass, but with more economy, because the solvents are really the expensive part of the operation; and in other respects, under the same circumstances, the crystal which contains the most silex, is the whitest and most brilliant. A well-constructed furnace, and combustibles of good quality, are essential elements for a good article. In order to obtain a good glass, the fusion should not be too rapid. If, for example, the mass should be completely melted at the end of from fifteen to eighteen hours, it would be best

17 *

to keep it in a state of fusion for five or six hours longer, so that the glass may refine, and free itself from the bubbles of gas which are found suspended in the mass.

The glass being once melted and refined, the foreign substances which mount to the surface in the melting-pot are removed; and by means of the blow-pipe, which is a tube of iron, the quantity of crystal required to make a particular article is taken up. By blowing into the pipe, and using a few simple tools, the workman fashions a great variety of articles.

Crystal glass may also be run into moulds of iron or copper. For some years past, in France, the Bohemian method has been practised, which consists in moulding the crystal in wooden moulds; this process has the advantage of leaving to the glass its own polish and purity, which is never the case when moulded in metal. To avoid the carbonization of the wooden moulds, the workman is careful occasionally to wet the mould with water, and constantly to turn the piece so that the red-hot crystal shall not remain too long in contact with the wood. By means of these precautions, about two hundred pieces may be formed in the same mould, without their dimensions being altered.

The glass is cut or ground, first roughly, by means of an iron wheel with sand, afterwards smoothed off on a grindstone, and then polished by means of a wooden wheel and pounce, or pumice-stone. The last polish is done with a wheel of cork and putty made with tin.

Colored Glasses.

The metallic oxides are generally used for coloring glass and crystal. For this purpose, they are prepared with great purity.

Colored glasses being intended for the most part to be *doubled*, that is to say, placed one over the other, ought to expand equally with heat. This result cannot be obtained except by experiment. The oxides to be used for coloring glass, should always be first tried either with common glass, or with lead glass (flint glass).

The principal colors are produced by the following bodies:

Sapphire blue . .	Oxide of cobalt.
Sky blue	Deutoxide of copper.
Purple red. . . .	Protoxide of copper.
Green	Oxide of chromium.
Canary yellow . .	Uranium.
Violet	Peroxide of manganese.
Red, or *rose* . . .	Gold.
Yellow.	Chloride of silver.

The name *doubled glass* is given to pieces formed of two glasses, placed one over the other; generally, white crystal covers colored glass. To obtain the effects of double glass, the workman first plunges his pipe into a melting-pot which contains colorless crystal, and afterwards into colored crystal. By blowing, or the ordinary processes of moulding, objects are obtained formed of two layers of different glass; then, by cutting or grinding off the colored glass where required, various designs may be produced on the colored glass over a white ground.

Trebled glass is formed of three layers of different kinds of glass; a layer of enamel, or opaque glass, is interposed between the colored and colorless glass. This glass is obtained in the same way as the double, by dipping the pipe successively into three pots, containing different kinds of glass.

Enamel.

Enamel is a white glass, holding in suspension in its substance, an opaque body. The bodies which may be employed to produce enamel, are stannic acid, arsenious acid, antimoniate of antimony, phosphate of lime, and sulphate of potash. Enamel is in general formed from a very fusible glass, so that the temperature used to melt it shall not be so high as to volatilize the body that opacates the glass. The mixture of materials used should be as perfect as possible.

To opacate with stannic acid, this acid is produced, together with the oxide of lead, by heating in the air a mixture of 15 parts of tin, and 100 parts of lead. There is thus formed a stannate of lead, which is cleared of all metallic particles which it may retain, by repeated washings; this is called *calcine*. This calcine is then fritted with sand and carbonate of potash, in the following proportions: 100 parts of sand, 200 of calcine, and 80 parts of carbonate of potash. This *frit* serves as a base for all enamels.

In Silesia and Bohemia, an opaline glass is made called *alabaster glass*, by introducing into melted glass a quantity of cold glass which has been first broken up by heat (*etonné*). The glass is then worked at as low a temperature as possible.

Venitian Glass—Filagree Glass.

The Venitian glass contains, in its thickness, various designs formed by threads of different-colored opaque enamel, of extreme tenuity. We will here explain the principle of this ingenious manufacture.

The fabrication is commenced by drawing out threads of enamel of one or two millimeters in diameter, and eight to ten centimetres long. These threads are placed in grooved moulds, in which is placed hot glass; this adheres to all the threads of enamel which preserve their parallelism and the position in which they are placed in the moulds. This form (*paraison*) is then introduced into the crystal; the threads of enamel are thus comprised between two layers of glass; the mass of glass procured by these operations is then drawn out, and at the same time turned between the fingers so as to form spirals. The variety of designs which these rods present, depends on the disposition in which the threads of enamel are first placed in the moulds. When a series of rods have thus been obtained, which are 15 to 20 metres long, they are divided into rods 30 to 35 centimetres long; these are placed side by side in an oven where the temperature can be raised so high that they will run together. A vitreous mass is thus obtained, which may be worked by the ordinary processes.

Millefiori (mosaic glass).

The glass called *millefiori* is similar, as to its preparation, to the Venitian glass. It is composed of small flowers, or stars, of differently-colored enamel, which are contained in a colorless glass.

The flowers or stars are made in moulds, like the rods of Venitian glass, and are united together by the same process.

Flint Glass.

This glass contains more oxide of lead than crystal.

It should be homogeneous, without bubbles, and but slightly colored.

It is used for optical purposes. Flint glass, fit for making object-glasses, of large size, is made by constantly stirring melted glass with an agitator of white clay, which may dissolve in the glass without coloring it; object-glasses of good quality and large diameter are rare, and bring a very high price.

Colorless Paste.—(Strass.)

Colorless paste approaches in its properties and composition to flint-glass. It is used in jewelry to imitate diamonds. The materials used in preparing it should be of a perfect purity. Their mixture should be as intimate as possible, their fusion should be slow; it should be prolonged at least for from twenty-five to thirty hours, and the mass should be so slowly cooled, as to undergo a complete annealing.

A paste is produced from rock crystal, which is harder than that produced from sand, but it is often too white, and throws less lustre than that which is slightly yellow. Different kinds of precious stones may be imitated by coloring paste with metallic oxides.

Avanturine.

For a long time, a glass has been manufactured at Venice by a secret process, which contains in its substance brilliant octahedric crystals of metallic copper.

Notwithstanding that numerous efforts have been made in France, they have not been able till within a short time to reproduce the *avanturine* of Venice, the value of which is very high.

The secret of this manufacture has lately been found out.

It is obtained by heating with a vitreous mass, a mixture of the silicate of the protoxide of iron and the protoxide of copper. In this reaction, the silicate of the protoxide of iron deprives the protoxide of copper of its oxygen, reduces it, and transforms itself into silicate of the peroxide of iron, which does not sensibly color the mass; the regenerated copper then crystallizes in perfectly regular octahedrons.

To obtain avanturine with all the qualities which are required for purposes of jewelry, circumstances of temperature are required which practice alone can point out, which renders this manufacture very difficult. (Clémandot and Fremy.)

PAINTING ON GLASS.

Two different processes are used for painting on glass.

In the first process, the glass is colored in its mass by metallic oxides, and then cut up, the fragments are put together again with strips of lead.

The second process consists in painting the glass, as porcelain is painted, and then baking it in a muffle furnace.

By combining these two processes, colored glass is obtained of beautiful appearance.

The colors used should have a transparency not required for painting on porcelain. So that, to paint on glass, the oxide of copper is preferred to the oxide of chromium to produce green colors; as the oxide of chromium gives opaque colors.

In painting on glass, both surfaces of the glass may be used; the surface, placed exteriorly, receives in general all the shades, which are thus more marked and clear.

POTTERIES.

The name *pottery* is given to different objects made of clay and then subjected to the action of fire.

Clay, which we have before spoken of as a silicate of alumine, forms the basis of all ceramic paste.* The potteries, however, never use clay alone, which, when calcined, cracks and shrinks considerably. They add a substance which impoverishes the clay, and forms in uniting with it, under the influence of heat, a homogeneous mixture, which, like stoneware and porcelain, contracts regularly in the fire, and undergoes a sort of demi-fusion.

Every potters' clay is then composed of a plastic clay, and a substance for reducing or impoverishing it.

The principal plastic matters are clay, marls, magnesite (silicate of magnesia), kaolin, and the talcs. The non-plastic or impoverishing matters are silex, sand, quartz, feldspar, chalk, calcined bones, and sulphate of barytes.

The nature and proportion of the bases which enter into the composition of potters' clays, have a great influence on the quality of the manufacture. Thus silex, united to pure alumine, would form the type of a completely infusible paste, which would answer for the manufacture of fire-bricks.

Lime, magnesia, and oxide of iron, added to silex and

---

* *Ceramic*, from κεραμος, an earthen pot, or *burned clay*.

alumine, would produce a ceramic clay, which by heat would undergo a sort of fritting, a demi-fusion.

Potash and soda make the mass more fusible, and render it suitable for the manufacture of porcelain, giving a composition approaching that of glass.

The preparation of the different fine potteries consists in a series of operations, of which we will give the details in a summary way.

Washing.

Clay is generally mixed with pebbles and silicious substances, rendering it unfit for use. These are taken out by washing it in water.

The pebbles being heavier than clay, immediately fall to the bottom in the water, which being rapidly decanted, and then left to settle, deposits pure clay.

Grinding.

The substances composing the potters' clay, such as quartz, silex, and feldspar, are often very hard.

To reduce them to powder, they are ground in a mill, after having been rendered more friable by being heated to redness, and suddenly cooled by immersion in cold water.

Intimate Mixture of Materials.

When the materials of the potters' clay are brought to the proper consistence, the mixture is worked by means of water. It should be brought to a clear, pulpy condition; too great a proportion of water would cause the separation of the solid materials according to the order of their density. The mixture, once made, must not be left to itself; first, because it cannot be handled; and also, because the substances composing it, being of different weights, would separate.

The operation which takes away from the paste its excess of moisture, is called *drying* or *stiffening the paste.* This is done by exposing it to the air, or placing it in porous pans of plaster, or in basins of earthen-ware slightly heated.

The paste being made sufficiently firm, by the drying process, to be worked, requires to be kneaded, beaten, and handled, to give it sufficient homogeneousness. The operation of *kneading*, essential to most pastes, is done by a workman walking over it with his bare feet, on a floor of wood or stone. He kneads the clay by tramping from the centre to the circumference.

In manufacturing common potteries, such as brick, tiles, common ware, &c., as soon as it has been subjected to the preceding operations, the clay is used; but for fine potteries, the paste is subjected to a preparatory operation called *ebauchage*, and then to a *battage* and *coupage* (a beating or slapping, and cutting). There is another operation which contributes to give the paste a perfect homogeneousness; it consists in leaving the paste for several months in wet pits, and is called *rotting of the paste.*

The organic matters contained in the paste being left in a wet place, undergo a sort of putrefaction, become black, and probably there is a disengagement of gas, which makes the mixture more homogeneous. We should say, however, that the benefit of this operation is not fully established; it often happens, in the manufacture of porcelain, that the demand for the article renders it necessary to use the paste shortly after its preparation, and it is known that articles made with this newly-mixed paste are not more defective than those made with the older paste.

When the potters' clay is made, the next process is to *shape it.*

We will not here describe the processes used to shape the pieces; we will only state that the form is given to it, either, by placing the wet paste in a potter's lathe, which is moved by the foot while the piece is fashioned with the hands, this is called *shaping*, or, by putting the paste into porous moulds mostly made of plaster: this operation is called *moulding*. The pieces are also shaped by *casting*, which consists in running into a porous mould a paste or pap "slip," of a clear, pulpy consistence; this, on account of the porosity of the mould, applies itself to its sides, and takes its form.

Glazing.

When the pieces are shaped and perfectly dry, sometimes they are immediately placed in an oven to give them a partial or complete baking; sometimes, before any baking, or after the partial baking, they receive a vitreous coating called *enamel-glazing*, intended to render them impermeable to liquids, to cover over their roughness, and their reddish color, as well as to give them tints agreeable to the eye. A good glazing should be spread uniformly over the surface, without penetrating too deeply into the ware, or it will be effaced, and will become what is called *tarnished, dried up*. The degree of fusibility of the glazing should be adapted to the paste; if too infusible, it will not spread.

An important condition, and one of the most difficult to fulfil in the composition of glazing, is to make it so that it will dilate with the paste, or else it *cracks*. These cracks injure the quality of the porcelain very much, especially if the clay is porous, permitting infiltrations of liquids and fatty matters.

However, if these fissures are arranged with symmetry, as in some China porcelain, it brings a high price, and is called *speckled porcelain* (porcelaine truitée).

The principal materials for glazing are feldspar, pounce, common salt, the alkalies, boracic acid, phosphate of lime, sulphate of barytes, the silicates of lead, stannic acid, the metallic sulphates, and the oxides of lead, manganese, iron, and copper.

Transparent glazing is made with alkaline and vitreous bodies, with feldspar and oxide of lead. Opaque glazing, with stannic acid or phosphate of lime. Colored glazing, with the metallic oxides and sulphurets.

Glazing is applied in different ways. While the pieces are still porous, they are dipped into water holding the enamel or glazing in suspension. If the paste has been baked, the glazing is applied by *sprinkling.*

Sometimes the glazing is applied by volatilization, by disengaging in the oven saline or metallic vapor, as common salt. This, by spreading itself over the pieces brought to incandescence, is decomposed by the action of silex and steam, forming a silicate of soda, which vitrifies their surface.

Often the glazing is done at the same temperature as the paste, as in common potteries; but often also it must be done at a much lower temperature: the piece then requires a second baking. The paste is first completely baked, and transformed into what is called *biscuit;* the glazing is then baked, being applied to the piece by sprinkling or watering.

Baking of Potters' Ware.

This is done to give the pieces solidity, so that they can be handled without breaking. It also renders them impermeable to liquids.

The scale of temperature used is extensive, being from 50° C. to 140° of Wedgwood's pyrometer; that is to say, to the point of fusion of iron.

The form of the ovens or kilns is very variable. For fine potteries, such as porcelain and delftware, they generally use kilns *à alandier*, from the name of the mouth at their base.

The term *encastage* (drying) is given to the operation which puts the pieces in a condition fit to be subjected to the action of the fire without being deformed.

To season the pieces, they are placed in a kind of support or case called *cazette* (saggers) made of fine clay, less fusible than the potters' clay.

The mode of seasoning varies according to the kind of ware. When the potteries are covered with the glazing or enamel which is to be vitrified by the fire, they are supported by the smallest and least numerous points possible.

The bottom of the *cazettes* is always covered with sand, to prevent the pieces from adhering to it.

The term *enfournement* is given to the manner in which the pieces are placed in the oven or kiln.

The fuel used in baking potteries is wood, coal and turf. These should burn with a flame. Wood is most generally used in fine porcelain. To judge of the temperature of the oven, small *proof-pieces* (watches) are placed within it; these are of the same material as the pottery to be baked. These *proof-pieces* are taken out from time to time, and by the changes they have undergone indicate the condition of the oven.

The action of the fire produces on the potters' clay the following modifications:—At first the water is driven off.

If the pieces have been first dried before baking, they remain porous and permeable; it is thus that refrigerators for water are made: these are called *alcarazas* (water-coolers.)

If the composition of the paste permits the molecules to approach each other in baking, the potteries will then become diminished in volume; this is called *shrinking*.

The shrinking varies with the temperature used in baking, the nature of the paste, and the mode of manufacture; it varies from a twelfth to a fifth in linear dimension.

It is not, however, the same in all directions, being generally greater in the vertical than in the horizontal direction.

The calculation of the amount of shrinking which the piece undergoes in baking, so that it shall lose none of its elegance and regularity of form, is one of the most delicate points in the art of pottery.

DECORATION OF POTTERIES.

The substances made use of to decorate potteries may be divided into four parts.

1st. The vitrifiable colors, properly so called.

2d. Les engobes, which are earthy matters fixed by a vitreous flux.

3d. Metals in the metallic state.

4th. Metallic lustres.

The vitrifiable substances used to decorate potteries should fulfil several indispensable conditions: they ought,

1st. To be fusible and unalterable at a red heat, which excludes all organic matter, volatilizing it and decomposing it.

2d. To adhere strongly to bodies to which they are applied.

3d. To preserve a vitreous appearance after baking.

4th. To resist the attacks of the air, moisture and the gases which may be in the air, and to be sufficiently hard to resist wear.

5th. To dilate properly with different potteries.

6th. To be more fusible than the potteries themselves.

The fluxes are colorless, vitrifiable matters, which are added to the metallic oxides and to the metals, to make them adhere to the potteries.

The materials composing the fluxes are sand, feldspar, borax or boracic acid, nitre, carbonate of potash, carbonate of soda, minium, litharge, and oxide of bismuth.

Some potteries are colored in the paste, others by the application of the vitrifiable colors to their surface on the enamel or glazing.

When it is proposed to color the paste, the color must resist the temperature of the baking without alteration; thus, those potteries which are baked at a very high temperature, such as hard porcelain, only admit of a limited number of colors.

When, on the contrary, the paste has been made more fusible by the addition of vitrifiable substances, as that intended for the tender porcelain, the paste may be variously colored.

Colors able to resist, without alteration, heat necessary to bake the enamel-glazing, or covering of potteries, are called *refractory colors*, (couleurs au grand feu.) Those which cannot support so high a temperature without altering, are called *muffle* or *reverberatory colors*.

The colors which resist a high temperature are but few. For the hard porcelains the blue of cobalt, chrome-green, the brown colors of iron, of manganese, and chromate of iron, the yellows obtained with the oxide of titanium, and the black of uranium are only used.

For tender porcelain, the violet reds and browns of manganese, of copper and iron: for the fine and common delft-wares the yellows of antimony, the browns of manganese, the greens of copper, and the blues of chrome are used.

The number of muffle colors is, on the contrary, great: at the manufactory of Sevres, seventy-five different compositions are used. These colors are ground in a porcelain mortar, with the essence of lavender or turpentine thickened in the air, and then applied to the pottery which is baked in a muffle furnace.

The painting of pottery is generally done at two bakings; first it is done roughly, then retouched, and baked over.

It is useless here to give the composition of all the colors used in the muffle. We will only speak of those which form the principal colors.

Blue.—Oxide of cobalt.

Red.—Protoxide of copper, purple powder of Cassius, and peroxide of iron.

Green.—Oxide of chrome, binoxide of copper, mixture

of the oxide of cobalt, antimonious acid, and oxide of lead.

Yellow.—Oxide of Uranium, chromate of lead, certain combinations of silver, subsulphate of iron, and a mixture of antimoniate of antimony and oxide of lead.

Violet.—Oxide of manganese, purple powders of Cassius.

Black.—Mixture of oxide of iron, oxide of manganese, and oxide of cobalt.

White.—Common enamel.

The name *metallic lustre* is given to a kind of decoration in which the colors partake of the brilliancy of the metals, or in which the metals appear during the baking with their natural brilliancy, without being subjected to *burnishing.*

The *lustre of gold* is obtained by applying a mixture of fulminating gold and essence of turpentine to the surface of the porcelain, and placing it in the heat of the muffle furnace.

The *cantharides lustre* has lively and brilliant tints with greenish reflections. This lustre is produced by the reduction of the chloride of silver under the influence of combustible vapors. It is obtained by applying first to the surface of the porcelain a mixture of vitrifiable glazing of oxide of bismuth and chloride of silver, heating the piece to redness in the fire of the muffle, and then exposing it in this condition to the smoke of a combustible.

To gild porcelain, a mixture of essence of turpentine, gold finely divided, and subnitrate of bismuth (for a flux) is applied to the pieces. The gold is obtained by precipitating the perchloride of gold by the sulphate of the protoxide of iron.

The metals subjected to the fire lose part of their brilliancy; the gold becomes dim. It is polished by rubbing it with a hard body. This operation is called *burnishing.* It is first smoothed off with a burnisher of agate, and finished with hematite.

After having presented general notions relative to the

properties of the ceramic pastes, it now remains for us to examine the principal kinds of potteries.

We will, as has been done by M. Brongniart, from whom we have borrowed these details on the ceramic arts, divide potteries into seven classes.

First Class.—Earthen wares, comprising bricks, paving-tiles, roofing-tiles, furnaces for laboratories, chafing-dishes, flower-pots, pipes for carrying off smoke, &c.

Second Class.—Common potteries.

Third Class.—Common delft, or Italian wares.

Fourth Class.—Fine delft, or English wares.

Fifth Class.—Stone ware.

Sixth Class.—Hard, or China porcelains.

Seventh Class.—Tender, or French porcelains.

Earthen Wares.

These products are not generally glazed; their paste is often heterogeneous, porous in texture, composed of common potters' clay, or clayey marls. This paste is worked or tramped, and sometimes washed; it is reduced either with sand, cement, or refuse coal.

The vitreous coating which covers some earthen ware, has lead, generally, for a base.

It is roughly formed, being generally done with the hand; though sometimes in moulds.

The *mode of baking* is variable, from the process of sun-drying, to that of the baking of stone-ware.

The *kiln* is often made with the articles to be baked.

The *fuel used* is coal, turf, or wood.

Bricks.

These are artificial materials, intended for the construction of houses or furnaces.

The properties of bricks should vary according to the purposes to which they are to be applied. A brick intended for building should be so solid as that it may be neatly cut, and baked at a temperature sufficiently high to prevent its being disintegrated by the atmospheric influences. A good brick, for an ordinary building, will support a considerable weight without being crushed; it ought not to crumble in water, nor to absorb too great a quantity of it; this is tried by weighing the brick before and after immersion in water.

Earths are often found, which, without preparation, are fit for the manufacture of bricks. Thus, at the mouths of large rivers, earths are often found fit for the manufacture of building-bricks; indeed, the commonest yellow vegetable earth, will generally answer for making bricks.

Bricks used for the construction of furnaces, ought to be refractory, and to resist for a long time the action of the ashes of the combustible. Fire-bricks are made with plastic clay, containing neither gypsum, lime, nor oxide of iron. This clay is washed to free it from the foreign substances which it contains. It is reduced with cement* of this clay, made expressly, and powdered. Even the purest sand, mixed with clay, would not make infusible bricks. A good fire-brick should have but little color; because the oxide of iron, which colors bricks red, renders them fusible.

Bricks are formed either with the hand, or by machinery. Two men, with the hand, can make from six to seven thousand bricks per day.

Bricks are burnt either with turf, coal, or wood. The kilns in which bricks are burnt, are built almost entirely with the bricks intended to be burnt, the base of the kilns being the only part made of old bricks. A kiln contains about four hundred thousand bricks, and it requires about twenty-five days to burn them.

The Flemish process, by which the bricks are burnt with coal, is the most economical.

---

* The cement here used is burnt clay.

Tiles — Paving-Brick.

The manufacture of tiles and paving-bricks, is very like that of building-brick.

Good tiles are impermeable to water; porous tiles are constantly moist, so that moss grows upon them, soon causing them to deteriorate.

To render them impervious to water, the density of the paste is increased, or they are covered with a glazing of lead, which is obtained with the sulphuret of lead, called *alquifou* (potters' ore).

Crucibles.

The principal quality of crucibles is to resist high temperatures.

The most refractory are made of a mixture of clay and graphite.

Sometimes they are made of porcelain, which has the advantage of being both refractory and impervious; but these crucibles, besides being high-priced, break easily with variations of temperature.

In laboratories, crucibles are used called *Hessian crucibles.* These have the inconvenience of being porous, and will not contain either nitre or common salt in fusion; but they will resist a high heat, and sudden changes of temperature. They are made of a mixture of fire-clay and quartz sand. The great quantity of silex which they contain, renders them easily attacked by the oxide of lead.

Crucibles are made in Paris, called *Paris crucibles.* They are of a good quality, contain less silex than the *Hessian crucibles*, and resist for a long time the action of litharge.

The paste of crucibles is made of raw clay, which forms the plastic part of it, and clay burnt at a red heat, used for reducing it. The burnt clay may be replaced by small fragments of coke, or old earthen-ware, reduced to powder.

The raw clay used in this manufacture is freed from foreign bodies by sifting and decantation (washing).

Water Coolers.

In some countries, vessels called *alcarazas* are used for cooling water. A small quantity of water oozes through to the outside, which in evaporating reduces the temperature of the liquid within. They are made of a clay, rendered porous by the introduction of a large quantity of sand; in baking, they are subjected to but a low degree of heat.

Common Pottery.

This ware has a homogeneous paste, tender, of an earthy fracture, and porous texture. It is opaque, and covered with a translucid lead glazing.

The material is composed of clay, clayey marl, and sand. The vitreous coating has principally lead for a base, and is obtained with galena (alquifou) or litharge. This glazing is colored with oxide of manganese, or oxide of copper.

These wares are in very general use, and are sold at moderate prices. Their porosity enables them to withstand variations of temperature. Their use has, however, some inconveniences; their coating is slight, and is easily scratched by table implements. They rapidly become impure; and more than that, all acids attack their glazing, which contains lead and copper, and form poisonous salts.

Common Delft, or Italian Ware, with Opaque Coating.

The paste of this pottery is opaque, slightly colored, of soft and open texture, and of an earthy fracture. It is covered with an opaque glazing, mostly with tin for a base.

This ware is composed of *figuline* (potters') clay, of clayey and calcareous marl, and sand; the clays are washed. The shaping is rude, and is done on a lathe.

The baking of the pieces is double. They are first baked *en biscuit* with a white heat; they are then coated with their glazing, and baked a second time. The same oven serves for both operations; the first is done in the upper part of the oven, the second below. The pieces are placed in cazettes.

This kind of ware is chiefly made at Paris, Rouen, Nevers, Luneville, &c. It is not strong, but will bear the fire sufficiently well for domestic uses.

The porosity and color of the material are corrected by a thick and opaque glazing obtained with tin, which is liable to fly off. This enamel often cracks, and comes off in scales.

M. Barral has examined, with the greatest care, the different causes which produce the cracks in the ware used for stoves, and the linings of fire-places. These wares, intended to support often a high heat, are generally made of a mixture of two parts of raw clay, and one part of burnt clay or sand. The grains of sand, or burnt clay, give elasticity to the paste; but the enamel does not dilate with the biscuit, and nearly always cracks. M. Barral has shown that by adding to the materials a small quantity of a flux, such as a frit of potash or soda, or even of carbonate of lime, the cracks will not be produced. This process is now followed in all the manufactories of chimney panels. It will even answer on old paste, to place, as an intermediary between the porcelain and enamel, a layer of the most fusible paste, to obtain a ware which will not crack in the fire.

Fine or English Ware, with transparent Glazing.

The paste of this ware is white, opaque, of very fine texture, dense, and sonorous; it is covered with a transparent coating, having lead for a base. It is principally formed of plastic clay washed, and of finely-ground silex; it sometimes contains a small quantity of chalk.

Its glazing is formed of silex, feldspar, soda, and oxide of lead. This covering, mixed with water to the consistence of thick fluid, is applied by immersion or sprinkling. It is formed with great care, and the pieces are thin and light. It is baked twice; first at a temperature of 100° of the pyrometer of Wedgwood; the glazing is then baked at a temperature not higher than 20° to 30° of the same pyrometer. The ovens are cylindrical; the number of "alandiers" (fire-places) varies from six to twelve. They are heated with wood or coal; the drying is done in closed cazettes (sagger-cases). This ware is esteemed, but it has many disadvantages for use. It will not bear the fire, its glazing is delicate, and is scratched by implements of iron and steel; the paste is porous, and soon becomes foul.

The fine wares are made at Choisy, Creil, Chantilly, and Montereau. Pipes are made of the same earths, but they are not baked at so high a temperature, and are without glazing.

Stone Ware.

The name *stone ware* is given to a ware of dense material, very hard, sonorous, opaque, rather fine in grain, and of various colors. It is sometimes covered with a vitreous, lead, or earthy glazing.

It is chiefly composed of a plastic clay, reduced with sand, silex, or cement of sand-stone.

The vitreous coating is sometimes saline, made by the volatilization of common salt; sometimes it has lead for a basis, and contains quartz, feldspar, and barytes. Often it is made with dross from the forge, pounce, and volcanic scoriæ.

The baking of stone ware is nearly always easy. It requires the high temperature of 120° of Wedgwood, and is often continued for eight days.

This ware is solid, hard, and impermeable without the aid of glazing; but it has the disadvantage of being easily broken by blows and sudden changes of temperature, and being hardly fit for use at the fire.

It is distinguished into common and fine ware.

The material of common ware is always yellowish, made

19

of plastic clay, reduced sometimes with quartz sand, and covered with chalk. The fine ware is very different in its composition from the common ware; it is more like porcelain. It is composed of the following bodies:

White plastic clay	25
Argillaceous kaolin	25
Feldspar	50
	100

This ware is colored with metallic oxides. Cobalt colors it blue; oxides of manganese and iron, black; oxide of antimony, yellow; oxide of nickel, pale yellow.

Hard, or China Porcelain.

This porcelain is made of a very fine material, though granular, hard, translucid; its coating is hard, and does not melt except at a high temperature.

The paste is formed of two elements, the one, clayey and infusible, is kaolin, or pure white plastic clay, or else magnesite; the other is fusible, it is feldspar alone, or else silicious sand, chalk and gypsum, taken separately or united together in different ways.

The coating consists of quartzy feldspar alone, or mixed with gypsum, or burnt and ground clay.

The paste is subjected to the same handling as that pointed out for the other wares, but is done with more care.

The shaping of this porcelain requires great care, because, in the baking, it shows the slightest irregularity of compression more than any other pottery.

It is twice fired; the first hardens the paste, so as to close it up sufficiently to receive the coating by immersion. This baking is generally done in the upper apartment of the oven, at a temperature of 60° of the pyrometer; this does not cause any sensible change in the size of the pieces, but the paste loses about one-eighth of its weight.

The second baking is performed in the interior apartment of the oven, and requires a temperature of 140° Wedgwood. The paste softens, becomes translucid, and shrinks considerably.

The pieces are placed in cases (*cazettes*) made of fire clay, sufficiently refractory to resist high temperature; the kiln is cylindrieal with five or six hearths (alandiers) at the most: it is heated with wood. It has been recently attempted in some factories to substitute coal for wood, but not with perfect success.

The temperature of the kiln is shown by *proof-pieces*, or *watches* (montres). Hard porcelain requires about thirty-six hours for baking; the oven ought to be from six to seven days in cooling. A well-made porcelain resists, without breaking, sudden changes of temperature varying from 0° to 100° Cent.

It should be milk-white, without spot, its glazing should be even and smooth. Porcelain too argillaceous is often colored yellow; the slightest negligence in forming it, or in mixing, makes defective pieces. A smoky fire discolors them, too great a heat deforms the pieces, causes the glazing to penetrate into the paste, renders them shrivelled and covered with small asperities. With too little baking they become uneven.

If the glazing and the paste do not dilate equally, they become fritted, and then crack, and are said to be *tressaillées.*

Tender French Porcelain.

The paste of this porcelain is fine, dense, with a texture almost vitreous, hard, translucid, and fusible at a high temperature. The glazing is vitreous, transparent, with lead for a base, and tender.

The distinctive mark of this porcelain is, that it contains a sufficient quantity of a substance to give it a fusibility almost like that of vitreous substances.

This substance is either soda, potash, alkaline salts, salts with an earthy base, feldspar, etc.

The glazing is a glass always containing lead.

The shaping of the old Sevres wares was always done by moulding, the paste not having any plasticity.

This porcelain is always twice fired; the pieces are first baked *en biscuit*, the glazing is applied by sprinkling, and it is then baked again. An oven of two floors is often used; the porcelain is baked below, and glazing above.

Tender porcelain may, like fine stone-ware, be colored variously in the paste. The glazing readily incorporates itself, and forms a brilliant, rich, and very *recherché* surface.

Tender English Porcelain.

This holds a middle place between the hard porcelain and the fine delft ware. It is distinguished from the first by the fusibility of its paste, and the nature of its glazing, which has lead for its base; and from the second by the heat at which it is baked, and the hardness of its glazing. The paste of the tender English porcelain contains kaolin; it is sufficiently plastic to be formed in the lathe.

Porcelain of Tournay.

A tender porcelain, highly esteemed, is made at Tournay, which is generally used in the restaurants of Paris. It is tougher and stronger than hard porcelain, but it does not bear changes of temperature so well.

BUILDING STONES.

Those stones are preferred for building, which are the least costly, strongest, most resisting to the weather, and at the same time are light, least porous, and bind the strongest with mortars.

We will examine in succession the different kinds of building stone.

Lime-Stone.

Lime-stone, called *pierre de taille*, is found in abundance in the tertiary and secondary formations. There are two principal varieties, the *silicious* and *shelly* stones.

The silicious stones are the most esteemed; they are hard, easy to polish, and are used in sculpture. They are not porous, and may be used when first taken out of the quarry.

The stones taken from Chateau-Landon are silicious lime-stone, very celebrated. The *shelly lime-stones* are also much sought after for building. They are porous, softer than the silicious stone, and contain a great deal of water; they ought to be left to dry in the yard before being used, in order to prevent them from cracking.

Some limestones are so porous, that the rain and atmospheric moisture penetrate them; and when they are exposed in winter to a temperature below 0° C., the water which they may contain freezes, expands, and breaks the stone; these stones are called *pierre gélive.* To know one of these stones, they may be plunged into a hot saturated solution of sulphate of soda. They will crack when the salt crystallizes.

In the article on *silicic acid*, we pointed out certain kinds of stones, such as *sand-stone*, *mill-stone*, &c., which are also used in building.

Granite is one of the hardest stones known; it constitutes the most ancient rocks; it is formed of feldspar, quartz, and mica. It is hard enough to strike fire with steel. It takes a beautiful polish, but its extreme hardness renders it difficult to work. It must be preserved a long time under water before working. It is used principally for making columns, obelisks, door-jambs, vases, polishing-tables, flags for foot-pavements, &c.

There are some volcanic stones which must be ranged

among building-stones, such as basalts, and certain lavas, which make building-stones which are light and very hard.

MORTARS.

The name *mortar* is given to a mixture used to bind together building materials.

These materials are divided into two classes, *regular* and *irregular.*

The term regular is applied to stones dressed on all their faces; these may be used without mortar, their cut faces being applied together. They will keep their position, if kept plumb; however, to make the contact more regular, a thin layer of mortar, which cannot become very hard, is applied between their faces.

In fact, liquid or gelatinous layers, placed between solid bodies, cause an adherence which it is difficult to destroy.

The term *irregular material* is applied to broken stones, pebbles, &c., which, on the contrary, require to be bound together by very consistent materials.

Common Mortar not Hydraulic.

Common mortar is made of a mixture of lime and coarse quartz sand; exposed for some time to the air, it acquires great hardness, and serves to bind together the irregular materials used in building.

The hardness which mortars acquire, is not attributed, as was formerly supposed, to a combination of the silex with the lime. For, if a hardened mortar is treated with an acid, gelatinous silex is never obtained, which ought to be the case if the silex had combined with the lime. These mortars solidify, by the action of the carbonic acid of the air on the lime, producing slowly a carbonate of lime, and adherence to the stones between which it is placed.

In order that mortar may bind together the materials, it

is necessary that the combination of the carbonic acid with the lime take place slowly. Mortar should not dry too rapidly; thus it is known that mortar used late in the season is better than that used early in the season.

The grain of the sand, the number of its rough points, and the quantity of water used, exercise a great influence on the solidification of common mortar.

The entire mass of the mortar used in buildings never becomes completely solid. It is well known that those parts of mortar placed in the interior of walls are often in the same moist state as when applied; the exterior solid parts preserving the interior parts from desiccation.

Hydraulic Lime and Mortars.

The common mortars of which we have just spoken solidify when they are exposed to the air, but crumble completely when placed in water.

The following article, on mortars which solidify in water, and which are called *hydraulic mortars*, is in part taken from the valuable works of M. Vicat on hydraulic lime.

It is known, that, in subjecting limestone to calcination, lime is obtained, which, in its contact with water, hydrates and swells considerably. This lime is called *fat lime.*

When limestone calcined contains a large proportion of magnesia, oxide of iron, or quartz sand, and but a small quantity of clay, but little heat is produced in slaking. When it is placed in water and swells but little, it is called *poor lime*, not hydraulic; after a time it hardens in the air.

But if the limestone contains a quantity of clay, it produces by calcination lime entirely of a different nature, which slakes but slowly in water, called *hydraulic lime.* By this calcination with clay, lime acquires a new property, of great service in building; mixed with water, it first forms a thick paste, and soon becomes so hard as to be comparable to the most resisting limestones.

The quality of hydraulic lime is determined by the proportion of clay which is contained in the limestone which produced it.

Stones *moderately hydraulic* contain 8 to 12 per cent. of clay, and give lime which hardens after fifteen or twenty days' immersion.

Hydraulic limestones contain 15 to 18 per cent. of clay; the lime which they give hardens in eight days.

Limestones *eminently hydraulic* contain 25 per cent. of clay; the lime which they produce hardens in three or four days.

When the proportion of clay increases to 30 or 40 per cent., the lime is called *Roman cement.* A good Roman cement becomes as hard as stone after an immersion of a quarter of an hour.

Theory of the Hardening of Hydraulic Lime.

We will now, still availing ourselves of the works of MM. Vicat and Berthier, give the theory of the solidification of hydraulic lime.

These savants have shown that the state of the silex exercises a great influence on the quality of hydraulic lime, and on its production. In fact, gelatinous silex, calcined with carbonate of lime, gives a hydraulic lime of a good quality. Rock crystal, on the contrary, reduced to powder, and calcined with carbonate of lime, produces a poor lime which is not hydraulic. The silex, such as it is found in clay, is in a condition favorable to the production of hydraulic lime.

The best hydraulic lime contains silex, lime, and magnesia, or alumine.

The solidification of hydraulic lime should be attributed to the formation of a silicate of alumine and lime, or of magnesia and lime, which combines with water and produces a hydrate excessively hard and insoluble in water. The hardening of hydraulic lime may then be compared to that of calcined plaster, which also combines with water to form a solid hydrate.

Care should be taken, in preparing hydraulic lime, not to calcine the stone at too high a temperature; the double silicate would in this case undergo a sort of *frit*, which would not hydrate in contact with water, and would give a poor lime not hydraulic.

Roman Cement.

Roman cement is produced by the calcination of limestones which are very argillaceous. It becomes exceedingly hard after having been plunged, for only a few minutes, in water. This remarkable property serves to distinguish this cement from the other varieties of hydraulic lime.

Roman cement was first made in London in 1796 by calcining a limestone which contained 30 per cent. of clay.

Afterwards it was found, that pebbles from Boulogne had a great resemblance to the limestone from which the Roman cement was made, and M. Lacordaire, mining engineer, found in Burgundy a cement which is equal to the Roman cement for hardness and resistance.

Artificial Hydraulic Lime.

In his experiments on hydraulic lime, M. Vicat was led to manufacture artificial hydraulic lime. He pointed out that artificial hydraulic lime could be obtained by calcining a mixture of carbonate of lime and clay.

It is prepared in the neighborhood of Paris by stirring into water a mixture of 1 part of clay of Passy and 4 parts of chalk; these materials are mixed by a vertical wheel which turns in a circular trough, and are then made to flow out into basins of masonry. A deposit soon takes place, which is formed into small bricks, which, after being dried in the air, are moderately calcined. Hydraulic lime, thus prepared, enlarges about two-thirds in volume when placed in water, while fat lime about triples its volume; it is completely dissolved in the acids, like the natural hydraulic lime.

The quality of the artificial hydraulic lime of M. Vicat is now established by trial on a large scale; for it has been used in all the hydraulic masonry of the St. Martin Canal.

M. Vicat, in enriching the arts with an excellent process for making artificial hydraulic lime, has at the same time proved, that France possesses in many localities clayey limestones, which will furnish by calcination hydraulic lime of good quality.

Hydraulic Mortars.

Hydraulic mortars are mixtures of lime and different substances, which, like hydraulic lime, have the property of solidifying in water.

Some solid bodies, mixed with hydraulic lime, exercise but little influence on its solidification; others, on the contrary, have the property of increasing this quality of lime moderately hydraulic, and even in some cases will make fat lime hydraulic.

The substances mixed with different kinds of lime, for making mortars, may then be divided into *inert* and *energetic materials.*

The inert materials are pebbles, sand, &c. Mixed with fat lime, these do not modify its action in water. M. Vichat thinks, however, that sand, added to hydraulic lime, may increase its cohesion.

Among the energetic substances, the mixture of which with lime will make hydraulic mortars, may be placed in the first rank, the volcanic products called *pouzzolanes.*

The pouzzolanes which were discovered by the Romans near Vesuvius, in the neighborhood of Pouzzoles, have the property of combining slowly with fat lime under the influence of water, and thus forming excellent hydraulic mortars. The Roman constructions owe their solidity to the use of mortars made of lime and pouzzolanes. Pseudo-volcanic substances, such as the materials resulting from burnt coal-mines, baked clay, tripoli, or lavas, act with fat lime like the pouzzolanes. Some artificial products may also render this lime hydraulic; such as refuse tiles, bricks, stone-ware, &c.

The pouzzolanes have the curious property of absorbing lime held in solution in water; and it may be said, generally, that pouzzolane is more energetic in proportion as it absorbs lime. The pouzzolanes which are best suited for making hydraulic mortars, are easily attacked by sulphuric acid.

CONCRETE.

The name concrete is given to mixtures of hydraulic mortars, and small pieces of stone. Concrete, so usefully applied in hydraulic structures, enables us to undertake works which were formerly considered impracticable. It solidifies in a short time, and takes the shape of the place in which it is applied. The composition of concrete varies with the purposes to which it is to be applied; it is ordinarily made of one volume of mortar, and two of small broken stones. The mixture of the materials is made either by hand, or by means of machinery.

MASTICS.

Mastic is used for covering terraces, lining basins, uniting stones, luting chemical or physical apparatus, &c. Mastic is of two kinds:

1st, such as is applied cold, in a pasty state, or dissolved in water, alcohol, ether, or oil.

2d, such as is applied by fusion.

Mastics applied cold.

Mastic of Dhil.—This is made of 9 parts of pounded brick or well-burnt clay, and one part of litharge; these two bodies are then mixed with linseed oil. It requires for solidification, from seven to eight days. Before applying it to stone, the stone is wet to prevent it from absorbing the oil. The mastic of Dhil is well adapted for jointing flags, and other cut stones.

Diamond Cement.—This mastic is used for mending porcelain and glass. It is obtained by making an aqueous solution of fish-glue, to which is added a little alcohol, gum ammoniac, or mastic resin dissolved in alcohol.

Mastic of White of Egg, is formed of the white of egg and lime powdered. It resists moisture, and answers for repairing marble.

Mastic of Cheese is a mixture of white cheese with powdered lime; this becomes very hard, and is hydraulic. A cement is obtained, which becomes as hard as stone by mixing 20 parts of sand, and 1 part of quick-lime, with dry linseed oil. By substituting for the quick-lime 10 parts of carbonate of lime, a *mastic cement* is formed, which may be used as Roman cement in some hydraulic constructions.

Mastic of iron, which is principally used to cement iron to iron, is made of 50 parts of iron filings with one part of sal-ammoniac. Formerly sulphur was added to this mixture.

Glaziers' mastic (putty) is made with a mixture of drying-oil and white lead, or even chalk (whiting).

Mastics for luting chemical apparatus are,

1st. A mixture of paste of almonds and flour paste.

2d. Iron filings, clay, and gum arabic.

3d. A mixture of fat clay, lime, and white of egg.

4th. A mixture of plaster and starch.

5th. A mixture of flour, clay, and melted caoutchouc; this lute resists acids.

6th. Melted caoutchouc, used for luting stop-cocks and ground-stoppers.

7th. Tallow, or a mixture of wax and essence of turpentine.

8th. Clay.

Mastics applied by Fusion.

The mastic used in the construction of electric machines is formed of 5 parts of rosin, 1 part of yellow wax, and 1 part of colcothar; to this is often added a small quantity of plaster in powder.

Common pitch is used as a fusible mastic; also sealing-wax, which is formed of a mixture of different resinous substances, generally colored red by vermilion.

IRON.

Iron is very abundant in nature; it is found in the state of oxide, sulphuret, and carbonate.

The iron of commerce is never pure; it always contains traces of carbon, silicium, and sometimes phosphorus.

As some difference is observed between the properties of iron of commerce and pure iron, we will separately examine iron in these two states.

Pure Iron.

To obtain iron in a state of absolute purity, one of its oxides must be reduced by dry hydrogen under the influence of heat.

The temperature at which the reduction is made exercises a great influence on the properties of the metal. If this is done at a bright red heat, the iron is of a silvery whiteness; it has most of the physical properties of the iron of commerce of the best quality; but it is a little more difficult to melt.

When, as M. Magnus has shown, pure peroxide of iron is reduced by hydrogen, at the temperature of an alcohol lamp, iron is obtained in the form of a black powder, very porous, which takes fire at the ordinary temperature when thrown into the air. In this state, iron bears the name of *pyrophoric iron of Magnus.*

To prepare pure iron in mass, iron wire with a fifth of its weight of oxide of iron is melted in a forge fire, in a refractory crucible luted with clay. The mixture ought to be covered with pure glass, free from metal; the oxygen of the oxide of iron burns the carbon contained in the iron of commerce, and oxidizes the silicium and the phosphorus, which pass in the glass, to the state of phosphates and alkaline silicates. We thus obtain iron in the form of residue of a silvery whiteness.

Iron of Commerce.

The iron of commerce is of a bluish grey color. It is ductile and the most tenacious of all metals; a wire of 2 millimetres diameter requires a weight of 250 kilogrammes to break it. When polished, it is very brilliant: it has a feeble taste and odor. It becomes brittle by cold hammering: its toughness is restored by *annealing.*

Iron is naturally grained, and its quality is better in proportion as its grain is fine and brilliant; it becomes fibrous (nerveux) by forging; the fracture of good iron presents a twisted fibre, fine and brilliant. If it is hammered in the cold, or if it is tempered, it becomes again grained. Polished iron often has brown spots, called *pailles,* owing to scoriæ, or oxide of iron.

The iron of commerce is known as *fersforts* (tough iron), which may be forged or bent either hot or cold, and into *fers rouverains* (brittle iron), which breaks in the cold, or at a more or less elevated temperature. These two principal kinds of iron comprise several varieties; thus, the *fers rouverains* are divided into two classes:

1st, *Red short* (*fers metis*), which break when hot. They owe this property to a certain quantity of sulphur and arsenic; the smallest proportion of sulphur will prevent iron from welding.

2d, *Cold short* (*fers tendres*), which contain a certain proportion of phosphorus, and break when cold; their fracture is in large brilliant grains. The *fers rouverains* which break hot and cold, cannot be applied to any use.

Melted iron which is slowly cooled, crystallizes in cubes and octahedrons. Iron may even crystallize without losing its solid state. This property has a great influence on its tenacity. When, in fact, fibrous iron is subjected for some time to frequent vibrations, a molecular movement is established in the mass, which causes the crystallization of the metal. It is not uncommon to see a bar of iron of good quality transform itself slowly, under the influence of vibrations, into iron crystallized in large facettes; it then becomes brittle, and loses a great part of its toughness.

This crystallization is frequently observed in the iron with which suspension-bridges are constructed, and in the axles of carriages and locomotives.

To restore to iron its fibrous texture and tenacity, it must be hammered hot (forged). The crystallization of iron may be seen in nails which have been exposed to the vibrations caused by the frequent passage of carriages. These nails become very brittle. In a few hours, a like change may be produced in bars of iron kept at a red heat in an annealing furnace, and then allowed to cool without hammering.

Iron fuses at a temperature estimated at 1500° C. It possesses a property of great advantage in the arts — it softens at a temperature a little below fusion; in this state, all the forms required in the arts may be given to it. It may be welded without the interposition of another metal, and the welded part is as solid as the rest of the bar, from which it cannot be distinguished.

Iron is highly magnetic. It shares this property with cobalt and nickel.

Pure iron is attracted by the magnet. Sustained by the action of this magnet, it has the power of attraction, but loses this property as soon as it is removed from the contact of the magnet.

Carburetted irons, as steel and cast iron, do not lose their magnetic properties when the action of the magnet has ceased. At a red heat, iron has no action on the magnetic needle.

Iron may be kept for an indefinite time at the ordinary temperature, in oxygen and dry air. Heated in the air, it absorbs oxygen, and becomes covered with a thin pellicle of oxide, which presents, as it increases, the phenomenon of colored rings. The different colors appear in the same order, and at the same temperature, as in steel. At a red heat, iron oxidizes rapidly, and gives rise to *scales of oxide*. At a white heat, iron burns with scintillations.

If iron first heated is introduced into a jar containing oxygen, it burns with brilliancy. When a piece of iron, heated in a forge, is placed before the bellows-pipe, it burns with the same intensity as in oxygen.

The combustion of a shred of iron is also very rapid, if when heated, it is whirled rapidly through the air attached to a wire. In working iron, it should then be preserved as much as possible from the oxidizing influence of the air. With this end, it is covered with a layer of fine sand, which forms a double silicate with the oxide of iron in preserving the metal from the further action of oxygen. It is well known, that when a piece of iron is sharply struck with a hard body, such as flint, it throws off sparks which will set fire to organic substances. This phenomenon is due to the combustion of the iron: thus, in striking the flint over a sheet of paper, it is seen that each fragment of the metal raised to a high temperature by the stroke, burns in the air, that is to say, aborbs the oxygen of the air, and is converted into peroxide of iron, or into an intermediate oxide.

Iron exposed to moist air is covered with a layer of hydrated oxide of iron, called *rust.* As soon as a spot of rust is produced on the surface of iron, the metal oxidizes with rapidity, because it forms an element of the pile, of which the rust is the negative pole, and iron the positive pole; the water is then decomposed and the iron becomes completely oxidized. The oxidation of iron is accelerated by the presence of the carbonic acid of the atmosphere. Rust generally contains ammonia. Iron is preserved from oxidation by covering it with a layer of a greasy substance or of varnish. It may also be preserved from rust by plunging it into water containing alkalies or alkaline salts in solution, such as potash, soda, lime, alkaline carbonates, borax, &c. Iron preserves its polish in water containing $\frac{1}{500}$th of its weight of carbonate of potash or carbonate of soda. Iron covered with zinc is preserved from rust; in this case, it is said to be *galvanized.*

When iron is heated to redness it decomposes steam, and gives rise to black and brilliant crystals of magnetic oxide of iron. This reaction has been described in the article on the *preparation of hydrogen.*

Most acids dissolve iron, disengaging hydrogen, and produce salts of iron.

PRINCIPAL COMPOUNDS OF IRON.

OXIDES OF IRON.

Iron combines with oxygen in several proportions: the first degree of combination is called *protoxide of iron;* the second, *sesquioxide* or *peroxide of iron.*

There are also other oxides, called the *magnetic oxide,* and *oxide of iron of blacksmiths' scales.*

The peroxide, which is that which is most highly oxygenated, is often called *English red, astringent crocus Martis, aperient crocus, colcothar.*

This oxide is abundant in nature; it may be obtained artificially by several methods, especially by calcining certain salts of iron, or by exposing to the air wet turnings of iron. Its color varies from yellow to a red brown; it is yellow when hydrated, and red when anhydrous.

It is generally oxide of iron which colors earth red or yellow; it is that also which produces the spots of rust on the surface of iron.

Oxide of iron calcined becomes excessively hard; that found in nature is often extremely hard. When polished, it is employed, under the name of *hematite*, to burnish gold and silver. *Colcothar* is used to polish glass, and many metals.

The peroxide of iron dissolves in some fluxes, such as glass and borax, and forms according to its proportion, glass but slightly colored, or that colored yellow and red.

The protoxide and magnetic oxide give to glass, on the contrary, a deep green color. Thus, it will be seen that to decolorize glass, the protoxide and magnetic oxide are to be transformed into sesquioxide, which colors much less. This oxidation is brought about by the peroxide of manganese.

Magnetic Oxide of Iron.

This oxide may be obtained by passing steam over iron heated to redness.

The *natural magnet*, or magnetic oxide of iron, forms deposits more or less considerable in the crystalline formations, but not in the sedimentary. It is generally in masses which form entire mountains, as at Taberg, in Sweden. Its aspect is metallic; it is very magnetic, and is found crystallized in regular octahedrons.

This body constitutes a valuable ore of iron, and is a great source of wealth to Sweden and Norway. The iron which it produces is nearly pure.

When a piece of iron, kept at a red heat for some time, is beaten with a hammer, a black oxide is detached in the form of scales.

Sulphurets of Iron.

Sulphur has a great affinity for iron; these two bodies combine together in different proportions. Sulphur and iron will react on each other at the ordinary temperature, under the influence of moisture, and give rise to a hydrated sulphuret of iron which is highly inflammable. This sulphuret is obtained by mixing in a flask 60 parts of iron filings, 40 parts of sulphur, and water sufficient to make a consistent paste; the iron and sulphur unite, and disengage sufficient heat to volatilize part of the water. If the product is then exposed to the air, it inflames, giving off sulphurous acid and vapor of water. When it is covered with earth, it produces some of the phenomena apparent in volcanoes. This preparation is called *volcano of Lémery.* The most important of all the sulphurets of iron, is that called *martial pyrites.*

Pyrites is of a brass-yellow, and sufficiently hard to strike fire with steel. When calcined, it loses part of its sulphur, and is transformed into magnetic pyrites. Roasted in the air, it disengages sulphurous acid, and is changed into peroxide of iron. Some varieties of pyrites will keep in the air without change; but others oxidize rapidly, and effloresce in absorbing the oxygen of the air, and are converted into sulphate of iron. That which effloresces the most readily is the *white pyrites.*

Pyrites is used in the manufacture of sulphuric acid; when roasted in the air, it gives off sulphurous acid, which is passed into chambers of lead.

It may also be used for making sulphur; by distillation, it changes into magnetic pyrites, giving off sulphur; the fixed residue, exposed to the air, is transformed into sulphate of iron (green vitriol).

Sulphate of Iron.

This is the most used of all the salts of iron. It is known as *green copperas* and *green vitriol;* it is the result of the combination of sulphuric acid with the protoxide of iron. This salt is obtained by exposing to the air certain varieties of pyrites called *efflorescent pyrites*, or common pyrites roasted. The pyrites then absorbs the oxygen of the air, and becomes changed into sulphate of iron. When the sulphatization is effected, the mass is treated with water which dissolves the sulphate of iron, and the salt is purified by crystallization.

Sulphate of iron has a greenish color, is soluble in water, its taste is styptic and astringent, and resembles that of ink.

This salt is decomposable by heat, producing fuming sulphuric acid, and peroxide of iron (colcothar). It is largely used in dyeing; when mixed with astringent solutions, it produces a black color, which is the basis of ink, and readily fixes itself in tissues.

The sulphate of iron renders indigo soluble, and for this purpose is used in dyeing blue, *to mount the indigo vats.*

The sulphate of iron is also used to precipitate gold from its solutions, &c.

The Extraction of Iron.

The name *iron ore* is given to every substance containing sufficient iron to make it worth working. As small quantities of phosphorus, sulphur, or arsenic, destroy the tenacity of iron, those ores are rejected in which the iron is united to either of these bodies.

The only ores worked are—

The magnetic oxide of iron; the anhydrous or hydrated peroxide of iron; the carbonate of the protoxide of iron (spathic iron, or carbonate of iron of the coal-measures).

All these bodies calcined in the air are brought to the

state of sesquioxide; thus it may be said, in general, that it is from the sesquioxide that iron is extracted.

The metallurgist divides the ores of iron into two principal kinds; into *earthy* and *stone ores.* The first are formed by the hydrate of the peroxide of iron, the second comprise all other kinds.

Charcoal is used for reducing the peroxide of iron. The principle of the metallurgy of iron, is the following:

The body which it is proposed to reduce is always a combination of iron and oxygen. When it is heated with charcoal, it is decomposed; its oxygen combines with the carbon, while the iron is set free. If the carbon is in excess, the reduced iron combines with a small quantity of carbon, to form a body more fusible than iron, called *cast-iron.* The processes of extraction vary with the state of purity of the ore to be worked. Suppose, for example, we had to treat a rich and very pure ore; in heating this, even at a moderate temperature, with an excess of charcoal, the iron would be reduced. This process is still in use in the Catalan forges.

If the ore is not pure, but contains carbonate of lime, silex, and other foreign matters, and is heated with charcoal in a catalan forge, the iron will be found so disseminated through the mass, that the workmen cannot collect it. They have then recourse to a reduction followed by a general melting; the fusion of the gang* is brought about, and that of the metal also, which passes to the state of cast-iron in uniting with carbon.

The process by which all the parts of the ore are melted, bears the name of the process by the *blast-furnace.*

Before treating an ore by the catalan method, or by that of the blast-furnace, it must be subjected to a series of operations called the *preparation of the ores.*

---

* *Gang* is the earthy, stony, saline, or combustible substance, which contains the ore of metals, or is only mingled with it without being chemically combined. If chemically combined it is called *gangue.*

Preparation of the Ores.—The earthy ores are not roasted; they are simply washed by being shovelled in a stream of water, or in a *buddle.*

The buddle is a wooden or iron box, the bottom of which is semi-cylindrical; arms of iron attached to the arbor of an hydraulic wheel kept in motion, agitate the ore in water, which is continually renewed; when the washing is completed, one of the sides of the box is opened, and the watei runs off.

The stone ores are frequently roasted. This operation is intended to soften the stone, and make it more porous and more easy to reduce, and also to drive off the water and carbonic acid which it contains.

The roasting may be carried on in heaps in the open air, in enclosures of masonry, or in continuous kilns like lime-kilns.

The Catalan Method.—The object of this method is to obtain iron directly from the ore, without making cast-iron. The ore is not reduced till it has first been roasted.

The furnace in which the reduction is effected consists in a rectangular hearth, lined with thick cast-iron plates, the bottom of which is formed of a refractory sand-stone. To increase the combustion, a current of air is driven into the hearth, through a copper *tuyère.*

The blowing-machine used is composed of a pipe pierced with holes, into which water rapidly falls, and, in its fall, carries along with it into a drum a quantity of air, which is then driven out by the water into the interior of the hearth. The fuel used is generally charcoal.

The operation is commenced by introducing lighted charcoal into the hearth, just over the tuyère. The hearth is divided into two parts by an iron plate; on one side of this the ore is placed, on the other, the charcoal. The fuel is always placed next the tuyère. When the hearth is charged, the plate of iron, which was placed there to prevent the ore and coal from mixing, is withdrawn. When the charge is made, the draft is put on at first with caution, afterwards as strong as possible.

The workman then stirs the mass with a hook; at a certain time in the operation, he allows the scoriæ to flow out, and, when he thinks the reduction complete, he collects together the lumps of iron disseminated through the agglutinated mass, forming out of them a mass called *loup*, or ball, which he takes out and places under the forge-hammer. The iron is beaten out by the hammer, and then divided by means of strong knives or chisels, into pieces, which are forged and drawn out into bars.

The process of the Catalan forges gives a tough iron of good quality; but it can only be applied to pure and rich ores. The iron thus obtained is often mixed with small grains of steel, which prevent it from being easily laminated.

We will now describe the process by the blast-furnace, in which iron is reduced into cast or pig iron, and which may be used for working very poor ores.*

Manufacture of Cast or Pig Iron.

The treatment of the ores of iron in blast-furnaces, requires a thorough melting. The iron reduced, unites to the carbon, and produces fusible cast-iron; the different substances forming the *gang*, ought themselves to enter into fusion to form the *slag* or *scoriæ*. When the gang of an ore is clayey, in order to cause it to fuse, some carbonate of lime is added as a flux, this forms, with the silex, a silicate of alumine and of lime, fusible at the high temperature of the blast-furnace.

If the gang is calcareous, the ore is mixed with some silicious substance as a flux.

The blast-furnaces are lined with fire-bricks, or sand-stone

---

* Every ferruginous clay-stone which contains more than 20 per cent. of metal, is regarded as an iron ore.—*Ure,* Vol. I., p. 1073

capable of supporting a high temperature without being fused.

Their form is that of two truncated cones, placed base to base, united with a slight curving, to avoid sharp angles in the interior of the furnace, which would interfere with the course of the flame, or that of the ores. Often, also, the two bases of the cones are placed together, without being curved at their junction.

The ordinary draught of a furnace would be altogether insufficient to produce the heat necessary to cause the fusion of the scoriæ and the metal in the blast furnace. Air is driven through two or three tuyères supplied from a blower worked by an hydraulic machine or steam-engine.

A blast-furnace is composed of different parts, each having its own name. We will here describe some of them.

The superior opening of the furnace is called *gueulard* or *furnace mouth;* this is circular: it is through this opening that the furnace is charged with the different layers of ore, fuel, and flux.

The parts of the blast furnace which are between the top or mouth and the *hearth*, are called *boshes*, *belly*, and *cavity*.

It is in the cylindrical part, called the *working area*, that the temperature is highest, and where the metal and scoriæ become fused, to pass then into the *hearth*, which is at the lower part. The opening of the hearth is closed by a large stone called the *dam stone;* over this is an opening through which the slag constantly pours out on an inclined plane.

At the side of the dam stone is a channel, which leads the metal from the furnace to the mould of the work-shop, when flowing out. During the reduction of the metal, the hole through which it flows is closed by a stopper made of clay and charcoal dust.

The melted iron is run out into moulds of sand, forming a mass called *pig*. This pig is then covered with sand, to prevent it cooling too rapidly.

The *tuyère* is the opening or pipe through which the air penetrates into the furnace; it is above the hearth.

The extremity of the pipe called *nose*, being required to support a high temperature, is surrounded with a casing through which a current of cold water is made to circulate; it then can be exposed to a white red heat without being fused.

The melted iron is run out every twelve to twenty-four hours. This time varies with the height of the furnace, and the size of the hearth.

The exterior wall of a blast furnace is provided with passages intended for the escape of moisture and to prevent cracking.

The furnace is charged at the top; an inclined way is built for transporting the ore and fuel to the platform.

The furnace is often built against the side of a hill, being careful to keep it off from the bank, so that the water will not soak through. It is common to have passages below to carry off the water.

The lining of the furnace is built of refractory stones, or brick; it is separated from the outside wall by a layer of sand or cinders, which prevents the escape of heat, and permits the expansion of the walls without cracking: by this arrangement, the lining can be repaired without deranging the outside walls.

The height of the furnace varies with the nature of the fuel used. For charcoal it is from 6 to 12 metres, and may be as high as 18 to 20 metres, and often more for those furnaces in which coke is used.

The fuel most commonly used is charcoal, coke and wood. Coke is preferred in all places where bituminous coals can be had at a moderate price.

Latterly, in blast furnaces, cold air has been replaced by air heated from 150° to 300° Cent. By using hot air, a much higher temperature can be obtained than it would be possible to produce with cold air: this is one of the greatest improvements in the history of the blast furnace. It may be conceived, that a marked difference in the intensity of the heat will produce a fusion or more easy reduction of the silicates, and increase the yield.

The advantage to be derived from the use of hot air, is chiefly in the quantity of heat which the air carries with it. The air may be heated in separate fires, or even by the waste heat of the blast-furnace. When the furnace is built, the fire is then made.

In order to dry it, the fire is first made in the chamber in front of the dam; a draft is established through the mouth, which carries off part of the moisture. After the lapse of a few days, when the moisture is driven off, some charcoal is put in the hearth, and lighted, when it is gradually filled to the working-chamber, and the furnace is entirely charged with coal, without adding the ore.

The drying of the furnace may last from twelve to fifteen days.

When the furnace is fully lighted, a small quantity of ore is added with each charge, which is gradually increased.

When the metal shows itself in the working area, the blast is given at first gently, and not with all its force till after two or three days.

The metal obtained in the first runnings is always white cast-iron, because the temperature is not sufficiently high; and it is only when the furnace gets to regular work that the grey metal is produced, if the nature of the ore will yield it.

Refining of Cast or Pig Iron.

Cast-iron, or pigs, are refined in work-shops called *forges*. The object is to decarbonize the cast-iron, and to transform its silicium into silicic acid, which then forms, with the oxide of iron, silicate of iron.

Cast-iron is refined by two different processes. The first is made with charcoal, in small hearths, or *refinery-fires*, often called the *procédé comtois*. The second, in *puddling-furnaces*, heated with pit-coal. We shall first describe the former process.

Before refining the iron, in some places it is first melted and run into flat trenches, and then broken into pieces. This preliminary operation bears the name of *mazéage*.

A hearth for a comtois refinery-fire is a prismatic cavity with a horizontal rectangular base, bounded by four vertical sides of iron, in which charcoal is burned, in order to produce, under the influence of heat, the decarbonization of the pig-iron, and a heat sufficient to weld all the parts of the metal, to forge it, and to draw it into bars.

The air which causes the combustion of the charcoal, is introduced into the fire-place by means of one or two tuyères which pass through one of the vertical sides of the hearth.

When the hearth is filled with burning charcoal, the pig-iron is placed over it, and soon melts and falls to the bottom of the fire. It is generally mixed with scoriæ, and oxide of iron.

The refining may be divided into two periods. In the first, the cast or pig iron being in contact with the oxide of iron, becomes decarbonized by its oxygen, and the oxide is reduced. All the efforts of the workman should then be directed to multiplying the contacts of the cast-iron and the oxide of iron. In the second period, the cast-iron is stirred about, in order to free it from the slag or scoriæ, which adhere to it in the bottom, in the angles of the hearth. The pig-iron, purified from the scoriæ, is presented to the blast of the tuyère, which oxidizes the silicium, and forms silicate of iron. When it is partly refined, it falls to the bottom of the fire, where the decarbonization is completed.

Then the workman *avale la loupe;* that is to say, he collects together all the parts of the refined iron, to form a *loupe*, or *ball*, which he then places under the large forge-hammer. This is called *cingler la loupe*, or forging the loupe.

The ball is divided into two pieces, which are reheated to a white welding-heat, and then drawn out under the large hammer, in two heatings. The drawing out with the small hammer is only used for small iron.

Refining with Pit-Coal, by the English Process.—The use of coke in blast-furnaces gave rise to the use of this fuel, or even pit-coal, for the refining of iron.

But this could not be performed in ordinary refining-furnaces, because the metal, being in contact with the coke, would rapidly combine with the sulphur, and become brittle.

The refining-furnaces have been replaced by furnaces in which the cast-iron is heated by the flame, alone, of the combustible.

In this process, the refining comprises three operations. The first is done in the *fineries*, very like the refining-hearths; the second in a reverberatory furnace, called a *puddling-furnace;* the third in another reverberatory furnace, called a *reheating-furnace.*

The *fineries* are composed of a hearth lined within with plates of cast-iron, covered with clay, and provided with a hole through which the scoriæ and cast-iron flow.

The hearth is surmounted with a chimney. Two tuyères, placed opposite each other, conduct the air to the surface of the bath.

The coke is introduced into the hearth, on which is placed from 1000 to 1200 kilogrammes of the pig-iron; and it is then raised to a temperature sufficient to render the metal very liquid.

At the end of two hours, it is run into a trench, and cooled with water, to make it brittle. It is thus *fine metal* is obtained.

In this first operation, the cast-iron is freed in a great measure from the foreign bodies which it contained.

To decarbonize the iron completely, it is heated in a puddling-furnace, mixing it constantly with scoriæ rich in iron, and with scales of iron; the object of this operation is to cause the oxide of iron to react on the cast-iron, in order to deprive it completely of its carbon.

The sole of the puddling-furnace is slightly inclined; it is made of very refractory bricks; it is generally covered with scoriæ, and pounded sand.

As the operation progresses, the cast-iron becomes pasty, and gives off carbonic oxide. When the decarbonization is complete, the furnace is carried to a white welding heat, and a ball or loupe is formed with the iron, which is first put under the forge-hammer, and then under the *reducing cylinders.*

These cylinders are channelled, and present grooves which successively diminish in size. The iron is passed five or six times through the grooves, according to the order of their decrease; the iron is thus reduced into common flat bars. In this operation, the pressure is so great that the scoriæ fly off with force, and the iron is compressed somewhat like a sponge.

To finish the refining of iron, it is cut while red-hot, and arranged in bundles, which are heated to a white welding heat in the *reheating-furnace,* and then subjected anew to the action of cylinders, the grooves of which progressively diminish, and vary in form to suit the patterns required.

This operation is called *forging.* Puddled iron is often subjected to a second, and sometimes to a third forging.

Different Kinds of Cast-Iron.

Iron, in combining in blast-furnaces with carbon, becomes fusible, and forms a substance known under the name of *cast-iron.*

Cast-iron is not formed exclusively of iron and carbon; it contains other foreign bodies, such as silicium, manganese, and phosphorus, which often greatly influence its properties.

There are three principal kinds of cast-iron; *black*, *grey*, and *white*. There is besides another kind, which is a mixture of white and grey, called *trout cast-iron*. There is also a peculiar kind of cast-iron, which is a product from the manganesian ores, called *white manganesian cast-iron.*

Cast-iron contains from two to four hundredths of carbon.

Black Cast-Iron.—This iron will take the mark of the hammer. It breaks easily, and shows large grains, in the midst of which graphite is distinctly seen; the presence of this body gives it its characteristic color. Cast-iron, then, has the property, under the influence of heat, of dissolving carbon, and depositing it in the form of graphite when slowly cooled.

Black cast-iron is more fusible than other cast-iron; when treated with acids, it disengages hydrogen, which is always accompanied with a fœtid carburet of hydrogen, and leaves an abundant residue of graphite. This cast-iron is produced in blast-furnaces, when an excess of carbon relative to the ore is used. It is much sought after for castings, from second fusion.

Grey Cast-Iron.—This is in general the product from ore of a good quality, when the furnace works regularly; its color is of a deep grey, and occasionally of clear grey. Its fracture is granular, it is always porous, and never takes a good polish, it may be filed, cut with the shears, and is easily drilled. Treated with the acids, it leaves a residue of graphite which is less considerable than from black cast-iron. Grey cast-iron always contains a considerable quantity of silicium. This iron, exposed to the air, oxidizes with more rapidity than white cast-iron, because it is more porous.

When, after having melted this iron, it is suddenly cooled by throwing it into cold water, it undergoes a sort of tempering, and is transformed into *white cast-iron*. This modification is always partially produced when grey cast-iron is too suddenly cooled; it then becomes more hard and brittle; it may, however, be softened by recasting it, and cooling it slowly. All white cast-iron is not softened by annealing; that which contains manganese always remains white.

In white cast-iron obtained by tempering grey cast-iron, the carbon is found probably in combination with the iron, and is not there in the state of graphite; for when this iron is dissolved in an acid, it leaves no black residue, and the greater part of the carbon is disengaged in the state of carburet of hydrogen. Black and grey iron, run into sleeves of thick iron, undergo a sort of liquation. The part which is cooled first, and which is probably less fusible, contains but about 1 or 1·5 per cent. of carbon; it is very hard, and has most of the properties of steel, while the part which is in the centre is rich in carbon, and much more fusible. This property is often made use of to harden the surface of certain pieces of cast-iron used in making rollers. The phosphorus contained in grey cast-iron diminishes its tenacity, but increases its fluidity, and renders it suitable for casting objects of art. Grey cast-iron may be used for casting from the first fusion, or for refining.

White Cast-Iron.—We have said that white cast-iron might be obtained by suddenly cooling grey cast iron, but the common white cast-iron is generally produced in the blast furnace, either by reducing manganesian ores, or by using too great a proportion of the ore for the carbon.

White cast-iron has a metallic brilliancy, it is sometimes of a silvery whiteness; when it is manganesian, it often crystallizes in large quadrangular pyramids.

It is very hard, cannot be cut with the file, and breaks under the hammer without receiving its mark. It resists crushing better than the grey; it is also more fusible, but remains in a pasty state of fusion, while the grey iron acquires great fluidity.

The carbon in it exists in a different state from that in the grey iron; when, in fact, it is heated with an acid, it leaves no residue of graphite.

According to M. Karsten, white cast-iron is harder in proportion as it contains more carbon. It is sometimes used for casting, but it is almost always refined. The manganesian cast-iron is generally used for the fabrication of the steels of the forge, or of iron for making steel.

Steel.

The name *steel* is given to a carburet of iron containing traces of silicium and phosphorus, and in which the proportion of carbon never exceeds one hundredth. Steel contains more carbon than the iron of commerce, and less than cast-iron.

Steel may, besides, contain small quantities of manganese, aluminium, and sometimes traces of arsenic. The ores of manganesian iron are eminently fitted for the manufacture of steel.

Steel is harder than iron; it will take a beautiful polish; it presents a very fine granular texture, even and close; it is sonorous, giving musical sounds.

Heated to a red temperature, and suddenly cooled, it undergoes the phenomenon of *tempering;* it becomes excessively hard and brittle, and will even scratch glass. It is this property which gives it so much value in the arts.

The hardness of its temper depends upon the temperature to which the steel has been heated, and on the nature of the body into which it has been plunged when red-hot.

To produce a very hard temper, the steel must be raised to a white heat, and plunged into very cold water, or, what is better, into mercury.

The soft tempers are produced by cooling the steel in a fat body or in melted rosin. Sometimes, in the arts, steel is tempered by raising it to a high temperature and suddenly cooling it. But, more frequently, steel is too highly tempered, and it is *annealed* at variable temperatures, so as to give it the degree of hardness required.

During the annealing, the steel loses its hardness, in proportion as it is heated to a high temperature. The workman judges of the temperature, in annealing, by the different colors which steel presents with the different temperatures to which it is exposed.

These colors are owing to the formation of very thin layers of oxide of iron, which reflect different colors according to their thickness. Heat gives to steel the following colors:

At 220° Cent. . . .	clear yellow.
" 245° "	golden yellow.
" 255° "	brown.
" 265° "	purple.
" 285° to 290° . . .	bluish.
" 300° "	indigo.
" 320° "	pale green.

Razors and pen-knives are annealed at a yellow, scissors and knives at a brown, watch-springs at a blue, carriage-springs at a red brown, etc.

The temperature for annealing is also judged of, by examining the changes produced on a layer of fat spread over the steel when heated. To temper at a yellow color, the moment the fat gives off white fumes the operation must stop: at a brown, when the vapors are abundant and colored; to temper at a blue color, the temperature must be raised till the fat is about to take fire.

Steel undergoes, by tempering, a modification somewhat like that of cast-iron. After the tempering, the carbon is not found in the same state as before. In fact, steel, not tempered, treated with an acid, is dissolved, leaving a very sensible residue of graphite, while tempered steel does not give graphite. The carbon disengages in the form of carburet of hydrogen.

Steel loses its sonorousness when it is tempered, and only gives dull and obscure sounds.

We will give the signs for distinguishing the best steels.

1st. When tempered at a low heat, it becomes very hard.

2d. Its hardness is uniform through its substance.

3d. After tempering, it resists a blow without breaking, and does not lose its hardness, except by annealing at an intense heat.

4th. It welds readily without cracking.

5th. It shows in its fracture a fine and equal grain; it is very dense and suitable for polishing.

There are four principal kinds of steel: *natural steel*, *steel of cementation*, *cast steel*, and *damask steel.*

Natural Steel.—This is also called *steel of the forge*, or *steel of cast-iron.* It is obtained by incompletely refining cast-iron in deep hearths in contact with the air, or under the influence of oxide of iron, which decarburets it.

Cast-iron contains more carbon than steel does. This explains why, by taking from cast-iron a part of its carbon, it can be transformed into steel.

In the preparation of natural steel, manganesian cast-iron is always used, for the reasons pointed out in speaking of Catalan forges.

This is operated in hearths like those for refinery fires, containing cast-iron in fusion, and some scales of oxide of iron.

Steel of the forge is chiefly used in the manufacture of agricultural instruments.

In the extraction of iron by the Catalan method, the iron is sometimes carburetted sufficiently to transform it into natural steel.

Steel of Cementation.—Cementation is an operation in which iron is made into steel by heating it for a long time in charcoal dust.

The iron then combines with about a hundredth of carbon, and is transformed into steel.

For cementation, crucibles or chests (troughs) made of fire-clay or bricks are used; these are so placed in the furnace that the flame envelopes them on all sides; the cnests are filled with alternate layers of a carbonaceous matter termed *cement* and bars of iron; these bars should not touch each other. Rods of iron called *trial-rods* are placed in the chests, which are from time to time taken out, in order to ascertain the degree of cementation. The temperature ought not to be high enough to melt the steel.

Various opinions have been formed as to the best cements for making steel; some use salt, ashes, etc., but it appears to be demonstrated that charcoal is the best.

Sometimes the surface only of small pieces of iron is cemented by a process which consists in heating the iron in chests of sheet-iron, cast-iron, or earth, with a cement composed of charcoal, fat, ashes, and sea-salt.

Several bodies will harden iron: it is sufficient, in fact, to rub a plate of iron heated to redness with a crystal of cyanoferride of potassium, to give to its surface the hardness of steel.

Steel of cementation is generally covered with blisters, which gives it the name in commerce of *blistered steel.*

When it is desired to diminish the hardness of the surface of steel, which is intended, for example, for engravers' work, it is kept for five or six hours at a white heat in iron filings.

Cast-Steel.—This steel is the most homogeneous, and most highly esteemed. It is obtained by fusing steel of cementation. It is very hard, will take a beautiful polish, and often has the valuable property of tempering by the action of air alone.

Damask Steel.—This name is given to a variety of steel which becomes covered with a kind of watering when treated with weak acids: this is often called *wootz* or *Indian steel.* Steel suitable for damasking is obtained by slowly cooling steel rich in carbon; there is then formed in the mass carburets of iron in definite proportions, which crystallize and then become visible under the action of acids.

Damask steel is likewise prepared, comparable to wootz steel, by melting iron of good quality with two-hundredths of lamp-black or coke.

When coke is used, the steel contains in general some thousandths of aluminium. The presence of this metal has been recognised in many damasked steels coming from India.

According to Messrs. Faraday, Bréant, Berthier, Fischer, Stodard, etc., damask steel may likewise be formed by alloying ordinary steel with chromium, platinum, and aluminium.

M. le duc de Luynes produced beautiful damask blades by alloying with steel small quantities of tungsten or molybdenum.

It results from the recent researches of a Russian engineer, (M. Anocoff,) that the surest method of obtaining steel fit for damasking, consists in melting in a refractory crucible 5 kilogrammes of very pure iron with $\frac{1}{12}$th of graphite, $\frac{1}{32}$d of scales of iron, $\frac{1}{24}$th of dolomite serving for a flux. To give the appearance of damasking, the steel is rubbed with sulphate of iron mixed with some sulphate of alumine.

Damask steel, thus prepared, appears to be much harder than the best cast-steel.

ZINC.

Zinc was known to the ancients, who used calamine for making brass. Paracelsus appears to have been the first chemist who described zinc as a metal; his researches date from the beginning of the thirteenth century.

Zinc was not regularly mined till about a century ago; it has, however, been very much developed of late years.

Properties.—Zinc is solid, of a bluish white color; its texture is lamellated. It has a peculiar softness; it adheres to files used in working it; it is said to *grease files*. It has not much sonorousness, and is rather soft, but less so than lead and tin.

When it is fine, it may be hammered into thin sheets, and does not crack along the edges. Zinc of commerce is not so malleable as pure zinc; in the cold it cracks under the hammer; raised to a temperature of 130° C. to 150° C., it becomes malleable, and may be forged, laminated, and even drawn into fine wire.

At 250° C., it becomes brittle. It may easily be pulverized in an iron mortar heated to this temperature.

This metal has but little tenacity. A wire of zinc, of two millimetres diameter, breaks with a weight of 12 kilogrammes.

Zinc fuses at a temperature of 412° C.; if allowed to cool slowly, it crystallizes. When it is melted, it is easily granulated by being poured into a vessel filled with water.

Zinc is volatile when heated to a red white heat; it enters into ebullition, and distils. This distillation may be effected in an earthen crucible, through the bottom of which an earthen tube rises into the interior a little more than one-half its height. This tube passes through the grate of the furnace, coming out under it into a vessel full of water. The zinc is placed in the crucible to the height of about half the tube, and the crucible is hermetically sealed. It is then heated to a red white heat; the metallic vapors are driven from above downwards, and condense in the tube, through which the zinc flows into the water.

Zinc may also be distilled in an earthen retort. To prevent the neck of the retort from being choked by the condensation of the metal, it is necessary from time to time to clear it with a rod of iron, and let it fall into a test or cupel.

The zinc of commerce is never pure; it contains about

one-hundredth of its weight of foreign bodies, which are iron, lead, and sometimes carbon, copper, cadmium, and arsenic.

Zinc does not oxidize in dry air; exposed to moist air, it is rapidly covered with a white and very thin coating of oxide of zinc, which is partly carbonated, and which preserves the rest of the metal from a subsequent alteration.

Zinc heated, in contact with the air, burns with a white flame at about 500°, the brightness of which is due especially to the presence of oxide of zinc, which is fixed and infusible. A crucible containing zinc heated to redness, is soon filled with downy flakes of the oxide of zinc.

Clippings of zinc, heated in the flame of a candle, take fire, and burn with a bright light.

Zinc easily decomposes steam under the influence of heat, giving hydrogen and oxide of zinc. The decomposition of water by zinc begins to be sensible at 190° C. This metal decomposes water in the cold, under the influence of even feeble acids, and gives off hydrogen; it is thus that hydrogen is prepared.

The zinc of commerce, which contains small quantities of iron or lead, dissolves rapidly in the acids; pure zinc, on the contrary, is slowly attacked by the acids, particularly in glass vessels.

Oxide of Zinc.

The oxide of zinc was formerly known under the name of *flowers of zinc*, *pompholyx*, *nihilum album*, *lana philosophica*, &c.

This oxide is white; it becomes yellow when calcined, but on cooling recovers its original color. It is completely fixed, and indecomposable by heat. The oxide is easily reduced by hydrogen and carbon; it attracts carbonic acid on exposure to the air, and thus acquires the property of effervescing with acids.

Crystallized oxide of zinc is obtained by submitting zinc to the action of steam. It is thus often found in the chimneys where zinc is volatilized; the crystals are then yellowish.

Anhydrous oxide of zinc is prepared by heating zinc in an open crucible: one part of the oxide escapes in the form of white flakes, and the remainder attaches itself to the sides of the crucible or to the surface of the metal: this is to be detached from time to time to give free access to the air.

The oxide of zinc may also be obtained by subjecting to calcination the carbonate of zinc prepared by double decomposition.

The oxide of zinc mixed with drying-oils, may be substituted for white lead, and is then called *white of zinc.* The oxide of zinc is produced in the large way by distilling the zinc of commerce in earthen cylinders, and burning the vapor by a current of air. An oxide of great whiteness is thus obtained, which is deposited in the chambers where it is collected. This oxide, mixed with 15 or 16 per cent. of linseed oil, first heated with a small quantity of bioxide of manganese, gives a paste which covers as well as white lead, and offers the advantage of not being blackened by sulphurous emanations.

The cost of manufacturing white of zinc is not greater than that of zinc, and its fabrication does not appear to be dangerous like that of white lead.

Extraction of Zinc.

Though the minerals which contain zinc are numerous, there are but three named as being sufficiently abundant to be used as ores, and one of those, the silicate, not being reducible by charcoal, cannot be used for the fabrication of zinc by the processes in use: the ores of zinc then may be reduced to two, viz.: calamine (carbonate of zinc) and blende (sulphuret of zinc).

Metallurgic Treatment.—The metallurgic treatment of the ores of zinc comprises two entirely distinct operations, the calcination or roasting of the ore, and the reduction of the oxide by charcoal, in properly arranged distilling apparatus.

Zinc is always extracted from its ores by distillation. The calamine is calcined to drive off the water and carbonic acid.

The roasting of blende is intended to bring this sulphuret to the state of oxide. It is seldom this can be done at a single operation; generally the blende in fragments is made to undergo a preliminary roasting to disintegrate it, and to drive off the most of the sulphur. This roasting is done either in reverberatory furnaces heated with wood, with coal, or waste flame, or in continuous kilns; the sulphur contained in the blende prevents the addition of fuel after the operation has commenced. The blende once roasted is finely pulverized, and roasted a second time in a reverberatory furnace, generally heated by the waste flame from the furnace for reduction.

The ore, being brought by the roasting or calcination to the state of oxide, is mixed with its bulk of fragments of coke, or dry coal coarsely broken, and then filled into earthen vessels of different forms which are heated to a bright red. The oxide of zinc is then reduced by the charcoal, the zinc distils, and condenses outside the furnace.

Lamination of Zinc.—Zinc intended for lamination is first purified by fusion; it is then drawn off and run into the ingot moulds. The superior layers are the purest; the inferior, where the lead and iron are concentrated, are improper for lamination, and produce zinc which is used for casting, and for the manufacture of colors, etc. This fusion is sometimes done in iron pots, but it is best done in reverberatory furnaces. The lamination can only be done at a temperature the limits of which are very narrow, and are comprised between 120° C. and 150° C. Above or below these limits, the zinc becomes brittle.

Uses.—Zinc is used for roofing, gutters, conduit-pipes, composition ornaments, castings (for voltaic batteries), for the manufacture of galvanized iron, brass, argental, white of zinc (for paint), etc. Zinc cannot be used for culinary vessels, because it is attacked by organic acids, forming salts of zinc which are poisonous.

TIN.

Tin is one of the most anciently known metals: it is almost as white as silver; its reflected color is however slightly yellow; it emits a disagreeable odor when rubbed between the fingers. It is very malleable, and may be reduced by beating into thin leaves. Tin is not tough; a wire of two millimetres breaks with a weight of about 34 kilogrammes. When bent, it gives rise to a peculiar cracking sound. Tin is one of the softest and least elastic of metals, and it has but little sonorousness.

It enters into fusion at the temperature of 228° C. It may be melted in a sheet of paper when it is in thin laminæ.

Tin does not volatilize at the highest temperatures; it has a great tendency to crystallize. Melted tin, slowly cooled, crystallizes in prisms with eight faces. When tin is cleaned off with an acid, its crystalline texture is recognised.

To reduce tin to powder, it is melted at as low a temperature as possible, and run into a box and agitated till it cools. A metallic powder is thus obtained, which is suspended in water, and the heavier parts separated by decantation. This powder, mixed with melted birdlime, is used in India to prepare a kind of metallic paint, which, on being burnished, looks like silver.

Tin does not act sensibly on air either dry or moist; so that it may be preserved a long time in the air without alteration; but when the temperature is raised, tin oxidizes rapidly. If a small quantity of tin is heated under the blowpipe to a white heat and then thrown on the ground, the metal breaks into small globules which burn with brilliancy.

The acids, and even alkalies, oxidize tin.

The tin of commerce generally contains a small quantity of lead, iron, copper and arsenic. The most highly esteemed tin is that of Malacca. The dealers in tin judge of its purity by melting it at a low heat, and examining the appearance of its surface at the moment of solidification; the purest tin is the whitest, most brilliant, and that which shows least marks of crystallization on its surface.

When tin is covered with crystalline ramifications on its surface after cooling, or by the action of acids, above all, when it presents a dull white surface, it is pretty certain that it is alloyed with other metals.

It was thought at one time that the tin of commerce contained sufficient arsenic to render it unsafe for domestic purposes.

In 1781 Bayen and Charlard demonstrated that Malacca and Banca tin did not contain arsenic in appreciable proportions, and that the other species of tin did not contain at the most the $\frac{1}{600}$th, a quantity altogether insufficient to give it poisonous properties.

Salt of Tin.

Tin dissolves in chlorhydric acid; on evaporating this solution, the protochloride is obtained, which is known in the arts under the name of *salt of tin.*

The salt of tin has a strong attraction for oxygen, and its solution is used in the arts, and especially in dyeing, as an energetic deoxidizer. When applied to stuffs which are colored yellow by the peroxide of iron, or *en solitaire* by the oxide of manganese, it carries off from these oxides a part of their oxygen, and decolorizes them.

The salt of tin is also used as a *mordant*, particularly for the violet colors, of which it heightens the brilliancy, or to modify the color of several other coloring matters.

Tin, dissolved in eau régale, produces the *mordant of tin*, which is used for the dyeing of woollens scarlet, and cottons Brazil red and yellow.

Tin Plate, or Fer Blanc.

This is sheet-iron, covered over with an alloy of iron and tin. The layers which are in contact with the iron are a true alloy of this metal with tin; the surface is metallic tin.

Tin plate is a very valuable alloy for family uses, for it has the toughness of iron, and, like tin, does not rust in moist air.

It is necessary to clean off the sheet-iron and free it from the scales of oxide of iron which generally cover it, or the tin would not adhere to it. With this view, it is immersed into water acidulated with a mixture of sulphuric and chlorhydric acids, until it becomes bright, and black spots are no longer seen on its surface. The sheet-iron is then passed through pure water and rubbed with hemp and sand. After these preparatory operations, it is immersed for about an hour in a bath of tallow, which dries it; it is then passed into a bath of tin, which is itself covered with tallow. The plates remain about an hour and a half in the tin. After this immersion, they are placed on an iron grating to drain.

The sheets which come out of this metallic bath always contain an excess of tin, which is removed by an operation called *washing*.

This operation consists in rapidly plunging the sheet of

tin into a very pure tin bath, which melts off the excess of tin on the surface of the tin-plate; the plate is brushed and passed into a new tin bath to efface the marks of the brush, and then plunged into a pot of tallow.

To terminate the preparation of the tin-plate, it is only necessary to clean it by means of bran.

Tin-plate has the appearance of tin; it preserves its brilliancy for a long time, unless there should be on the surface a crack exposing the iron; there are then formed spots of rust which spread rapidly.

MOIRÉ METALLIQUE, OR CRYSTALLIZED TIN PLATE.

When tin-plate is dipped into an acid solution, the layers of tin on the surface are dissolved, and expose to view the inferior layers, which then show a crystalline appearance of different colors.

To crystallize tin-plate, the following mixtures may be used:

1st. 8 parts of water, 4 of sea-salt, and 2 of nitric acid.

2d. 8 parts of water, 2 of nitric acid, and 3 of chlorhydric acid.

3d. 8 parts of water, 1 of sulphuric acid, and 2 of chlorhydric acid.

The best tin-plate for this purpose is that made of pure tin.

Tin-plate for crystallization ought to be covered over with a thicker layer of tin than is commonly used; for if it is too thin, the crystals will be very small.

To prepare moiré, tin-plate is slightly heated, and by means of a sponge it is evenly covered with the acid solution, and the metallic crystals are soon seen to appear. The action of the acid is arrested by dipping the plate in water; if the acid should act too long, the sheet-iron would be exposed, and there would be black spots.

The appearance of the moiré may be modified at will by changing the crystallization of the tin. In the ordinary state of the tin covering, the plate is slowly cooled, and its crystals are large. But if the plate is heated so as to melt the tin, and the sheet is sprinkled with sal-ammoniac, so as to reduce the oxide formed, and then plunged rapidly into cold water, the tin takes the form of small radiated crystals.

If such a plate is treated with acidulated water, a moiré is obtained which looks like granite. To preserve the moiré, it must be rapidly dried and then coated with varnish. By using different colored varnishes, moirés of beautiful effect are obtained.

Uses of Tin.—This being a metal, alterable with difficulty, is largely used in the mauufacture of a great number of vessels and utensils for domestic use. It answers well for coating copper. Thin sheets of tin are used for preserving a great number of articles from the action of the air and moisture. The tinning of glasses, the manufacture of bronze, the solder of plumbers, and tin-plate, consume a large quantity of tin. The two chlorides are mordants, very useful in dyeing.

It is used also for making mosaic gold, purple powder of Cassius, mineral lake, etc.

LEAD.

Lead is of a bluish grey color, generally dull when exposed to the air, but bright when recently cut: it is very soft. Plates of lead, of great thickness, can be bent, and it can be scratched even by the nail: rubbed on paper, it leaves a mark of a metallic grey.

Lead may be reduced into thin leaves; it has but little tenacity: a wire of 2 millimetres in diameter breaks with a weight of 9 kilogrammes.

Lead fuses at 334°. At a high temperature it volatilizes sensibly, and gives off very visible fumes.

It has the property of dissolving a proportion of oxide of lead, which makes it hard. This solubility of the oxide in the metal explains the modifications which lead undergoes in some of its physical properties, and especially in its malleability, when it has been kept in a state of fusion for a long time in contact with the air. To restore to lead its softness, it should be agitated, when it is melted, with a small quantity of carbon, which reduces the oxide of lead.

Lead, exposed in distilled water to the contact of the air, oxidizes rapidly; the presence of a foreign salt, particularly the sulphate of lime, prevents this oxidation: thus lead only oxidizes superficially when immersed in common water, which always contains salts in solution. Diluted sulphuric acid, and chlorhydric acid, do not readily attack lead.

The oxides and salts of lead are very poisonous; they produce a peculiar poisoning, known as *painters' colic* or *lead colic.* The workmen who use white lead, which is the carbonate, are liable to this dreadful disease, which occasions great pains and nervous trembling.

Workmen attacked with this colic are treated with drinks containing sulphuric acid; this acid combines with the oxide and forms sulphate of lead, a salt insoluble in water, and which does not appear to have poisonous properties.

It would be dangerous to keep drinks of wine or vinegar in vessels of lead, as the metal would form with these liquids, which are always acid, or might become so, soluble salts of lead, which would then act on the animal economy.

The Protoxide of Lead.

The oxide of lead, prepared in the dry way, and which has not been fused, is called *massicot;* when it has been fused, it is called litharge.

Properties. — The protoxide of lead is solid; it varies in color from a yellow citron to a yellow red; it fuses a little below a red heat, and crystallizes on cooling in micaceous scales.

When litharge is fused in an earthen crucible, it acts on the silex of the crucible forming a fusible silicate of lead, and the crucible is thus destroyed.

The protoxide of lead is sensibly soluble in pure water, to which it gives an alkaline reaction.

This oxide is easily reduced by carbon and by hydrogen. It combines with all the acids, and will even attract carbonic acid from the air: it is considered as an energetic base.

The protoxide of lead, heated in contact with the air, absorbs oxygen, and becomes transformed at about 300° C. into *minium*.

The oxide combines in the wet and dry way with alkalies and earths, and forms salts which are called *plumbites*.

The plumbite of lime crystallizes: it is obtained by boiling the oxide of lead with milk of lime.

This compound is sometimes used for dyeing the hair black: in this case, the lead of the plumbite of lime reacts on the sulphur contained in the organic substance constituting the hair, and forms sulphuret of lead, which is black.

The anhydrous protoxide of lead is obtained under different circumstances. It is often prepared by heating lead in the air, or by calcining the carbonate of lead; if the temperature is sufficiently high, the oxide fuses, and crystallizes on cooling.

Minium.

This is an oxide of lead of a beautiful red color. To prepare minium, lead is calcined in a reverberatory furnace, until it is entirely transformed into yellow oxide; the oxide should not be allowed to fuse.

The product of this calcination is ground between two stones and subjected to the action of a current of water, which carries off the oxide into vessels where it is allowed to settle. In this way is prepared the oxide called *massicot.*

As the lead used in the preparation of massicot is not pure, and as the different metals contained as impurities are differently oxidizable, the purity of the massicot varies during the course of the operation; so that, by carefully selecting the products, different qualities of minium are obtained.

The massicot intended for the preparation of minium is introduced into sheet-iron boxes, which hold about 25 kilogrammes of the oxide, and subjected in a reverberatory furnace to a heat which ought not to exceed 300° C. A stronger heat would decompose the minium. A single fire is often not sufficient to convert the massicot into minium; it is subjected to a second, and often to a third fire.

Lead as pure as possible should be used for making minium.

A highly esteemed minium is prepared by calcining in the air carbonate of lead, called *orange mine.*

Uses.—Minium, on account of its beautiful color, is used for coloring paper-hangings, sealing-wax, etc.: it is used largely in the manufacture of crystal glass. In this last application it is preferred to litharge, because it contains neither silver nor oxide of copper, which always color the glass, nor metallic lead, which in reacting on traces of carbonate of potash existing in it, would produce bubbles of carbonic oxide which would remain in the vitreous material, and render its refining almost impossible. The excess of oxygen which the minium loses in forming a silicate of lead is useful, besides, for burning the organic matters contained in the potash.

Sulphuret of Lead.

The sulphuret of lead, known as *Galena,* is the commonest ore of lead, and the only one worked for extracting the metal.

Galena is of a metallic bluish gray, brilliant and brittle. Its crystalline forms are derived from the cube. It is less fusible than lead; it cannot be kept in a state of fusion in earthen crucibles, because it penetrates the crucible, forming the silicate of lead. Galena is decomposed in part by heat, a part volatilizes, and a sub-sulphuret remains; it volatilizes in a current of gas.

Hydrogen takes off its sulphur under the influence of heat; steam decomposes it, forming sulphurous acid and sulphuretted hydrogen, leaving the metallic lead.

Roasting transforms Galena into a mixture of oxide and sulphate of lead; sulphurous acid is disengaged.

Many metals, such as iron, copper, zinc, tin, react under the influence of heat on Galena, and decompose it: iron separates the lead from it in a state of purity.

Litharge easily decomposes the sulphuret of lead under the influence of heat, producing sulphurous acid and lead.

A mixture of Galena and sulphate of lead, exposed to a red-heat, gives sulphurous acid and lead.

These two last reactions are the basis of the metallurgy of lead.

Uses. — The sulphuret of lead is chiefly used for the extraction of metallic lead. It is used by potters under the name of *alquifou* (or potters' ore) to glaze their wares.

There are two principal varieties of Galena, the lamellated sulphuret with large and small facettes, and the compact sulphuret.

The Galena with small facettes is always richer in silver than that with large crystals.

Carbonate of Lead.

The carbonate of lead is white, pulverulent, insoluble in water, sensibly soluble in carbonic acid: it is decomposed by heat, giving off carbonic acid, and leaving a residue of protoxide of lead. When it is heated in the air at a temperature too low to fuse the protoxide of lead, it is transformed into a kind of minium, of a very bright red, called *orange mine.*

The carbonate of lead is used in painting, and is called *white lead, ceruse, blanc d'argent:* it forms the base of all paintings in oil.

For this purpose, it is mixed with drying-oils, such as linseed oil. Ceruse ground with small quantities of oil produces *putty.*

A mixture of equal parts of white lead, linseed oil, and minium, forms a mastic which in time acquires the hardness of stone.

Manufacture.—White lead is made at Clichy in the factory of M. Roard, by a method due to M. Thenard: it consists in passing a current of carbonic acid into a solution of sub-acetate of lead: the carbonate of lead, which is insoluble, precipitates.

The *Holland process* is a method used in Holland, and in Lille and Valenciennes.

The process consists in putting into glazed earthen pots a small quantity of vinegar of poor quality (that from beer or fermented barley,) etc.

These pots have a ledge in the inside, upon which is placed a spiral roll of thin sheet-lead; these are ranged side by side in large wooden boxes, and are covered with sheets of lead. These pots are placed over a thick body of horse manure, and thus are arranged alternately for a height of 5 to 6 metres, beds of manure and rows of pots.

The manure soon ferments; the temperature gradually rises to 100° C.; a considerable quantity of carbonic acid is given off. Currents of air are carefully led into the angles of the cases, and the sheets of lead are soon found exposed to the action of the air, of the carbonic acid, and of the vapors of the vinegar which are disengaged from the pots by the elevated temperature.

The metal absorbs the oxygen of the air, and the carbonic

acid of the manure, under the influence of acetic acid, and in a few weeks the lead is almost completely transformed into ceruse.

The scales of ceruse are easily knocked off. The ceruse is ground either dry or with water; a soft, homogeneous paste is made, which is dried first by putting it into small pots of porous earth, and then into a stove.

Alloys of Lead.

Lead alloys with nearly all the metals: we will speak only of the alloys which are used in the arts.

Lead forms different alloys with tin, which are less brilliant, but harder and more fusible than this last metal. The alloy which contains equal parts of tin and lead is used for solder, called *plumbers' solder.* This is more oxidizable than either of the metals which it contains; it is also used to make *tin putty*, (mixture of oxide of lead and tin) used in the manufacture of *delft-ware.*

The different objects made with pure tin, or its alloys with lead, may be reduced to three classes.

1st. Pure tin used for kitchen utensils.

2d. The alloy formed of 8 parts of lead and 92 of tin, used for water-cisterns, plates, and dishes.

3d. The alloy containing 20 of lead and 80 of tin, with which spoons, lamps, and inkstands are made.

The alloy of lead and antimony, which is made of 4 parts of lead and 1 of antimony, is used for making type. It oxidizes easily when heated in the air.

The physical properties of this alloy are of great importance for the use for which it is intended. It should be very fusible, so as to mould with precision; if it is too soft, it is deformed by the action of the press; if too hard, it cuts the paper.

COPPER.

Copper was known to the ancients. It is found in the native state and crystallized in different forms derived from the cube; but it is mostly in amorphous masses, in fragments, in leaves, or in grains.

Copper is of a brilliant reddish brown; it acquires a disagreeable odor by friction between the fingers; it is very malleable and very ductile. Copper occupies the third rank among the metals for malleability, and the fifth for ductility; it is harder than either gold or silver. After iron, it is the most tenacious of all metals: a wire of copper 2 millimetres in diameter requires a weight of 137 kilogrammes to break it.

Copper fuses at a temperatue of 27° of thė pyrometer, which corresponds to about 788° of Centigrade.

Copper, raised to a very high temperature, volatilizes sensibly, and produces vapors which burn with a beautiful color; it is not, however, very volatile.

Copper has but small affinity for oxygen; it may be kept for an indefinite time in dry air and oxygen without alteration. But when it is kept in moist air, it becomes covered with a green layer, called improperly *vert-de-gris*, which is a hydrocarbonate of copper.

When copper is heated in the air at a moderate temperature, there is formed on the surface of the metal a reddish layer of protoxide of copper.

If the action of the oxygen is prolonged, the protoxide is changed into binoxide, which is black.

Copper decomposes water slowly, and only at a very high temperature; it does not decompose it in the cold, even under the influence of the most energetic acids.

Nitric acid attacks copper, and produces the nitrate of the binoxide of copper and the deutoxide of nitrogen.

Diluted sulphuric acid does not act on copper, but if it is

concentrated, and the temperature is raised, it forms the sulphate of copper and sulphurous acid. When sheets of copper are moistened with diluted sulphuric acid, and exposed to the air, they absorb oxygen, and the sulphate of copper is produced.

Organic acids in a short time cause the oxidation of copper; fat oils, and fat, act in the same way, and become colored green.

Ammonia dissolves copper under the influence of the oxygen of the air: the deutoxide of copper is formed which is dissolved by the ammonia, coloring it blue.

Weak solutions of sea-salt dissolve copper rapidly; those which are concentrated do not attack it sensibly.

All the salts of copper are poisonous; to combat their action on the animal economy, iron filings are used, as recommended by M. H. Edwards; in this case, the iron causes the precipitation of the metallic copper.

To avoid danger from the use of copper utensils, it is usual to cover the inside of those vessels with tin; but as tin is a soft metal and soon wears off, to avoid accident, the tinning of copper vessels should be often renewed.

Copper, in combining with oxygen, forms two oxides.

The protoxide is red; it is used for coloring glass red.

The deutoxide is black, and is also used for coloring glass green or blue.

Among the different salts of copper, we will name:

The sulphate of copper, often called *blue copperas.*

This salt enters into the composition of ink, and black dye for wool and silk; it serves to soften such colors as violet and lilac; it is used in medicine and galvanoplastie (the electrotype.)

Malachite, which is much sought after in jewelry, and which is used for making objects of art, is a natural carbonate of copper,

Scheele's Green, a combination of oxide of copper and

arsenious acid, which is of a beautiful green color, is used in the decoration of paper-hangings.

Alloys of Copper.

Copper combines with nearly all the metals, and forms many alloys, which are much used in the arts.

Copper and iron combine together with difficulty. The brown product, however, which proceeds from the reduction of the double sulphuret of copper and iron, called *black copper*, ought to be considered as an alloy of copper and iron.

Alloys of Copper and Zinc.

These are largely employed in the arts; their price is lower than that of copper. These are called *brass*, *yellow copper*, *gold of Manheim*, *tombac*, (*prince's metal*), *pinchbeck*, *metal of Prince Robert*, *similor*.

Zinc, in alloying with copper, gives a paler color; and in certain proportions it communicates a golden tint: in larger proportions the color is yellowish green; when it forms more than half the alloy, it makes a bluish grey.

These alloys have, in general, greater densities than the mean densities of the two metals which constitute them. They are more fusible than copper.

When brass is melted in the air, a part of the zinc oxidizes; by taking off the film of oxide, which is formed on the surface of the bath, as it forms, all the zinc can be thus oxidized.

The alloys which contain one-third of their weight of zinc are very ductile and malleable in the cold, but they become very brittle when hot.

Sometimes a certain quantity of lead is added to these alloys to give them dryness, and to prevent them from greasing or sticking to the file: the addition of tin hardens them.

Bronze.

Bronze, or airain, is almost always an alloy of copper and tin; but often a small quantity of iron, zinc, and lead is introduced into it, and thus are obtained alloys which resemble common brass.

Bronze was used by the ancients for making agricultural implements, arms, etc., before iron and steel were known.

This alloy is now used for making cannons, clocks, statues, moulded objects, clock bells, reflectors of telescopes, etc.

Bronze is harder and more fusible than copper. It oxidizes in the air less easily; its density is greater than the mean density of the metals which form it.

When melted bronze is kept in contact with the air, the tin oxidizes much more rapidly than the copper, and this last metal remains in a state of purity. The copper may also be separated from the bronze by mixing it with a certain quantity of the same alloy first oxidized.

These alloys of copper and tin have a great tendency to decompose themselves by *liquation.*

They separate themselves, even during fusion, into two alloys, one with an excess of tin which floats on the top, and which is very fusible, the other heavier, which is very rich in copper.

The liquation which takes place during the cooling of alloys of copper and tin, proves that it is impossible to obtain large pieces of bronze homogeneous. This is a very serious inconvenience in casting bronze cannons; for the piece being formed of alloys of different degrees of fusibility, after a certain number of discharges, there are formed in it inequalities, which injure the solidity of the piece, and the accuracy of the aim.

M. D'Arcet has shown that bronze has the curious property of acquiring, by tempering, sufficient malleability to be worked by the hammer. If, on heating it again, it is allowed to cool slowly, it becomes hard, brittle, and very sonorous. This property is used in the arts for making tam-tams, cymbals, medals, and money. The articles of bronze being cast, are tempered, and can then be subjected to the action of the hammer, of the lathe, or the die; their hardness is afterwards restored by reheating.

MERCURY.

Mercury is the only metal liquid at ordinary temperatures. It is almost as white and as brilliant as silver. Subjected to a cold of —40° C., it solidifies, and crystallizes in octahedrons. During the expedition of Captain Parry in the northern seas, they were able to examine the physical properties of solid mercury; and it was shown that this metal takes its place by the side of lead and tin, for malleability, ductility, and tenacity. The experiments of M. Thilorier have confirmed and extended the results observed by Captain Parry. By submitting several kilogrammes of mercury to the cold produced by a mixture of solid carbonic acid and ether, M. Thilorier has shown that this metal may be laminated, and that it is easy to make medals of it, some of which were struck by the die.

Solidified mercury produces, when placed on the skin, the same sensation as a hot body, and disorganizes it almost immediately. This metal has neither taste nor smell.

Mercury absorbs both air and water, from which it is freed by continued ebullition. It boils at about 360° C. The volatility of mercury is taken advantage of to distil it, and free it from foreign metals, which it often holds in solution.

Mercury is generally distilled in the wrought-iron bottles in which the metal is transported, and which, in this case, take the place of retorts; the bottle communicates with a recipient by means of a curved gun-barrel. In laboratories, this distillation is performed in common glass retorts.

Mercury exercises a slow but deleterious action on the animal economy, producing tremors and salivation, which are often seen with workmen exposed to the direct contact of mercury, or to its vaporous emanations.

When mercury is pure, it soils scarcely anything; this property tests its purity. But when it holds foreign metals in solution, such as copper, tin, or lead, it soils glass vessels. When impure mercury is thrown on a plane surface, the globules, instead of being spherical, are seen to take an elongated form.

Mercury, amalgamated with $\frac{1}{4000}$th of lead, forms a plane surface in tubes and may be used for graduating glass instruments.

Mercury, exposed to the air, tarnishes slowly by oxidizing; when mixed with a fat body, it takes a grey color more or less deep, becomes dull, and converted into a black body, which some chemists have considered as protoxide of mercury, but it appears to be mercury finely divided.

Mercury alloys with a large number of metals, producing *amalgams.*

It combines with oxygen in two proportions. The protoxide is hardly known; the binoxide is red. This last oxide is obtained perfectly pure by introducing mercury into a matrass with a long and finely-drawn neck, and raising its temperature to the boiling-point; the length of the neck of the matrass prevents the evaporation of the metal; the mercury absorbs the oxygen of the air, and transforms it by degrees into small crystalline scales of a beautiful deep red color, which was formerly called *precipitate per se.*

The red and crystalline oxide is prepared by subjecting to a regulated calcination, the nitrate of mercury.

The most important compounds of mercury are the following:

Vermillion is the sulphuret of mercury reduced to a fine powder.

Protochloride of Mercury is a medicine much used in practice, called sometimes *calomel;* this body is remarkable for its insolubility in water.

The *bichloride* is, on the contrary, soluble in water; it is exceedingly poisonous. Taken into the stomach, it rapidly corrodes the membranes, and soon causes death; thus it is often called *corrosive sublimate.* Its antidote is white of egg, which forms with it an insoluble compound.

Corrosive sublimate is a powerful antiseptic; it is used for preserving anatomical preparations, and objects of natural history, which it preserves from the attacks of insects. It is used dissolved in water, or better, in alcohol.

AMALGAMS OR ALLOYS OF MERCURY.

Mercury does not alloy, in general, with metals whose point of fusion is high, such as iron, manganese, &c. It combines, however, with platina, when this last metal is very much divided.

Amalgams are liquid when the mercury is in great excess, and solid when the alloyed metal predominates. They may crystallize, and form combinations in constant proportions.

All amalgams are decomposed by heat; the mercury volatilizes.

Mercury amalgamates readily with potassium and sodium, and forms alloys which decompose water.

Amalgams of Tin.

The amalgam which is formed of one part of tin and ten parts of mercury, is liquid, but much less fluid than mercury. That which contains one part of tin, and three of mercury, is soft, and crystallizes readily. When it is formed of equal parts of mercury and tin, it is solid. Generally, a contraction is noticed in the combination of tin with mercury, except in the amalgam which is formed of one part of tin and two of mercury.

The amalgams of tin are brilliant, and do not alter in the air; they are used for coating mirrors.

This operation is performed by spreading a sheet of tin on a horizontal table, and pouring over it some mercury, forming thus a thin layer. A glass is then pushed over it so as to cut the layer, and it is then loaded with weights; the amalgam in a short time adheres to the glass.

Amalgams of Bismuth.

Bismuth amalgamates readily. When the mercury is in excess, the amalgam is liquid, and has the property of dissolving lead without solidifying; in this way, by introducing lead and bismuth, mercury is often adulterated; but this mercury always soils glass (*fait la queue.*)

The amalgam formed of one part of bismuth, and four parts of mercury, has the curious property of adhering strongly to bodies with which it is placed in contact. When this amalgam is poured into a flask quite dry and warm, and applied over the surface, a beautiful coating is often produced.

SILVER.

Silver is the whitest of all metals, and is susceptible of a most beautiful polish. When precipitated from a solution by another metal, it presents itself in the form of white sponge, composed of crystalline grains, which acquire great cohesion by pressure and hammering. When melted, and allowed to cool slowly, it crystallizes in large octahedrons, or cubes.

Silver has neither taste nor smell. It is harder than gold, and softer than copper. After gold, it is the most ductile and malleable of metals.

Silver fuses at 22° of Wedgwood's pyrometer; this temperature corresponds to about 1000° C.; it is slightly volatile. Its volatility increases considerably in presence of a current of gas; it is very rapid under the influence of heat produced with a lens, or with the oxyhydrogen blowpipe.

In the work-shops for refining the precious metals, where large masses of silver are melted daily, they avoid heavy losses which would result from the volatilization and mechanical escape of the metal, by making the furnace to communicate with passages of masonry of 25 to 30 metres long, which, before entering the chimney of the factory, pass into large chambers, where the silver is deposited.

When silver is melted in earthen crucibles, metallic globules are often seen adhering to the lid, which proceed from the volatilization of the silver.

Silver oxidizes neither in wet nor dry air; it does not tarnish in the air, except under the influence of sulphurous vapors. Silver in sponge becomes incandescent in a current of hydrogen, when heated to 130° C.

M. Lucas has shown that pure silver, when melted, has the property of dissolving oxygen, and of giving it up on solidifying. Gay-Lussac has demonstrated that the quantity of oxygen absorbed may amount to twenty-two times the volume of the silver.

The presence of a small quantity of copper prevents silver from absorbing oxygen. Silver does not dissolve nitrogen.

When silver which has absorbed oxygen is left to cool in the air, at the moment of the solidification of the metal, the gas, in disengaging, often causes a projection of the silver, which forms a sort of metallic vegetation.

Silver is attacked by but few acids. Nitric acid is its best dissolvent; it produces nitrate of silver. Dilute sulphuric acid does not attack silver, but when concentrated dissolves it with the aid of heat, giving off sulphurous acid.

Sulphydric acid is decomposed by silver; a plate of silver,

dipped in a solution of this acid, immediately becomes black, covering itself with sulphuret of silver.

The vegetable acids have no action on pure silver.

Nitrate of Silver.

This is the most important and best known salt of silver. It crystallizes in square plates colorless and unalterable in the air. It is soluble in its own weight of cold water, and much more soluble in boiling water. Alcohol dissolves one fourth of its own weight when warm, and a tenth only when cold. Light appears to decompose it only in presence of organic matters: it corrodes the skin and makes black spots, which the iodide of potassium will remove.

The nitrate of silver fuses without decomposing, and forms in cooling a crystalline mass designated by the name of *lapis infernalis*, or lunar caustic.

Lapis infernalis, when pure, is white; but it is often black on its surface, and sometimes through its entire mass, owing to the presence of a small quantity of silver reduced by the metal of the mould, or by traces of organic substances. This bronze color may also proceed from the binoxide of copper resulting from the decomposition by heat, of the nitrate of copper, which is found in the nitrate of silver. Nitrate of silver may be obtained by attacking pure silver, or silver coin, with nitric acid.

To prepare this salt from coin, which is an alloy of silver and copper, it is dissolved in nitric acid, the solution is evaporated to dryness in a porcelain capsule; the residue is melted, and kept at a heat a little below a dark red. The nitrate of copper is decomposed, and an insoluble oxide of copper is left.

When the melted mass which at first was blue, becomes colorless by the separation of the black oxide of copper, and when, besides, it ceases to give off yellow fumes, it is then evident that all the nitrate of copper has been decomposed, and the calcination may be stopped.

The nitrate of silver being thus separated from the nitrate of copper, the mass is treated with water, which takes up the pure nitrate of silver and leaves the oxide of copper.

Nitrate of silver blackens most organic substances; its dilute solution is used for dyeing the hair and marking linen. The ink used for this last purpose is formed of 2 parts of nitrate of silver dissolved in 7 parts of distilled water, to which is added a small quantity of gum-arabic. The linen is first rendered firm by the carbonate of soda and soap, and smoothed with a hot iron. Soon the salt of silver is reduced, and the black mark of metallic silver, which is entirely indelible, is left on the linen.

Alloys of Silver.

Silver alloys with a great number of metals; but the only really important alloys are those which silver forms with copper and with some unoxidizable metals, as gold and platinum.

Alloys of Silver and Copper.

Copper alloys with silver, in a state of fusion, in every proportion. Those alloys are less ductile, harder, and more elastic than silver. They are in general white, and do not take a red color, except when the proportion of copper is very considerable.

As the color of the alloys of silver is not so beautiful as that of pure silver, they undergo the operation of *blanchiment* or *whitening*. This is done by heating the alloy in the air, so as to oxidize the copper, and then dipping it in water acidulated with nitric and sulphuric acids; these acids dissolve the oxide of copper, and bare the silver, and consequently raise its standard, and bring it, so to speak, to a state of purity. Articles of silver thus whitened, have a dull color, which is rendered brilliant by rubbing.

Alloys of silver and copper, when the proportion of copper exceeds 100 thousandths, become rapidly changed in a moist air, and particularly when in contact with organic matters. When subjected to roasting, the copper oxidizes together with a small quantity of silver. The oxidation becomes slower in proportion as the silver predominates: it is always very difficult entirely to free silver from copper by roasting.

Money of low standard is known as base coin.

The pieces of six liards, the two sous pieces with the letter N, which are now out of circulation, were alloys of $\frac{200}{1000}$ strongly whitened. The pieces of 15 and 30 sous, also drawn out of circulation, were of the standard of $\frac{666}{1000}$.

The money which has been preserved, and that now coined in France, is of the standard of $\frac{900}{1000}$, some being $\frac{3}{1000}$ above, and some $\frac{3}{1000}$ below the legal standard.

Medals contain more silver; they are of the standard of $\frac{950}{1000}$, with the same variation as in coin.

The standard for silver ware is of two qualities; that most used for plates and dishes is $\frac{950}{1000}$. The allowance being $\frac{5}{1000}$; a dish, for example, which contains $\frac{945}{1000}$ of pure silver, is within the limits of the law.

There are no fixed limits for standards above $\frac{950}{1000}$; the manufacturer is interested in not passing these limits.

The second standard is $\frac{800}{1000}$. The variation allowed being $\frac{5}{1000}$ below this.

To solder silver, alloys of lower standard are used, and a small quantity of zinc is sometimes added.

The alloy mostly used for soldering silver of $\frac{950}{1000}$ is made of 666·67 of silver, 233·33 of copper, and 100·00 of zinc.

Plate.

The name *plate* is given to sheets of copper covered with sheets of silver.

To plate copper, it is thoroughly dressed on its surface to remove all defects and render it perfectly uniform: it is then rolled out to about double its original surface, and dressed again; it is then ready to receive the silver.

When it is proposed to plate one-twentieth; an ingot of fine silver one-twentieth the weight of the copper is taken; it is so rolled out that its surface will be a little larger than that of the copper.

The two plates being thus prepared, a solution of concentrated nitrate of silver is passed over the plate of copper; the workmen then say that the plate is *amorcée*, that is, put in train for the operation.

The plate of silver being spread upon the bench, the surface of the plate of copper which has been prepared is placed upon it, and the edge of the silver-plate is turned up over its thickness: the two plates are then heated to a low red, and they are passed through rollers, so as to reduce their thickness to about 1 millimetre. The two plates are in this way soldered together so that they cannot be separated.

GOLD.

Gold is tasteless, inodorous, of a yellowish-red color. When reduced to thin leaves, it appears green by transmission, and yellowish red by reflection of light. When in fine powder, it is of a violet-yellow color.

Gold crystallizes in quadrangular pyramids or in octahedrons; it is found in nature in different forms, which are derived from the cube. It is less hard than silver, and almost as soft as lead: it is the most malleable and ductile of metals.

It may be reduced into leaves the ten thousandth of a millimetre thick; five centigrammes of gold may be drawn into a wire $162^{m}\cdot419$ in length.

Its tenacity is less than that of iron, copper, platinum

and silver. This metal contracts much more than the others in passing from the liquid to the solid state.

Gold fuses at 32° of the pyrometer, corresponding to about 1100° of Cent. In the liquid state, gold appears to be green.

When heated in an ordinary furnace, it may be said to be fixed, but it is sensibly volatilized in the focus of a powerful glass, in the flame of the oxyhydrogen blow-pipe, or when exposed in thin leaves to the action of a powerful battery or voltaic pile.

Gold finely divided becomes incandescent in a current of hydrogen gas, when heated to 50°.

It partakes, in common with platina, silver, iron, &c., of the property of being welded without being first fused. If, after being precipitated from its solutions by iron, it is washed and strongly compressed by means of a hydraulic press, it adheres together, produces a ductile and malleable mass, which can be forged, laminated, or drawn into wire.

Gold is one of the least alterable metals known. It resists the action of air, oxygen, water, sulphuric, nitric, and muriatic acids.

Thus gold can immediately be distinguished from copper, by treating these metals with nitric acid: if the metal dissolves in the acid it is copper, if it resists it is gold. This operation is performed on a hard black stone called *touch-stone*, on which the alloy to be assayed is rubbed. This method is employed to appreciate the quantity of gold which exists in an alloy of copper and gold.

Nitric acid mixed with muriatic acid forms *aqua regia* which dissolves gold by converting it into a chloride.

In commerce gold is dissolved in aqua regia formed of one part of nitric and four of chlorhydric acid (muriatic acid). Thus is obtained chloride of gold, which is used to prepare all the other compounds of gold.

Gold is not tarnished by sulphuretted hydrogen. It gives glass a rose color.

Gold, finely divided, is used to gild glass or porcelain. Gold may be obtained in powder, by precipitating it from its solution by the sulphate of the protoxide of iron, or better, by grinding gold leaf with honey. Gold prepared by this last method is generally placed in shells in thin layers, and is called *gold in shells*, *painter's gold.*

Purple Powder of Cassius.

This name is given to a red purple precipitate obtained by mixing chloride of gold with a salt of tin.

The purple of Cassius has been repeatedly investigated, and yet its true nature is not known: it appears to contain gold finely divided.

The purple of Cassius is used for coloring glasses, crystal, and porcelain, rose-color and purple.

Alloys of Gold and Copper.

Gold alloys in every proportion with copper. Copper heightens the color of gold, increases its hardness, and renders it more fusible, but diminishes its malleability and its ductility.

The density of the alloys of gold and copper, is less than the mean densities of the metals composing it.

The presence of an extremely small quantity of lead in the alloys of copper and gold, and particularly in the money alloys, renders them brittle.

The alloys of copper and gold being much more fusible than gold (their fusibility augments with the proportion of copper which they contain), are used to solder gold.

The solder known as *red gold*, is formed of 5 parts of gold, and 1 part of copper.

Sometimes a small quantity of silver is added to the alloys of copper and gold intended for soldering.

Gold of $\frac{750}{1000}$ths, is generally soldered with an alloy of 4 parts of gold, 1 part of copper, and 1 of silver.

Gold not being a very hard metal, cannot be used pure for making coin, medals, or jewelry. Coin of pure gold would soon become deformed; its stamp would not be durable; it is hardened by adding copper. The standard of the gold coins of France is $\frac{900}{1000}$ths. The law allows a variation of $\frac{2}{1000}$ths, either above or below this; thus, coin between 898 and 902 thousandths, is of the legal standard.

Medals are of a higher standard than coin; they contain $\frac{916}{1000}$ths of gold: $\frac{2}{1000}$ths above or below this standard, is allowed.

The standard alloys for jewelry are also determined by law.

There are for this, three standards. That most used is $\frac{750}{1000}$ths; $\frac{3}{1000}$ below is allowed; thus, jewelry of $\frac{747}{1000}$ths, is of the legal standard. There is no limit above the standard.

The two other standards for jewelry are $\frac{840}{1000}$ths, and $\frac{920}{1000}$ths; but these alloys are not much used. The alloys of copper and gold tarnish in the air in proportion as their standard is low; they can be rendered bright by rubbing with caustic ammonia, and then washing in plenty of water.

To give brilliancy to alloys of gold, they undergo an operation called *mise en couleur*, which consists in dissolving a part of the copper and silver contained in the alloy, in order to leave the gold nearly pure at the surface.

Amalgams of Gold.

Gold unites very readily with mercury, even at the ordinary temperature. It suffices to expose a leaf of gold to the most feeble mercurial emanation, to whiten it. This sometimes serves for recognising traces of mercury. A piece of gold rubbed with mercury becomes very brittle, and may easily be broken between the fingers.

Mercury dissolves a large quantity of gold without ceasing to be liquid. The amalgam is white like silver; when saturated with gold, it becomes slightly yellow, and then has the consistence of wax. When the liquid amalgam is placed in a bag of chamois skin, the mercury containing a small quantity of gold filters out, and a white amalgam remains in the skin, of a pasty consistence, formed of about two parts of gold, and one of mercury.

Gold in scales, sometimes called powder of gold, used in painting, is obtained by alloying 1 part of gold and 8 of mercury, and separating the last metal by distillation.

All the amalgams of gold leave a residue of pure gold when heated to a bright red: they are used for making what is called *gilding with mercury.*

Alloys of Gold and Silver.

Gold and silver may be united in all proportions.

These alloys left to cool very slowly, undergo a sort of liquation.

The density of these alloys is about the same as that of the mean densities of the metals composing them.

The alloys of gold and silver are more fusible than gold.

They are harder and more elastic than gold and silver.

They are much used by goldsmiths, and are called *yellow gold, pale gold, green gold, electrum.*

Green gold, which is the alloy most used, is formed of about 70 parts of gold and 30 parts of silver.

Electrum is composed of 4 parts of gold, and 1 part of silver.

Vermeil is silver-gilt.

There are natural alloys of gold and silver which have various compositions.

The Extraction of Gold.

Gold is always found in the native state pure, or combined with certain metals, mostly silver.

It is generally crystallized in cubes, or in octahedra, or forms derived from it; it is also met with in scales, in gold sand, or in fragments.

It is found, though rarely, in isolated masses called *pepitas*.

M. Humboldt mentions a pepita from the mines of Peru which weighed 12 kilogrammes.

The richest mines of gold are the veins of auriferous sulphuret of silver which pass through certain intermediary formations: such are the mines of Mexico, Peru, Hungary, and Transylvania, and those of the Ural mountains in Siberia.

Gold scattered in grains through the argillaceous and ferruginous sands, forms the auriferous sands which are drifted by a great number of rivers; these are often worked to advantage.

The richest sands are those of California. The auriferous sands of Brazil cover an immense space, and contain platina, diamonds, etc.: they also exist in Chili, New Granada, Mexico, and Peru.

There are great numbers of auriferous sands in Europe, but much less rich than those of America.

In France, auriferous veins are numerous: we will mention the Arriége, the Gardon, the Rhone in the neighborhood of Geneva, the Rhine near Strasburg, the Garonne near Toulouse, the Herault near Montpelier, etc.: these sands are not sufficiently rich to be worked.

Gold is extracted from alluvial formations, or from veins, by the following processes:

Gold from Alluvial Formations.

The auriferous alluvion is subjected to the action of a sufficiently rapid current of water in a narrow canal.

The earthy matters are carried off by the water. As soon as there remains nothing but gravel, the washing is finished in a large wooden bowl of a conical form: ferruginous sand is first obtained, which, subjected to a second washing, gives the gold in powder.

When gold contains grains of platinum, it is amalgamated by rubbing it under water with mercury: in this case the gold alone dissolves in the mercury; the amalgam is then distilled.

The ore of Choco contains about 12 per cent. of platinum in grains.

Gold in Veins.

Gold in veins is generally found mixed with iron pyrites, oxide of iron, sulphuret of zinc, sulphuret of antimony, etc.

In America every ore which contains $\frac{1}{30000}$th of gold is worked.

The gold is extracted 1st by fusion, 2d by washing, 3d by amalgamation.

The treatment by fusion consists in melting the ore either alone or with plumbiferous substances, so as to form *mattes*, which are submitted to the action of melted lead which alloys with the gold.

The gold is then separated by cupellation: this operation causes the oxidation of the lead, and leaves the gold in a state of purity.

The extraction of gold by washing is an operation in which the ore, after having been first roasted in a reverberatory furnace, is washed in wooden bowls, and thus freed from foreign matters lighter than the gold, which are washed away.

Amalgamation may be applied to all the ores. This method consists in grinding the ore with mercury in a mill, very much like that used for grinding sand for porcelain.

A continuous current of water passes over the ore to free it from foreign bodies, the amalgam is taken away as it forms, this is filtered to free it from an excess of mercury, and it is then subjected to distillation. Thus argentiferous gold is obtained.

To separate the gold from the silver, the alloy is heated to a dull redness in a porous vessel, for 24 to 30 hours with a cement composed of common salt and powdered brick; the gold then frees itself from nearly all the silver which it contains, the silver passes in the cement to the state of chloride, from which it is extracted by amalgamation.

PLATINUM.

Platinum was not introduced into Europe till about the year 1740; its name comes from the Spanish word *platina*, which signifies *little silver*. This metal has been a long time known in America, but was not put to any use: it is only since the beginning of this century that it has been used in chemical laboratories, and in the arts.

Forged platinum is nearly as white as silver; it may be polished to great brilliancy; it has no taste nor odor; it is very ductile and very malleable; it occupies the fifth rank among metals for malleability, and the third for ductility.

A wire of 2 millimetres breaks with a weight of 124 kilogrammes.

Platinum is softer than silver. It is harder than copper, and less hard than iron. It is the least expansible of all the metals.

It is infusible in the fire of the forge, but it easily melts under the oxyhydrogen blow-pipe. At a white heat it softens, and may be forged and welded like iron.

This property is of great value, for by means of it platina utensils may be made. Platinum appears to be volatile when carried to a very high temperature; it produces brilliant scintillations when exposed to the hydrogen blow-pipe.

Platinum is not oxidized in the air, either hot or cold; it does not decompose water under any circumstances, and is attacked by but few acids.

Nitric acid has no action on pure platinum, but dissolves it when alloyed with a sufficient quantity of silver and gold. Its true solvent is aqua regia, 100 parts of which, formed of 75 parts of muriatic acid at 15°, and 25 parts of nitric acid at 35°, will dissolve 14 parts of platinum.

Sulphuric and muriatic acids do not dissolve it. Spongy platinum unites with sulphur when sufficiently heated; forged platinum combines with it with great difficulty.

Phosphorus and arsenic unite with platinum in sponge, and form a very fusible phosphoret and sulphuret. When an organic matter containing phosphorus is burned in a platinum crucible, it produces, by the reduction of the phosphoric acid, phosphoret of platinum, which is very fusible; thus, in a very short time the crucible is pierced.

Chlorine is slowly absorbed by platinum.

Platinum alloys with nearly all the metals, at a temperature more or less elevated; when very much divided, it alloys with mercury.

Metallic platinum presents itself under different aspects.

When it proceeds from the calcination of the ammoniacal chloride of platinum, it is spongy, dull, and of an ashy grey; it is then called *spongy platinum*, or *en mousse*. This may be polished by rubbing.

When platinum is precipitated from one of its solutions, it sometimes appears in the form of a black powder, called the *black of platinum*. The sponge and the black will, like carbon, condense gases producing heat, and cause, under the influence of the atmosphere, the inflammation of combustible gases and vapors.

The black of platinum absorbs as much as 745 times its volume of hydrogen.

Platinum, divided, possesses the property of causing combinations by its presence alone. We will mention the principal reactions which are brought about by the influence of black and spongy platinum.

Oxygen and hydrogen unite in the presence of divided platinum, and produce water. A finely-drawn wire of platinum, or very thin leaves of this metal, become incandescent in a detonating mixture, and cause the combination of the two gases.

Sulphurous acid and oxygen unite under the influence of spongy platinum slightly heated, and form anhydrous sulphuric acid. Alcohol absorbs oxygen in presence of black of platinum, and is changed into acetic acid.

The solution of platinum in aqua regia, is called *chloride of platinum;* it is of a brown color, and is used in the manufacture of porcelain to produce *platina lustre.* When a mixture of chloride of platinum and essence of lavender, is applied to a piece of pottery, which is then put in the fire, a very brilliant lustre is obtained.

The chloride of platinum is used in laboratories as a test for the salts of potash. It produces in their solutions a yellow precipitate, slightly soluble in water.

In finishing the history of the metals, we will give some generalities on gilding, galvanoplastie, and photography.

GILDING.

A great many objects are gilded by means of leaves of gold, which are fixed with different *mordants*, or *size.* It is thus are gilded wood, pasteboard, leather, iron, &c. For a long time, copper and its alloys were gilded by means of an amalgam of gold applied to the pieces perfectly cleaned. The application of heat volatilized the mercury, and an adherent pellicle of gold was obtained, the thickness of which could not exceed a certain limit.

This method is almost entirely replaced by gilding, by washing, and galvanic gilding.

Gilding by washing, is employed by jewellers for covering copper with a light coating of gold. A solution of perchloride of gold, and bicarbonate of potassa, is used; the articles being first thoroughly cleaned, are dipped into this liquor for half a minute, and then carefully washed in water, and dried.

By the galvanic processes, a deposit of gold may be made on any metal, of any thickness required. The solution of gold best adapted to this purpose, is the double cyanide of gold and potassium.

The articles ought to be cleaned, and subjected to preparations varying with the nature of the metal; they are plunged into the auriferous bath, putting them in communication with the two poles of a pile, the intensity of which must be suitable. The thickness of the layer of gold depends on the duration of the immersion, and the intensity of the electric current. Articles are silvered by analogous processes, replacing the auriferous bath by a solution of the cyanide of potassium and silver, or better, the cyanide of calcium and silver.

By the galvanic process are also obtained deposits of platinum, copper, zinc, &c. Sheet-iron, covered with zinc, is called *galvanized iron*.

GALVANOPLASTIE.

This is a process for precipitating on an object a continued and consistent layer, but not adherent, of a metal, from its solution, which is decomposed by an electric current. The metal mostly used is copper; a saturated and slightly acid solution of sulphate of copper, is decomposed by a weak electric current. The object to be covered with copper, is placed at the negative pole of the pile, and plunged in the solution; if it is a non-conductor, it is covered with plumbagine; this is most generally the case, for mostly a plaster cast is made of the object to be reproduced. This non-conductor cast is *metallized* with plumbagine, and is soon covered over with a deposit of copper, which exactly represents the object proposed.

Of late years, bas-reliefs of silver, in great perfection, have been obtained by the galvanoplastic processes.

PHOTOGRAPHY.

The art of photography is based on the property which the salts of silver possess, of blackening under the influence of light. A plate of silver, or silver plate, is covered with a thin coating of iodide of silver, by exposing it to the vapor of iodine; the plate, thus prepared, is placed in the focus of a camera obscura, and the iodide is decomposed in the clear parts of the image; its change is incomplete in the demi-tints, and it is not altered in the dark parts. To render the image visible, it is exposed to the vapors of mercury; this forms an amalgam of silver, which produces the bright parts of the image. The plate is then washed with a feeble solution of the hyposulphite of soda, in order to take away the unaltered iodide of silver, which would blacken in the light.

This method, due to Daguerre, has been improved by the use of certain substances called accelerators, vapor of bromine, for example, which makes the layer of iodide of silver more sensible, and permits the images to be taken in a very short time. Further, different processes have been discovered to render the images more exact and unalterable; they are fixed by subjecting them to the influence of chloride of gold.

Of late years, photographic images have been obtained in great perfection, by using *photogenic papers*, prepared with different compounds of silver. The paper blackens in the luminous parts of the image, and remains white in the dark parts; this *negative* proof is exposed to the light, placing behind it another photogenic paper, which becomes black in the parts which correspond to the white parts of the first, and thus is formed the *positive* proof. This proof is then washed with the hyposulphite of soda.

ORGANIC CHEMISTRY.

GENERALITIES ON ORGANIC BODIES.

The term *organic bodies* is given to those which enter into the constitution of vegetables or animals. Most organic substances composing vegetables are formed of three bodies, which are carbon, hydrogen, and oxygen.

The organic substances which are produced by the animal organization, often contain nitrogen; they may then be formed of carbon, hydrogen, oxygen, and nitrogen. In fact, a number of nitrogenous bodies are likewise obtained from vegetables. The most different organic substances, and those which are most widely separated from each other by their properties, such as sugar, vinegar, wood, starch, gum, resin, and fat, are formed of the same elements. In this relation, organic bodies differ essentially from inorganic bodies, which contain different elements. Thus sulphuric, nitric, and carbonic acids, have indeed a common element, which is oxygen, but they contain the simple bodies, sulphur, nitrogen, and carbon, which give to these acids their special properties. Organic bodies, then, do not differ from each other in the nature of these elements, but rather in the proportion of these elements. Thus, sugar and fat are formed, it is true, of carbon, oxygen, and hydrogen; but the proportion of these elements is very different in the two cases. Fat contains much more carbon and hydrogen than sugar.

The action of heat often serves to distinguish an organic from an inorganic body. It may be said, indeed, that all organic substances are decomposed when they are subjected to the action of a temperature sufficiently elevated; they give off water, acid vapors, tar, and inflammable gases, leaving often a residue of carbon, while most inorganic bodies, which are not volatile, resist the action of heat.

Again, organic bodies have the characteristic property of decomposing spontaneously when exposed to the influence of moist air: they undergo a sort of slow combustion, and become changed into water and carbonic acid.

When an organic body formed of oxygen, hydrogen, and carbon, is heated with the oxide of copper, under the influence of the oxygen contained in this oxide, the elements of the organic body change into water and carbonic acid. By estimating the proportions of water and carbonic acid which are produced during this decomposition, the quantity of each of the elements which constitute the organic matter may be determined: *elementary organic analysis* is founded on this principle.

When the body, subjected to the influence of the oxide of copper, contains nitrogen, the nitrogen disengages in a state of purity if care has been taken to use a certain quantity of metallic copper, which decomposes the binoxide of nitrogen which has a tendency to form. By collecting the nitrogen in a graduated bell-glass, the quantity contained in the substance analyzed can be easily determined.

Organic bodies may be divided into three great classes, which comprise, 1st, acids, 2d, bases, 3d, neutral bodies.

To facilitate the study of neutral organic bodies, which are very numerous, they have been separated into a certain number of groups, which comprise; woody substances, starchy substances, the gums, sugars, alcohol, essential oils, fat bodies, coloring matters and animal substances.

We will select from these different classes, those which

are most interesting, and will explain their principal properties: we will commence with the study of the organic acids.

ACETIC ACID.

This is one of the most important acids in organic chemistry; it is abundant in the vegetable organization: it may be artificially produced by a great number of different processes. Acetic acid is set free where the acetates are used in many of the arts.

Acetic acid is met with in the sap of all plants; it is generally combined with potash, soda, or lime. The acetates are produced by the calcination of the carbonates which are found in the ashes of vegetables; acetic acid is also met with in many of the animal secretions. Most liquids which have undergone putrefaction or fermentation contain this acid.

Organic substances heated with potash, sulphuric, nitric acids, etc., give rise to acetic acid: it may, besides, be considered as one of the constant products in the distillation of organic matters. Acetic acid is known to form almost constantly, when the equilibrium in the elements of an organic substance is disturbed.

Acetification.—It is known that wine may become acid and transformed into vinegar.

Wine may be looked upon as a liquor holding in solution alcohol and an azotized substance.

It is well demonstrated, that in the acetification of wine, it is the alcohol that absorbs the oxygen of the air, and becomes converted into vinegar; for when wine has become sour, it is no longer alcoholic. But this transformation does not take place without the influence of azotized matter, which causes the absorption of the oxygen of the air by the alcohol.

To make acetic acid with wine, it is seen, then, that it is necessary to bring the alcohol in contact with the oxygen of the air, and besides, to cause a third body to intervene, which acts by its presence, causing the action of the oxygen on the alcohol. The number of organic bodies fit for converting spirituous liquids into vinegar, is considerable; but that one which possesses this property in a high degree, is the *mother of vinegar*, a gelatinous matter which is deposited in the vats where wine is fermented.

Wine may be prevented from becoming sour, either by precipitating the azotized matter by boiling, or by preventing the access of air.

All the circumstances favorable to the transformation of alcohol into acetic acid, have been realized in a method of acetification due to M. Schuzenbach. Alcohol is made to circulate repeatedly through shavings of beech-wood, mixed with the must of beer, placed in vats whose sides are pierced with many holes.

Generally, alcohol of $\frac{80}{100}$ths is used, mixed with six parts of water. The operation is generally terminated in thirty-six hours. M. Schuzenbach remarks that the acetification is much more rapid when the shavings have been first sprinkled with strong vinegar.

It is easy to explain the theory of this operation. The shavings are intended to divide the liquid, and increase the contact of the alcohol with the air; the wood contains, besides, an azotized matter which determines the oxidation of the alcohol. The must of beer acts also as a ferment, and renders the acetification more active. It is noticed that during this operation, the chemical reaction produces sufficient heat to raise the temperature to 40° C.

Table vinegars are usually made with wine. White vinegars are generally preferred. The following process is used in Orleans, and in most wine countries, to prepare vinegar.

Several rows of casks are placed on end in the factory, where the temperature is maintained at 25° to 30° C. Those casks are preferred which have been before used for this purpose, as their sides are covered with ferment, commonly known as *mother of vinegar*. These casks are pierced with two holes on their upper head; one used for the introduction of wine, the other for the escape of the air. A certain quantity of boiling vinegar is first poured into each cask, then, every eight days, 10 to 12 litres of wine, which have been run over shavings of beech-wood. In fifteen days, the acetification is finished; one-half of the vinegar contained in each cask is then racked off, and the operation is recommenced with new wine. Beer without hops, cider, perry, and a great number of saccharine and alcoholic liquors, may be acetified by analogous means. Acetic acid may be made by another process, which consists in subjecting wood to distillation. The idea of extracting acetic acid from the products of the calcination of organic matters, and particularly wood, is due to Lebon.

The manufacture of *pyroligneous acid*, was first practised by the Mollerat brothers.

About five cubic metres of wood are introduced into large sheet-iron cylinders, which communicate with a series of condensers. The gases produced are carried by pipes into the hearth, to burn and keep up the distillation. The liquid which condenses is composed in part of water, tar, spirit of wood, and acetic acid.

It is freed from the tar which floats on the surface, and then introduced into a copper alembic.

The spirit of wood is found in the first products of the distillation; acetic acid, which is called pyroligneous acid, in the second.

This pyroligneous acid is always colored, and tastes of tar, which a new distillation does not entirely remove. In order to be removed completely, it must be combined with a base, so as to fix it. Heat, properly applied, will then decompose the tar without altering the acetate. In some factories, the pyroligneous acid is saturated by the carbonate of soda. But generally it is found to be more economical first to heat the pyroligneous acid with carbonate of lime, and then to decompose the acetate of lime by the sulphate of soda: sulphate of lime precipitates, and the acetate of soda which remains in the liquor may be purified by crystallization.

100 kilogrammes of this salt are roasted in a large shallow cauldron of iron for twenty-four hours, breaking up the mass constantly with an iron poker. This operation requires great caution, for too high a temperature would immediately bring about the combustion of the acetate of soda. The acetate roasted is dissolved in water: this solution then deposits crystals of pure salt.

The acetate of soda is then treated with sulphuric acid diluted with one-half its weight of water. 100 parts of the acetate require about 36 parts of concentrated sulphuric acid. The acetic acid which disengages is received in condensers, and may be purified by a second distillation.

Different woods yield by distillation very variable proportions of acetic acid. According to M. Stolze, ash wood gives the most, while fir gives the least.

In the distillation of wood, an inflammable liquid is produced, soluble in water, called the *spirit of wood:* this may be substituted for alcohol for some purposes; it dissolves the resins, and answers for making varnish.

Spirit of wood is a considerable article of commerce.

Acetic acid may also be obtained by distilling the metallic acetates: the acetate of the deutoxide of copper is generally used for this purpose.

This salt is introduced into an earthen retort or one of glass well luted, to which a condenser is fitted. It is heated cautiously. In a short time a colored liquid distils over, which is formed of acetic acid, a small quantity of acetone and acetate of copper carried over. This liquid rectified, gives acetic acid perfectly white, called *radical vinegar.*

Properties of Acetic Acid.—This acid, as concentrated as possible, is solid below the temperature of 17° C. At this temperature it fuses, and forms a colorless liquid: its odor is characteristic, its taste is biting; it is as corrosive as the most energetic mineral acids; it is soluble in water in all proportions; it boils at 120° C. Its vapor is inflammable, and burns with a blue flame.

Acetic acid has a property which distinguishes it from some vegetable acids, viz., it does not precipitate albumen and dissolves fibrine.

Acetates.—Several acetates are used in the arts.

The acetate of alumine is used in dyeing, and in calico-printing; it is known as *red mordant of the calico-printers.*

The acetates of iron in the arts, are known as pyrolignite of iron, and liquor of old iron. These salts are used in dyeing, and in the process for the preservation of wood, due to M. Boucherie.

The neutral acetate of lead, known as salt of Saturn, is used in the arts, and in medicine; it is one of the most valuable chemical reagents. The basic acetate of lead is called, in medicine, *extract of Saturn*, and *Goulard's solution;* it is used in the manufacture of white lead.

The oxide of copper combines with acetic acid in several proportions. The body called *vert-de-gris*, is a basic acetate of copper; it is generally obtained by exposing to the air sheets of copper, covered with skins of grapes. The alcohol contained in the skins acetifies; the acetic acid determines the oxidation of the copper by the oxygen of the air, and forms, with the metallic oxide, a bibasic acetate.

The sheets of copper covered with the acetate of copper are scraped, and a green powder is procured, called vert-de-gris of commerce.

The oxidation of the copper in the preparation of vert-de-gris, resembles that of lead in the manufacture of Holland white lead.

The readiness with which the acetate of copper forms, when copper with acetic acid is placed in the air, proves that great caution is necessary in the domestic use of copper, to preserve it from contact with liquids containing acetic acid, or from those which will acidify in the air.

The acetates of copper are used as green colors, in dyeing green on wool.

OXALIC ACID.

Oxalic acid crystallizes; its crystals are colorless, and perfectly transparent. They dissolve in water, giving rise to a sort of decrepitation, which appears to be produced by the disengagement of a small quantity of gas, shut up in the crystals at the moment of their formation.

Eight parts of cold water dissolve one part of oxalic acid; boiling water dissolves its own weight of it. This acid is also soluble in alcohol.

Its taste is sharp and pungent; it is poisonous in the dose of 15 to 20 grammes.

It fuses at about the temperature of 100° C.

Natural State.—Oxalic acid is very widely spread in organic nature. It is met with in a free state in the beard of chick pea; it is generally combined with bases, most frequently with potash, soda, or lime. It is thus that it is found in sorrel, and other kinds of *oxalis*, in the state of binoxalate of potassa (*salt of sorrel*); in barilla and marine plants, in the state of neutral oxalate of soda. The lichens which grow on calcareous rocks, all contain oxalate of lime; and the proportion of this salt often attains even to three-fourths of the weight of some species of lichens.

The oxalate of lime constitutes nearly entirely the urinary calculus, called, on account of their form and color, *mulberry calculus.*

Sometimes, in the lignites, a basic oxalate of the sesquioxide of iron is found, which mineralogists designate under the name of *humboldtite.*

Preparation.—Oxalic acid is developed in a great number of reactions. All bodies which retain their oxygen feebly, such as nitric acid, plombic acid, permanganate of potassa, &c., transform organic matters into oxalic acid; those which appear best suited to this transformation, are in general substances which contain oxygen and hydrogen in the proportions to form water. Oxalic acid is generally prepared by making nitric acid react on starch, or on sugars of inferior quality.

One part of starch, treated with 8 parts of nitric acid of a density of 1·42, diluted with 10 parts of water, gives, after prolonged ebullition, a liquor from which are separated by evaporation, beautiful crystals of oxalic acid.

Starch, subjected to the action of nitric acid, gives about the eighth part of its weight of oxalic acid. Sugar produces about one-half of its own weight of it.

The oxalic acid of commerce is sometimes impaired by traces of nitric acid, from which it is easily freed by crystallization. It ought not to leave any residue by calcination.

In Switzerland, oxalic acid is extracted from the salt of sorrel which the *rumex acetosa*, or great sorrel, contains. This plant is bruised, and the juice expressed, which is moderately heated to clarify it, and after being filtered, it is concentrated. The juice is also sometimes clarified by boiling with a little white clay, which deprives it of its coloring matter. At the end of about six weeks, there is deposited from the liquors a considerable quantity of greenish crystals, which are purified by repeated crystallizations.

100 kilogrammes of fresh leaves produce about 320 grammes of binoxalate of potassa.

The excess of acid in this salt is saturated by carbonate of potassa or soda, and the neutral oxalate which results is precipitated by the acetate of lead. The oxalate of lead, well washed, is then decomposed by dilute sulphuric acid. The liquor filtered and concentrated, deposits oxalic acid on cooling.

Gay-Lussac has recommended to extract the oxalic acid from barilla, which grows in abundance on the coast of Spain. The juice of this plant contains oxalate of soda, which may be precipitated with the chloride of barium. It then remains to wash the oxalate of barytes, and extract oxalic acid by means of sulphuric acid, diluted with water.

Uses.— Oxalic acid is used in large quantities in the manufacture of painted calicoes. It is used as *a discharge* to destroy the mordant on the parts of the stuff which ought to preserve their whiteness. It is useful for the displacement of some colors.

It is also used for scouring copper utensils, and taking out rust-spots from linen. These applications of free oxalic acid, and of the acid oxalates, are based on the property which this acid possesses of forming soluble salts with the oxides of copper and iron.

Oxalic acid is used in the state of oxalate of ammonia, as a test for lime

TARTARIC ACID.

Tartaric acid exists in the juice of the grape. It is that which, combined with potash, forms the *tartar* of wines, and precipitates in the casks, carrying with it a part of the coloring substance of the wine.

Tartaric acid is solid, and soluble in water; its taste is distinctly acid. It crystallizes in voluminous prisms.

This acid has a great tendency to form acid salts, or rather, double salts.

Cream of tartar is the acid tartrate of potash. This salt is but slightly soluble in cold water; thus, tartaric acid in excess is often used as a test for potash, forming with it the acid tartrate of potash. This salt precipitates in granular crystals.

Tartar emetic is a double salt, formed by the combination of tartaric acid with potash and oxide of antimony. Its action on the animal economy is very energetic; it vomits in the dose of 5 to 10 centigrammes; in the dose of one gramme, it may produce death.

Tartaric acid is used in the manufacture of printed calicoes. Its uses are the same as those of oxalic acid.

It is used in a free state, but oftener in the state of cream of tartar, as a mordant on wool. Mixed with chalk and alum, tartaric acid is used for cleaning silver-ware. It is often used as a substitute for citric acid, in the preparation of lemonade.

CITRIC ACID.

Citric acid exists in a great number of fruits; it is extracted mostly from the juice of lemons and gooseberries. This acid is solid, and crystallizes readily; its taste is acid and pleasant; it is generally used for making lemonade.

Citric acid is largely used in dyeing; it is used to prepare, and particularly to brighten, the red of carthamine. It forms, with tin and cochineal, beautiful scarlets, which are applied upon silk and morocco. The calico-printers use it for decoloring, and reserves. It is also used to prepare a solution of iron, employed by binders to give leather a marbled appearance.

LACTIC ACID.

Lactic acid exists abundantly in the animal organization. It is found in a free state or in combination, in the muscles, in blood, in wine, in milk, in the gastric juice, and the yolk of egg. It is also found in the vegetable organization. It exists in nearly all the vegetable juices which are subjected to fermentation.

Before describing the properties of lactic acid, we will examine in detail the phenomena which constitute *lactic fermentation.*

Lactic acid which is found in the vegetable and animal organization, appears to be the result of a transformation brought about under the influence of a sort of fermentation.

Most nitrogenous organic substances, such as fibrine, albumine, and casein, which are exposed for some time to the air, and undergo the commencement of change, become transformed into *lactic ferments*, and acquire the property of changing into lactic acid neutral matters, such as sugar, gum, starch, sugar of milk, &c. This transformation is easily understood. If, in fact, we compare the composition of the neutral substances which we have named, with lactic acid, it will be seen that these bodies have the same composition with it, or differ from it only by the elements of water. It may then be said that lactic fermentation is for the most part, an isomeric modification or transformation, which fixes the elements of water.

These generalities on the lactic formation, enable us to explain the formation of lactic acid in the most important cases.

Milk contains in suspension a fat body, to which is given the name of butter; and in solution, two neutral bodies; one, resembling sugar, has been called the *sugar of milk*, the other is nitrogenous and albuminous, and is called *casein*.

If milk is kept from the air, the casein does not become transformed into lactic ferment, and it will keep indefinitely. If, on the contrary, it is exposed to the air, the casein which it contains alters, and becomes an active ferment, which acts on the sugar of milk, and transforms it into lactic acid. Thus milk kept for some time in the air becomes sensibly acid.

When the casein does not thus change into lactic acid all the sugar of milk contained in the milk, the fermentation soon stops. When the lactic acid is in sufficient quantity, it causes the precipitation of the casein; the milk coagulates, and the casein, rendered insoluble, does not act on the sugar of milk. If, at this moment, the lactic acid is saturated by a base like potash, soda or lime, the casein recovers its solubility, and acts anew on the sugar of milk. By thus saturating lactic acid as it is produced, not only is all the sugar of milk contained in the milk transformed into lactic acid, but even sugar of milk which may have been added previously.

The lactic acid which is found in the animal organization, and especially in the fluids of the stomach, is formed under circumstances similar to the preceding; there are the azotized substances of every kind introduced into the stomach, which react on the neutral materials, such as starch and sugar, and transform them into lactic acid.

The presence of lactic acid in the vegetable organization is explained in the same way. If sprouted barley is taken and exposed to the air for some days after having been moistened, the grain becomes strongly acid, and contains lactic acid. In this case, it is the albuminous matter of the grain which is changed in the air into lactic ferment, which acidifies the starch.

The rapid change which beet-root undergoes in the manufacture of sugar, is due in part to the lactic fermentation. The beet-roots contain an albuminous matter, which is transformed into lactic ferment when these roots are exposed to the air: this ferment acts on the sugar, and changes it rapidly into lactic acid.

A lactic fermentation may be arrested by coagulating the ferment by heat, or by precipitating it by tannin.

It may also be opposed by preserving from the air the animal substance, which might be transformed into ferment. The processes for the preservation of animal matters are founded on these principles.

The juices of onions, carrots, celery, turnips, as well as all of the saccharine plants, acidify in the air, forming lactic acid.

The soured water of the starch factories, yeast, sour beer, cider kept in the air, alcoholic fermentations which have been retarded, all saccharine liquors extracted from vegetables which have been exposed to the air, contain lactic acid. Difficult digestions develop large quantities of it in the stomach.

Lactic acid is liquid, inodorous, colorless, uncrystallizable, of a syrupy consistence: it is soluble in all proportions, in water, alcohol, and ether. Its taste is pleasant and agreeably acid. Poured in small quantities into milk, whether hot or cold, it causes its coagulation; thus milk, curdled naturally, is always acid.

THE ASTRINGENT PRINCIPLES OF VEGETABLES.

TANNINS.

Several vegetable substances, and principally gall-nut, the bark of oak, horse-chestnut, elm, willow, the leaves of trees, the skin of several fleshy fruits, grape-stones, sumac, catechu, some kinds of sap, etc., contain certain peculiar astringent principles, which appear to differ among themselves in their composition and properties. These matters are considered as feeble acids, and they have received the generic name of *tannins*.

We shall here only speak of the tannin which exists in gall-nut and in oak bark; it is the most important in its applications: the other astringent principles have, besides, a great analogy to gall-nut.

TANNIC ACID, OR TANNIN.

It would be difficult to obtain tannin pure, by precipitating it from an aqueous solution of gall-nut or oak bark, by means of reagents: tannin would always retain the coloring matters, and besides would undergo during precipitation a more or less important alteration. To prepare pure tannin, a very simple process is followed, founded chiefly on the solubility of tannin in ether.

The gall-nut is introduced in powder into an apparatus formed of a glass tube fitted in a bottle. The gall-nut is kept in the tube by means of a grooved cork fitted in its end covered with cotton, so that the liquid may filter through.

Ether of commerce, containing 10 per cent. of water, is poured over the gall-nut.

The tannin is first dissolved by the ether, and is then found precipitated in the form of thick syrup, by the water which the ether contains. The ether and the syrupy solution of tannin pass through into the bottle.

The syrupy layer, after having been repeatedy washed with pure ether, is evaporated in vacuo, or at a temperature which should not exceed 100° C. It leaves tannin pure, or at least containing but traces of foreign substances (Pelouze).

When tannin is to be prepared in the large way economically, gall-nuts are macerated for twenty-four hours in aqueous ether; this is filtered to separate the ethereal solution, and submitted to evaporation, which leaves then a considerable quantity of tannin.

Thus we are enabled to extract from the gall-nut 66 per cent. of tannin.

Expression does not give as pure a tannin as the process by displacement.

Tannic acid is solid, white, and without smell; its taste is powerfully astringent; it is soluble in water, and uncrystallizable; it dissolves in alcohol and ether, but ether dissolves it less readily than alcohol.

The aqueous solution feebly reddens the tincture of litmus. It decomposes the alkaline carbonates with effervescence.

Tannin in solution in water readily absorbs the oxygen of the air, and transforms itself into gallic acid, giving off carbonic acid.

This transformation is favored by the presence of animal matter, and constitutes *tannic fermentation.*

Tannin pure and dry is unalterable in the air; thus it is always preserved in powder. It is only dissolved at the time it is to be used.

Tannin forms with bases compounds slightly soluble; thus solutions of potash, lime and barytes are precipitated by tannin: tannin precipitates nearly all the salts which contain organic alkalies.

This property is made use of in some cases of poisoning, to stop the action of the organic alkalies.

Tannin in solution in water is entirely absorbed by an animal skin, there is formed an insoluble combination of tannin and animal substance, and the water no longer contains astringent matter.

We can thus analyse a solution of tannin by weighing the skin before and after absorption. Skin which is combined with tannin, is called *leather;* it becomes almost imputrescent and impermeable.

Solutions of gelatine are entirely precipitated by tannin; there is thus formed an insoluble white precipitate.

Tannin precipitates nearly all animal matters; it is used in the manufacture of white wines, to coagulate a substance called *glaïadine.* This substance will excite in white wines the viscous fermentation, and make them *muddy.* Tannin is sometimes used in medicine to stop hemorrhages.

INK.

Ink is a combination of tannin and peroxide of iron.

Ink is prepared by making tannin or nut-galls react on a salt of iron at the maximum: 1 kilogramme of nut-gall is boiled in 15 litres of water; the liquor is filtered, and mixed with 500 grammes of sulphate of iron, and 500 grammes of gum; sugar and sulphate of copper are often added. The liquor is exposed to the air until it shall become of a deep black color.

The gum which is added to ink is intended to prevent the tannate of the peroxide of iron from separating, in the form of a black precipitate. Ink is not unalterable; chlorine and oxalic acid decolorize it readily; it even alters spontaneously in the air.

The instability of ink has led persons, for some time, to search for a less alterable composition than common ink, which might be used with safety. This problem has been solved, to a certain extent, by composing an ink formed of lamp-black, held in suspension in gum-water, to which has been added small quantities of chlorhydric acid, or soda.

TANNING.

Astringent vegetable substances are used in the arts to render skins imputrescent, to *tan them.*

Tanning is, then, an art, which has for its end the combination of tannin with the animal matter forming skin, in order to render it impermeable, elastic, and imputrescent.

The tannin which is generally used in tanning, exists in the bark of the oak; this bark, ground, constitutes *tan.*

When the skins, properly prepared, are left in vats for a sufficient time with wet tan, the animal substance constituting the skin forms a true combination with the tannin, becomes impermeable and imputrescent, and changes into *leather.*

This operation, to be well done, requires a long time, which varies from six months to two years.

ACIDS OF FRUITS.

Most green fruits owe their acidity to three acids, which are *tartaric*, *citric*, and *malic.* Tartaric acid exists particularly in the juice of the grape, while citric and malic acids are found in nearly all fruits, such as apples, cherries, plumbs, peaches, apricots, strawberries, pears, &c.

Some fruits, such as apples, contain a gelatinous acid, called *pectic acid.* It is this acid which gives to some fruits the property of becoming jelly.

ORGANIC ALKALIES.

Those compounds, which, like the metallic oxides or ammonia, combine with acids to form true salts, are called *bases* or organic *alkalies.*

For a long time it was thought that only neutral bodies or acids could be extracted from the vegetable and animal organization. About the year 1803, M. Derosne extracted from opium a crystalline substance which was alkaline in its character. In 1804, MM. Seguin and Sertuerner discovered simultaneously a new body in opium, whose alkalinity they proved; this was at first attributed to ammonia, or some mineral base which had been used in the preparation.

In 1817, Sertuerner repeated his experiments on opium, and proved that this body contained a true basic substance, whose alkalinity is not due to the bases used in its preparation. It is then to Sertuerner that the discovery of the organic alkalies belongs.

Several chemists, among whom must be cited in the first rank MM. Pelletier and Caventou, soon began to enrich science by the discovery of several organic alkalies.

At this time, the number of bases extracted from the animal and vegetable organizations is considerable, and a great number of organic bases have even been artificially produced.

The bases extracted from the vegetable organization render the syrup of violets green; like the mineral bases, they saturate the most energetic acids and form crystallizable salts, which are subject to the ordinary laws of the decomposition of salts.

The organic alkalies are generally solid and fixed: some form perfectly defined crystals. Some organic bases are liquid and volatile; such are the alkalies of tobacco and cicuta.

The solid organic alkalies are, in general, inodorous and fixed.

The organic bases are slightly soluble in water; their solvents are alcohol and ether.

Some are amorphous; others, as codeine,* crystallize with remarkable regularity.

Their taste is generally sharp and bitter, their action on the animal economy is energetic. Used in small doses, they are often valuable medicines, but if administered in moderately large quantities they become true poisons. As the organic alkalies are precipitated by astringent solutions, their true antidote is tannin, which forms with them a precipitate insoluble in water, which has no action on the animal economy.

THE NATURAL STATE AND EXTRACTION OF THE ORGANIC ALKALIES.

The organic bases pre-exist ordinarily in plants; they are seldom found in them in a free state; they are found combined with muriatic, malic, lactic, acetic acids, or with peculiar acids, such as the meconic or quinic, which are only found in opium and the cinchonas.

The pre-existence of the organic alkalies in the organs of vegetables, has been for a long time denied by divers chemists, who attributed their formation to the reagents used in their extraction; but all the facts are contrary to this opinion.

We shall only relate that M. Dupuy has withdrawn from

* Extracted from opium.

opium sulphate of morphine, perfectly pure, by the contact alone of this extract with distilled water.

To extract the organic bases of vegetables, different processes are used, according as the base is soluble or insoluble in water, or volatile.

When it is insoluble, which is most frequently the case, the vegetable which contains the base is exhausted by acidulated water, and the salt which is formed is decomposed by ammonia, lime, or magnesia.

The organic alkali is separated from the mineral bases which are used for precipitating it, by using ether or alcohol, which dissolve the organic alkali, and sometimes let it crystallize in a state of absolute purity. Sometimes, to separate the organic base from the foreign matters which color it, a salt is made with muriatic or sulphuric acids; this salt is treated with animal charcoal, and the organic base is precipitated anew by an alkali. A second crystallization in alcohol, then gives the base in a state of purity.

To purify the alkalies which are insoluble, or slightly soluble in water, they may also be precipitated from their solutions by a great excess of potash, or caustic soda, the mixture being brought to ebullition. The coloring and resinous matters which are almost always combined with the alkalies, with which they comport themselves like acids, dissolve in the potash or soda, and the organic alkali is often decolorized with great facility.

When the organic bases are soluble in water, their extraction has many difficulties. Crystallizable salts are formed which are purified, and the acid which is found united to the base, is precipitated. When the organic base is volatile, the vegetable is distilled with an excess of potash or lime.

The base which passes in the distillation, is purified by engaging it in saline combinations. The volatile base thus obtained, is generally mixed with ammonia. To separate it, the mixture is saturated with oxalic or sulphuric acid, and evaporated to dryness. The deposit is taken up by alcohol, which only dissolves the sulphate or oxalate of the organic alkali, which may thus be obtained pure by evaporation and crystallization. The salt is then mixed with a solution of potash, and an equal volume of ether; two layers are formed; that which is above is the ethereal solution of the organic alkali. This liquid, distilled in a retort, disengages the ether, and traces of ammonia. The alkali remains in the retort, and may itself be distilled afterwards, to be obtained perfectly pure. The same plant may sometimes contain several different organic alkalies; thus, several bases are extracted from opium; the principal are *morphine*, *narcotine*, and *codeine*.

In nux vomica, two very poisonous organic alkalies are obtained, *strychnine* and *brucine*. The bark of cinchona, which is used with so much advantage to combat fevers, contains two alkalies, which are *quinine* and *cinchonine*. The sulphate of quinine is the only salt with organic alkaline base, which is prepared in the large way for medical supply. To obtain it, yellow cinchona is boiled with water acidulated with sulphuric acid; there is thus formed soluble sulphate of quinine; an excess of lime is added, which precipitates the quinine. The precipitate thus obtained is yellow; it is dried, and then treated with boiling alcohol, which redissolves the quinine. The alcoholic liquor is then neutralized with sulphuric acid; it is evaporated and decolorized by animal charcoal, and on cooling, the sulphate of quinine deposits in small fine needles, which are purified by a second crystallization. Two grammes of sulphate of quinine, in doses, are generally sufficient to cut short a fever.

NEUTRAL BODIES.

LIGNINE.

THE fibrous tissue of wood, deprived by the action of solvents of the minute quantities of foreign matter which it contains, was for a long time thought to be a pure body. But M. Payen has shown that the substance called *lignine* is a mixture of different bodies, and that it is composed of elongated cellules in juxtaposition, lined in their interior with a hard and amorphous matter, which is found in more or less irregular layers. It may then be said that each cellule of wood is formed principally of an exterior substance, which M. Payen calls *cellulose*, and of *an incrusting matter*, the composition of which appears to be very complex. There is besides found in the cellules, an azotized matter which is often rather abundant, and which appears to precede the formation of cellules in the cambium.*

CELLULOSE.

Cellulose† may be represented in its composition by carbon and water.

Cellulose constitutes, in great part, old linen, lint, paper, and elder-pith. Swedish filtering-paper, called *paper of Berzelius*, is cellulose nearly pure.

To obtain it absolutely pure, these different bodies may be treated with water, alcohol, ether, feeble acids, and dilute

* Cambium is a mucous, viscid layer, interposed beneath the bark, and above the wood, found to contain minute transparent granules, and to exhibit faint traces of a delicate cellular organization. It appears to be exuded by both the bark and wood. (Lindley.)

† Cellulose is the base of all the organs of vegetables, cellules, and vessels. (M. Payen.)

alkalies. Cellulose may also be prepared with common cotton, or the pith of the plant which furnishes rice-paper.

The spongioles of the radicles of young plants, also easily give pure cellulose; but its purification presents great difficulties when the organic tissues which contain it are strongly organized, as in leaves. The cellulose contained in lignine is found in large quantity in the excrements of ruminating animals; it does not digest, while the other parts of lignine are completely absorbed.

Pure cellulose is white, solid, transparent, and insoluble in water, alcohol, and ether, and the fixed and volatile oils. Water disintegrates it readily, when its organization is not far advanced. Sulphuric acid dissolves it without coloring it, and transforms it into dextrine, and then into glucose, as M. Braconnot has shown.

It is admitted that cellulose, under the influence of acids, changes first into starch. M. Payen satisfied himself, in fact, that cellulose, treated with dilute sulphuric acid, acquires the property of being colored blue by iodine.

In fact, cellulose which is found in the rudimentary state in certain cryptogamous plants, is colored blue with iodine; when cellulose is strongly aggregated, it is not colored by iodine. The acids, even when dilute, act in the end on cellulose, color it brown, and make it friable; it is thus that paper pulp, which has not been thoroughly freed from acid by washing, produces a paper which is without substance, and which becomes of a yellow color on long exposure to the air.

Cellulose, dipped in fuming nitric acid, combines with the acid without changing form, and constitutes a very inflammable body, called *pyroxylin*, or *gun-cotton*.

It is seen that all these actions tend to bring together cellulose and amidon, and seem to show that these two bodies are isomeric modifications of the same substance. Finally, in observing cellulose in its different natural states, it is seen to offer very different degrees of cohesion, from the resisting fibres of the ligneous and textile plants, to the delicate membranes which form the tissue of the cryptogames: cellulose then resembles closely amylaceous substance.

Dilute alkaline solutions do not act sensibly on cellulose; it is the same with chlorine. It is on those properties are founded the operations of bleaching the tissues of hemp, linen and cotton.

However, chlorine or the alkalies used in excess, in the end disaggregate cellulose, and completely destroy it. The hypochlorite of lime gives the same results.

Cellulose, when feebly aggregated, as in the parenchyma of young leaves, lichens, the perisperms of certain fruits, may be used for aliment as amylaceous matter.

In the form of long tubes, more or less thick and strongly aggregated, it constitutes the filaments of different textile plants, flax, hemp, cotton, aloe *phormium tenax* (New Zealand flax) *urtica nivea*, (nettle) which are used for making thread, cordage, cloth, paper, and gun-cotton.

INCRUSTING MATTER.

Incrusting matter predominates in hard wood; it is abundant in the stones of fruits; it is that which forms the stony concretions in some pears. It is deposited in layers more or less thick and irregular in the elongated cellules of the woody tissue: it is more abundant in the heart than in the sap-wood. It is often of a yellow or brown color. It is found in greater quantity in heavy and hard wood, than in white and light wood.

The incrusting matter of wood may be extracted by rubbing the wood for a long time in a mortar.

The incrusting matter, which is friable, is separated by a sieve, and then purified by alcohol. It differs in its composition from cellulose, and cannot be represented in its composition by carbon and water.

The incrusting matter containing an excess of carbon and hydrogen, in proportion to oxygen, gives out more heat in combustion than cellulose; thus hard woods which are rich in incrusting matter have a greater heating power than soft woods.

The composition of wood varies with the proportions of cellulose and incrusting matter which they contain.

The incrusting matter is colored black by sulphuric acid, and is dissolved in chlorine; these are the distinguishing properties of cellulose; thus that part of wood which colors black with concentrated sulphuric acid is not cellulose, but incrusting matter.

GENERAL PROPERTIES OF WOOD AND THE COMBUSTIBLES.

Wood is denser than water; it swims on this liquid only in consequence of the air which it contains in its pores. Finally, the age, climate, and nature of the soil, exercise a great influence on the density of the same kind of wood.

Green wood contains on an average 40 per cent. of water; when dried at a temperature of 10° C., it loses but 25 per cent.

It is in this state that wood is ordinarily found, containing, consequently, the fourth of its weight in water.

This water is evidently injurious to combustion, requiring a great quantity of heat to drive it off. In many factories, wood, before being used, is first dried. Wood dried at 100° C., and exposed anew to the air, at the ordinary temperature, takes back from 10 to 12 per cent. of water.

Wood contains, besides carbon, oxygen and hydrogen, which constitute its combustible parts, a certain number of fixed mineral substances which form the ashes. The proportion of ashes is always small, and varies, indeed, in the same vegetable. Thus, the leaves and bark of wood give more ashes than the branches, and the branches more than the trunk.

The ashes of wood are composed of soluble alkaline salts, which are formed by the combination of potash and soda with carbonic, muriatic, and sulphuric acids, and also of insoluble matters, which contain carbonic acid, phosphoric acid, lime, magnesia, and the oxides of iron, manganese, and silex.

Wood subjected to the action of heat, is decomposed; it begins to alter at about 140°.

The products from the distillation of wood vary with the nature of the lignine, and with the temperature at which it is decomposed. A disengagement of inflammable gases is always noticed in this distillation, also of water, which contains in solution acetic acid and spirit of wood; there are formed, besides, different tarry substances.

The residue of the distillation is charcoal, which preserves the form of the wood; the quantities of the coal produced vary with the rapidity of the distillation; there is obtained from 13 to 28 per cent. of it.

Wood of a compact and hard tissue, like oak, beech, and elm, carbonize more slowly than the white woods, and leave a more considerable deposit of carbon. Charcoal produced by the light woods, burns rapidly, and with flame. In some of the blast-furnaces of Ardennes, a combustible known as *roasted wood*, is used. This is prepared by placing common wood in iron cylinders, which are heated by the waste flame from the furnace; the wood thus loses as much as 50 per cent. of its weight, and produces, in burning, more heat than common wood.

Wood changes readily when exposed to the joint action of moisture and air; it becomes colored, disaggregates, and transforms the oxygen of the air into carbonic acid. Peat and mould are formed in these conditions.

PEAT.

Peat proceeds from the alteration which aquatic herbaceous plants undergo, in marshy places.

Peat is often found in horizontal banks, frequently very thick. The remains of earthen utensils are often met with in peat, proving its modern origin.

There are two kinds of peat, which correspond to different states of decomposition.

1st, *compact peat*, which is brown, and in which are seen some vegetable remains; 2d, *herbaceous peat*, which is spongy, and formed of vegetable remains easy to distinguish.

Peat is got out in the spring. It is extracted in the form of small bricks, which are left during the summer to dry in the air; they shrink from three to four fifths. The weight of a cubic metre is generally from 250 to 400 kilogrammes.

In chemical composition, it is very like mould. The composition of the ashes of peat, varies with the nature of the ground which is in the vicinity of the peat-beds.

Charcoal, made from peat, is in general very friable and very light; it is not much used.

FOSSIL COMBUSTIBLES.

Lignite, Pit-Coal, Anthracite.*—The fossil combustibles are generally divided into three great classes, comprising lignite, pit-coal, and anthracite. Anthracite and pit-coal are found in the transition and secondary formations; the lignites, in the tertiary.

The fossil combustibles evidently proceed from the alteration of vegetable substances. In fact, there are found in the lignites which approach nearest the present epoch,

---

* Bituminous coal is here called *coal*, or *pit-coal*.

portions which present traces of vegetable organization, and which unite the lignites to fossil wood and peat. There are also found in the lignites, portions which have a great analogy to the coals. By a single inspection of the fossil combustibles, a gradual passage may be shown of the ligneous substances into anthracite, formed almost exclusively of carbon.

The calorific power of anthracite and pit-coal is at least equal to that of charcoal. As we approach the combustibles of the present epoch, their calorific power diminishes rapidly. The bitumens alone form an exception to this rule.

In proportion as the decomposition of the organic substances advances, the proportion of oxygen and hydrogen diminishes, and the quantity of carbon increases.

Properties of Coal.—Pit-coal is formed by a mixture of different bodies insoluble in all the dissolvents, and which consequently cannot be separated one from the other. The *caking property* of coal, that is to say, the facility they have of softening and caking in the fire, depends in general on the relation between the oxygen and the hydrogen which they contain. Coal cakes in proportion to the excess of hydrogen to oxygen which it contains. When the proportion of hydrogen becomes very considerable, as in the bitumens, these combustibles do not leave a sensible quantity of coke; nearly all the carbon then passes into the state of carburets of hydrogen, which are volatile.

The fat and hard pit-coals are chiefly used for the manufacture of coke, which is slightly swelled or puffed out, dense, and having great power of cohesion, it is valuable for the fusion of iron ore.

Smith's Coal (coal with a long flame) is used in forges or reverberatory furnaces. It is used for making illuminating gas, because it gives large gaseous products. The coke which it produces is puffed out, and is not in general suited to metallurgic applications.

The *poor coals* give a coke which has but little consistence. They are used for heating steam-boilers and purposes which do not require a high temperature.

Iron pyrites is frequently met with in pit-coals, and is a great injury to their quality. In fact, the sulphur which they contain, alters steam-boilers rapidly, and modifies the properties of metals with which the combustible comes in contact.

Pit-coal gives by distillation water, combustible gases, ammonia, empyreumatic oils, and tar containing a solid carburet of hydrogen, which has received the name of *naphthaliue.*

PROCESSES FOR THE PRESERVATION OF WOOD.

The preservation of wood is one of the most important chemical questions applied to industry. Wood contains an azotized substance, which, acting as a ferment, causes its decomposition.

Many causes concur in the destruction of wood; the chief of which are the alternating influence of air and moisture, the boring of insects, certain cryptogamous plants which grow on the surface of wood, and often penetrate into its interior. The azotized matter contained in the woody tissue serves both as nourishment for the insects, and as manure for the mushrooms which grow on the wood.

It has been long known that wood, which contains a resinous principle, such as ebony and quince, are very enduring. It was at first proposed to cover wood with resinous substances, to preserve it from the contact of the air and moisture; it was afterwards thought that the essences, creosote, pyroligneous acid, sulphate or acetate of iron, bichloride of mercury, and arsenious acid, might be advantageously used for preserving wood.

Latterly, Messrs. Bréant and Mohl, and M. Boucherie in particular, conceived the idea of causing preservatives to penetrate the capillary ducts of the wood.

M. Bréant first proposed to introduce oil into the interior of the wood, by great pressure; from this cause, the planks impregnated with *dry linseed oil*, used as beams on a bridge over the Seine, have resisted decay for ten years, while planks of the same wood, not prepared, were completely rotted in a few years. The pressure used by M. Bréant is sufficiently powerful to cause the fusible alloy of D'Arcet to penetrate the central parts of the thickest pieces of wood.

The process of M. Bréant is, however, so expensive, that it has not yet been applied in the arts. M. Mohl has proposed the introduction of steam into the tissue of the wood, which leaves a vacuum on cooling, and thus causes the absorption of the preservative liquids.

M. Boucherie, to preserve wood, in the first place made use of different saline solutions, which he introduced into the sap-vessels by means of the ascending force, which causes the sap to flow through the ligneous tissues, from the root of the tree to the summit clothed with foliage.

The tree, cut down, is plunged into a vessel containing the liquid to be absorbed; in order that the absorption should take place, it is not necessary that the tree should be standing: the experiment succeeds with a tree lying down, provided it is in contact with the liquid; and even by making a cut around the base of the tree with a saw, while still supported by its roots, and surrounding it with a kind of basin to contain the fluid, this will soon be completely absorbed, and will penetrate all the tissues.

For this operation, all the lateral branches and leaves of the tree may be cut off, leaving only a tuft of leaves at its summit, which causes the ascent of the fluid. The liquids used vary with the results desired to be obtained. If it is desired to preserve the wood from the dry or wet rot, to increase its hardness, and insure its preservation, M. Boucherie proposed to use the pyrolignite of iron, or the sulphate of copper: the earthy chlorides are especially used when it is desirable to preserve the pliantness of the wood.

The introduction of saline substances into the interior of wood, is now done by displacement, and has the great advantage of preserving it from *voilage* (the rot), from shrinking, and renders it in some degree incombustible. The displacement of the sap by saline solutions, is very prompt. Thus, a poplar of 40 centimetres at its base, has absorbed in six days, 3 hectolitres of pyrolignite of iron; a plane tree of 30 centimetres has absorbed 2 hectolitres of chloride of calcium, in seven days. In 1843, it was shown in the forest of Compiègne, by the forest agents, that in a beech cubing 294 metres, there was displaced in twenty-four hours, 3·060 litres of pure sap, which was replaced by 3·210 litres of pyrolignite. This experiment is interesting in a physiological point of view, and permits us to establish the relation which exists between the solid parts of the wood, and the fluids which circulate in the sap-ducts.

M. Boucherie conceived that cabinet-work might also profit by these processes, by introducing into the pores of wood, substances having the power of giving rise to coloring-matters by their mutual decomposition. Thus, with a salt of iron and a tanning substance, prussiate of potash, acetate of lead, and chromate of potash, are produced in wood, veins of black, grey, blue, yellow, brown, green, and other colors, which may be varied infinitely. Experience has already pronounced on these processes of M. Boucherie, applied to the preservation of wood. Thus, the cross-ties intended for rail-roads have been buried in the ground, after having been prepared by the method of M. Boucherie, and after many years they have been found in a state of perfect preservation; while the cross-ties of the same wood, not prepared, and placed in the same circumstances, were entirely decomposed.

Some chemists have proposed to introduce into wood, coloring and preservative matters, by a process which consisted in cutting the trunk of the tree at its base and forks, and placing one of its extremities in communication, with the aid of impermeable cloths, with a reservoir containing the liquid; an apparatus is adapted to the other extremity, in which a partial vacuum is made by dilating the air by burning a small quantity of alcohol in a cylinder; the liquid passes through the tree longitudinally, driving the sap before it.

The solution of the sulphate of copper has recently been used with great success for preserving sails. (M. Boucherie.)

PYROXYLIN, PYROXYLE, GUN-COTTON.

Towards the end of the year 1846, M. Schœnbein announced that he had made a new powder, much more energetic than gunpowder; but he neither pointed out its nature, nor mode of preparation; he contented himself with showing the projectile effects of this inflammable material, to which he gave the name of *gun-cotton.*

Several chemists thought they saw, in the discovery of M. Schœnbein, the application of a material already known and previously described (Pelouze). They announced that the product of the impregnation of ligneous matters (cotton, paper, &c.,) with nitric acid, burnt in fire-arms like real powder; and in consequence of this, uttered the opinion that the gun-cotton of M. Schœnbein was nothing but this very material.

Some months afterwards, M. Schœnbein made public the preparation of gun-cotton. His process consists in plunging for some moments, carded cotton into a mixture of concentrated nitric and sulphuric acid.

To prepare gun-cotton, a mixture is made of 3 volumes of nitric acid, and 5 volumes of concentrated sulphuric acid; this mixture is allowed to cool, and carded cotton, such as is found in commerce, or rather after being dried in an oven, is dipped into it. To avoid an elevation of temperature, and combustion which would follow, only a small quantity of the cotton is dipped into the bath at a time, and besides, the weight of the acid should always be considerable relatively to that of the organic material. After fifteen or twenty minutes of contact with the acid, the cotton is withdrawn, and as much as possible of the liquid is pressed out; it is then washed in plenty of water, until there is neither odor nor taste, nor action on litmus paper.

The water for washing may be either cold, warm, or boiling.

The inflammable cotton, pressed in a cloth or the hand, divided between the fingers or carded, dries readily at the ordinary temperature. The drying may, however, be hastened by subjecting it to the action of a current of air at 30° or 40°, or by placing it in a vessel with a substance having an attraction for moisture, as lime for example.

Paper and tissues are made inflammable in the same way as gun-cotton; and it is useless to add, that the materials properly prepared all give an identical product.

100 parts of pure cellulose give on an average 175 parts of pyroxylin.

Properties of Pyroxylin. — Cotton and ligneous matters properly called do not change, so to say, either form or aspect when they are transformed into pyroxylin. Cotton rendered inflammable is, however, a little less soft to tho touch, and its fibres break more readily.

Pyroxylin is completely insoluble in water, whether cold or hot. Concentrated alcohol and ether do not dissolve it, but it is slightly soluble in a mixture of these two liquids; it then becomes transparent, and is called *collodiun*. The acetate of méthylène and acetic ether dissolve it entirely.

Pyroxylin, subjected to a temperature slightly elevated, detonates. The inflammation manifests itself in general from 140° to 150° C. But when pyroxylin is kept for a certain time at 100° C., and even between 60° and 80°, it gradually alters, gives off a nitric odor, becomes friable, and then in a short time it suddenly detonates at a temperature below 100° C.

Pyroxylin, when burnt on a cloth, on a piece of white paper, or on a porcelain dish, leaves no trace of residue when it is quite pure, and the products of the combustion have, in general, no sensible odor. It however sometimes gives off yellow fumes and gases slightly prussic. Nitrous vapors are easily recognised by burning a few milligrammes of gun-cotton in a tube closed at one end. Looking lengthwise into the tube, the atmosphere appears of an orange-red, and besides the gases of the combustion smell of hyponitric acid.

The nitrous and prussic products do not appear to be produced in appreciable quantities, when the pyroxylin burns in the ordinary way of powder in guns or in mining. The detonation, about as strong as that of powder, is not accompanied with smoke.

In the experiments made by Messrs. Combes and Flandin with gun-cotton used in mining, the combustion of many kilogrammes of pyroxylin left no perceptible nitrous gas nor prussic odor. The ordinary and most abundant products of the inflammation of pyroxylin are carbonic oxide, carbonic acid, nitrogen, and the vapor of water.

If, instead of burning gun-cotton with a body in a flame, or by raising its temperature, it is twisted into threads and placed on a body which is a good conductor, like a metal, and touched with a coal, it burns slowly and almost without flame, giving off a strong nitrous odor.

Exposed to the air, gun-cotton attracts but little moisture. Its weight hardly increases 2 or 3 hundredths in the space of several months, and its projectile powers are not sensibly modified.

Common cotton, in the same conditions, is much more hygrometric. Pyroxylin, kept in water for two years, does not alter. It may be immersed for a long time without inconvenience, and it is equally true that it will behave with sea-water as with common water.

MM. Flandin and Comtes have made numerous experiments in blasting in mines, which have established in the clearest manner that this powder produces effects much greater than those of common powder, and that the advantage of pyroxyle over powder increases with the hardness and resistance of the rocks. However, in some circumstances, the carbonic oxide which escapes through the cracks may give rise to accidents and interrupt the work, for this gas is poisonous, and besides inflammable. The cost of pyroxyle, whichis 4 to 5 francs the kilogramme, renders doubtful the advantages which may be derived from it in the industrial operations.

In 1840, M. Courbes thought to render complete the combustion of pyroxyle, by the use of different oxydising salts. New experiments on the use, in mines, of a mixture of 10 parts of pyroxyle, and 8 to 9 of nitre, have much interest. When fire is communicated to this mixture, the pyroxyle which it contains is completely reduced into aqueous vapor, carbonic acid, and azote; while pyroxyle by itself gives a great quantity of carbonic oxide. Experiment has shown that in thus furnishing to gun-cotton all the oxygen which it requires to transfer its carbon into carbonic acid, and its hydrogen into water, a material is made whose power on hard and brittle rocks, is at least seven times greater than an equal weight of mining-powder, and five or six times greater than that of gunpowder. From this time, then, pyroxyle may economically replace mining-powder. It is difficult to say whether the day will come when it may be usefully employed in fire-arms, and in the art of war. It is very certain that it burns too quickly, constituting a true *poudre brisante* (bursting); but it is probable that this combustibility may be moderated and regulated, as has been done with gunpowder itself, which, when it is not sufficiently dense, or when its grain is too fine or too porous, causes guns to burst.

All the tissues, paper more or less thick, pulp for paper, wood, saw-dust, and many other organic matters formed of cellulose, give inflammable pyroxylines. The degree of compression of carded gun-cotton influences the rapidity of its inflammation, and its projectile effects. It may be brought, by compression, to occupy a volume equal to that of an equivalent weight of gunpowder.

Pyroxylin does not moisten fire-arms, as was at first supposed; the high temperature produced by the combustion, drives out of the gun the large quantity of vapor of water produced by the explosion.

AMIDON.

When potato is rasped, and the pulp washed on a sieve, the water which passes is milky, and deposits a white substance called *fecula*. The general name of *amidon* is given to the amylaceous substance extracted from the cereals. Chemically, fecula and amidon are identical. Leeuwenhoeck, on examining amidon with a microscope, discovered that this body is of a globular form, and that the internal part of the globule is different from the external. This important observation was for a long time overlooked. Its chemical properties were examined, while the microscopical observations on amidon were neglected; but as amidon is organised, it cannot be properly studied without the microscope.

From 1825 to 1830, M. Raspail published a series of very important microscopic observations on amidon, and repeated those of Leeuwenhoeck. After him, many chemists, among whom we will mention Messrs. Gay-Lussac, Chevreul, Biot, and Dumas, completed the chemical study of amidon; and since the last works of M. Payen, from whom we have borrowed nearly all the details we give on amidon, this body is perhaps one of those best studied in organic chemistry.

PHYSIOLOGICAL NOTIONS ON AMIDON.

Amidon is never found in the tissues which are in a rudimentary state: thus, the spongioles of the radicals, and the rudiments of the buds, do not contain it. Amidon, on the contrary, is met with in the epidermis of vegetables; it is often enclosed in the cellules, in the form of grains, which increase in volume and in quantity as we approach the central parts of vegetables.

The grains of amidon first present themselves in the vegetable organization, in the form of almost imperceptible granules, which have a peculiar duct called *hile* (umbilicus of seeds.) It is by this duct that the granule receives its nourishment and increases in volume. This augmentation appears to be intermittent, for the granules are composed of concentric layers different in density and cohesion.

The dimension of the grain seems to depend on that of the cellules which enclose it, and on the extensibility of the exterior layers of this grain. It is indeed very variable, as the following table demonstrates:

Large Rohan potatoes,	185	thousandths	of a	millimetre.
Other potatoes,	140	"	" "	"
Sago,	70	"	" "	"
Large beans,	75	"	" "	"
Lentils,	67	"	" "	"
Large peas,	50	"	" "	"
White wheat,	50	"	" "	"
Haricots,	36	"	" "	"
Indian corn,	30	"	" "	"
Root of parsnip,	7·5	"	" "	"
Seed of beets,	4	"	" "	"

Not only do the dimensions vary in different kinds of amidon, but also the shape, and the microscopic inspection of a grain of amidon will often serve to identify different kinds of fecula.

The existence of the duct (hile) is often difficult to prove: to render it evident, M. Payen advises to submit the grain of amidon to a powerful desiccation, which overcomes the cohesiveness. In fact, the parts of the amidon which were distended with water, diminish more in volume than the others. The *hile* then opens, and exposes to view in the interior of the amidon, the concentric layers which constitute it.

This observation now demonstrates that the internal part of the amidon is consistent and not liquid, as was maintained formerly.

The concentric layers may also be seen by compressing the grain of amidon between two plates of glass: the grain opens in breaking, and the inside is thus seen.

M. Payen, in order to demonstrate in an evident manner the internal structure of a grain of amidon, has contrived to exfoliate completely the fecula.

To effect a local solution of the exterior layer of the amidon, it is sufficient to plunge the amidon, first heated to 180° C., into some aqueous alcohol; the alcohol evaporates faster than the water, and there remains on each grain of amidon a small drop of water, which perforates the outside layer. If the amidon thus perforated is put into alcoholized water, the internal layers of the amidon dilate under the influence of the water more readily than the external layers, and the grain of amidon opens somewhat like a flower.

It results from the microscopic observations which have been made on amidon, that this substance is not crystallized, as was at one time supposed, but that it is truly organized.

Properties of Amidon.—When amidon is heated to 200°

C., it undergoes a very remarkable isomeric change, and is transformed into a soluble body called *dextrine.*

If moist fecula is introduced into a copper tube closed at both ends, and the tube is heated to 170°C., the fecula is transformed into dextrine under the simultaneous influence of water and pressure.

Warm water exercises a rapid action on amidon. If 1 part of fecula is put into 15 parts of water, and the temperature of the liquid is slowly raised, as soon as it reaches about 55° C., the consistence of the liquid is seen to change; it becomes thick and mucilaginous, *starch* begins to form at this temperature, and increases particularly from 72° to 100° C.

Examining starch with a microscope, the grains of the fecula are seen to be split; the interior layers in hydrating are considerably developed; the grains of fecula have increased to thirty times their volume.

The starch which heat has produced may be destroyed by cold: when, in fact, the starch is subjected to a temperature of 10° C., the internal parts of the amidon which were developed in the boiling water, contract under the influence of cold water, re-enter their envelope, the starch loses its consistence, and the liquor resumes its original fluidity.

Several bodies have the property of converting amidon into starch; we will mention soda particularly, which, in the proportion of 0·02, increases amidon to seventy-five times its volume.

It results from what precedes, that when some grains of amidon are so placed that they cannot swell freely, they will adhere to each other and form a gelatinous starch.

Even boiling alcohol is without action on amidon, and only dissolves the most feeble traces of it.

Amidon, which is generally considered as a neutral body,

may, however, combine with certain bases, as lime, barytes, and oxide of lead.

Iodine exercises a characteristic action on amidon. It colors it a deep blue: the color obtained in this case depends on the state of aggregation of the amidon: it is generally blue or violet, and in some cases becomes red when the amidon has undergone a partial disaggregation.

The iodide of amidon is destroyed when exposed to light; the iodine which it contains is converted into iodhydric acid, as M. Guibourt has demonstrated.

When the iodide of amidon, held in suspension in water, is exposed to a temperature of 66° C., it becomes colorless and resumes its color on cooling.

Dry amidon may be kept indefinitely without change; it is not the same with starch, which in warm weather alters, acidifies, and is transformed into dextrine and water. The azotized matter which amidon often contains appears to influence this change.

According to M. Braconnot, when amidon is treated with concentrated nitric acid, it is entirely dissolved; the liquor diluted with water deposits a substance called *xyloïdine* or *azotate of amidon.*

All weak acids appear to act on amidon; they first disaggregate it, and then transform it into dextrine and sugar.

Sulphuric acid is mostly used for modifying amidon. When 500 parts of fecula are treated with 1000 parts of water and 10 parts of sulphuric acid, and steam is made to pass through the liquor to heat it in a uniform manner, the amidon is rapidly dissolved; if the acid is saturated with carbonate of lime, sugar or dextrine is found in the liquor, according as the reaction has been more or less prolonged.

DIASTASE.

It has been long known that sprouted barley digested in water at about 70° C., gives a gummy and dense liquid which afterwards becomes saccharine; this liquid, aromatised with hops, and fermented, forms beer. The theory of the fermentation of beer was for a long time unknown, but has been completely cleared up by the observations of M. Dubrunfaut on the property which the infusion of malt possesses, of converting at 60° C., amidon into sugar, and, above all, by the important discovery of diastase, due to Messrs. Payen and Persoz.

These chemists have proved, that, in the germination of the seeds of barley, oats, wheat, etc., there is developed near the sprouts and roots, a substance which has the characteristic property of disaggregating the amidon, and transforming it first into dextrine and then into sugar; it is this property which has given it the name of *diastase.*

Diastase disorganizes amidon, when it is made to act at a temperature of 70° C., and transforms it first into dextrine and then into glucose. Messrs. Persoz and Payen have shown that 1 part of diastase metamorphoses 2000 parts of amidon.

In the manufacture of beer, when the germinated barley is dissolved in water at 70° C., the diastase which is found there transforms the amidon into glucose; this sugar then fermenting, gives to beer its alcoholic principle.

DEXTRINE.

Dextrine has the same composition as amidon: it is solid, soluble in water, and uncrystallizable. Dextrine is insoluble in alcohol.

Dextrine is manufactured for use in the arts, by three different processes.

The first is founded on the transformation of amidon into dextrine by the action of acids.

According to this process, to transform for example 1000 kilogrammes of dry fecula into dextrine, 2 kilogrammes of nitric acid at 36° are diluted with 300 kilogrammes of water; the fecula is then mixed with this acidulated water, and afterwards placed in the open air in a dryer.

When, by drying, the cakes crack spontaneously, they are crushed with a shovel, and spread out in layers of 3 or 4 centimetres, in sheet-iron drawers in a stove, where the temperature is kept up to between 110° and 120° C. In an hour, or an hour and a half, the transformation is complete.

The second process depends on the transformation of fecula into dextrine by diastase.

A mixture of water and malt is heated to the temperature of 75° C., and fecula is gradually added. When the solution of the fecula is complete, the temperature is rapidly carried to 100° C., to stop the action of the diastase, and prevent the formation of glucose. The liquor is filtered and evaporated to the consistence of syrup in a boiler provided with an agitator.

Impure dextrine, or *roasted amidon*, is also prepared in the arts by a third method, by pulverizing the amidon of the cereals, and gradually heating it to 140° or 160° in an oven or cylinder like that used for roasting coffee. The operation is finished, when the material has become of a clear brown, and when it gives off an odor like overdone bread.

Dextrine may replace gum in nearly all its applications: it is used for dressing chintzes, and other cotton stuffs, in the application of mordants in prints, the sizing of certain kinds of paper, &c.

It is used in surgery for giving firmness to bandages, which may afterwards be removed by hot water.

EXTRACTION OF AMIDON.

Amidon is found in the cereals mixed with an azotized substance called *gluten*. The hardest grain is richest in gluten; it always contains large proportions of azotized substances, and is generally richer, containing more inorganic salts, and less amidon than the soft grain.

There are two processes for separating amidon from gluten. The first, which is the oldest and still most used, consists in exposing the farina to a long fermentation; the gluten is then destroyed, becomes soluble, and leaves the amidon. In the second process, the farina is exposed to the action of a weak current of water, which carries off the amidon, and leaves the gluten in the form of a viscous and insoluble body. We will first examine the former method.

The grain, well malted, is soaked in water which has been used in previous operations, called *sour waters*. These waters contain alcohol, acetate of ammonia, phosphate of lime, sulphuric acid, lactic acid, dextrine, and besides an azotized matter in a state of decomposition, which shortly causes the fermentation of the gluten contained in the farina.

In from fifteen to thirty days, the fermentation determines the solution of the gluten in the sour waters, while amidon, being insoluble, is deposited in the bottom of the vessel.

This deposit is repeatedly washed, until the water is clear and colorless; it is then thrown on a sieve, which retains the bran and part of the foreign substances.

The water in which it was washed, is milky, and deposits the amidon. There is often found on the surface of the deposit some colored points, which are taken off with a wooden shovel.

Finally, the amidon is placed on a silk sieve; it is

then poured into a box with holes, or baskets lined with cloth, where it takes the form of loaves.

When the water is drained out, these loaves are placed in a loft on a floor of thick plaster.

When they have acquired a suitable consistence, they are divided into four equal pieces, which are exposed for twenty-four or thirty-six hours, if the weather is good, in a dryer in the open air; after this, the desiccation is completed in a stove, the temperature of which is gradually raised from 40° to 80° C.

If the temperature is rapidly raised to 80° C., the portions of the amidon still moist might be changed into starch.

This mode of manufacture is most followed.

It has, however, many inconveniences. The decomposition of the gluten is accompanied with fetid emanations which are absorbed by the sour waters and the water of the washings, so that these establishments must be by themselves; and besides, the yield in amidon is always ten per cent. below what it ought to be: the grain, instead of giving 50 per cent., gives but 40. This loss is certainly due to the transformation of part of the amidon into dextrine, caused by the action of the azotized substances and lactic acid.

By the following process, a much larger quantity of amidon may be extracted from the farina.

A paste is made, containing 40 to 50 parts of water for 100 of farina. This is kneaded to render it homogeneous, and then left to stand for about a half hour in summer, and an hour in winter. It is then subjected to a mechanical washing, which is performed in a kind of semi-cylidrical kneading-trough called *amidonniere,* furnished laterally with two wire cloths through which the amidon can escape. The paste is subjected to the action of fine streams of water, forming a sort of fountain, whilst a grooved cylinder, with a backward and forward movement, rolls the paste against its sides. About 38 kilogrammes of paste may be placed in the kneading-trough (amidonniere), which will take from four to five times its weight of water to wash it.

The gluten which remains in the amidonniere becomes tenacious when the amidon has been completely carried off. As some of the gluten is carried off with the water, the amidon is passed over a silk sieve; in this way a very beautiful amidon is obtained, which is dried by the ordinary process.

However, as the amidon retains some portions of the gluten, the crude amidon is subjected to a fermentation for twenty-four hours, which is brought about by the use of the foam which is collected in the washing vessels, and which probably contains a small quantity of ferment.

The yield of amidon by this method is, as we have said before, very considerable, and has the further advantage of leaving the gluten in the hands of the manufacturer, which will certainly be found useful. It is, in fact, already ascertained, that fresh gluten, mixed with fecula and cooked potatoes, makes bread of good quality.

Gluten, mixed with farina, makes a very good paste for vermicelli and maccaroni. Mixed with bran, it makes bread which can be baked in an oven, and is also fit for feeding cattle, particularly when a little salt is added. Finally, gluten may be used to saccharify the feculas, and for bringing about promptly the fermentation of molasses.

In commerce, amidon of the first quality is required to be of a peculiar form; it is called *amidon in needles.* This characteristic is indicative of great purity; it is owing to the lenticular form of the amidon of wheat, which gives to the grains, when they are in juxtaposition, a certain degree of adherence. The shrinking caused by the desiccation breaks this adherence with uniformity, so that the cakes of amidon on coming from the stove, are composed of needles extending from the circumference to the centre, to a depth of 6 to 8 centimetres.

By the process of washing called *process Martin*, 100 kilogrammes of farina yield 40 to 42 of amidon of first quality, and 18 to 20 of amidon of second quality.

By the process of fermenting called the *old process*, 100 kilogrammes of farina yield from 28 to 30 kilogrammes of first quality, and 12 to 15 of the second quality, and all the gluten is lost.

EXTRACTION OF FECULA.

Fecula is extracted from potatoes, as the *patraque jaune*,* *shaw d'Ecosse*, *la tardise d'Ireland*, *Siberian* and *Ségonzac*.

Before being used, the potatoes should be thoroughly washed, either with the hand, or better, in a cylinder like that used for washing beet-roots. The potatoes are then rasped with a rasp like that used for beets, except that the teeth are shorter, in order that they may more readily tear the cellules which contain the fecula. The rasp is moved with great speed: it reduces into pulp 15 hectolitres in an hour. The pulp falls directly on a sieve, where, by means of a continual jet of water, a separation of the fecula from the skin of the potatoes is effected.

The fecula, carried off by the water, is received into basins, where it is deposited; the deposit is stirred and washed, till the water runs colorless; the fecula is then agitated with a small quantity of water, and passed repeatedly over sieves, sufficiently fine to retain the gravel and earth which may have escaped the washing of the tubers.

After remaining some time, it is decanted; the fecula is put into baskets, of a somewhat conical form lined with

* The names of potatoes as known in France. The patraque is a round potato, with many eyes or buds. The shaw of Scotland and Ségonzac, a round yellow potato, of excellent quality; a variety most extensively grown in France and very productive.

cloth, to drain. In this state, it is taken out of the baskets in the form of loaves, and placed in a dryer in the open air, the plaster floor of which absorbs a still further portion of water. At the end of 6 or 12 hours, the loaves are divided, generally into eight or a dozen pieces; these are placed on shelves made of strips of wood, where they are subjected for three or four days to a strong current of air; after this, the loaves are broken with a wooden roller, and the fecula thus divided is placed in an oven heated by a pipe, bringing in a current of hot air, which completes the desiccation. In this state it is put under an iron roller, which pulverizes it, and fits it for the bolting-cloth.

The fecula, when it is first put in the oven, contains a considerable quantity of water, so that it should be cautiously heated.

Dry fecula, thus obtained, does not contain more than 18 per cent. of water.

GUM.

The term *gum* is given to neutral substances, which exude or flow out from some trees: they are tasteless, insoluble in alcohol, but soluble in water. The most important gum is gum-arabic, used in medicine and for dressing fabrics.

SUGARS.

Saccharine matter is abundant in vegetables. We will describe principally:

1st. The sugar which is found in all saccharine fruits, and which may be artificially produced by different processes, especially by the action of dilute acids on neutral substances, the composition of which is represented by carbon and water: the name *glucose* has been given to this sugar.

2d. The crystallizable sugar met with in sugar-cane, beet root, maple, pine-apple, pumpkin, chestnuts, corn-stalks, and nearly all tropical fruits.

3d. Sugar of milk, which, in its properties, ranks between gums and sugars.

4th. Uncrystallizable sugar.

The distinctive character of sugar is that of undergoing alcoholic fermentation, that is to say, of transforming itself, by the aid of a ferment, into alcohol and carbonic acid.

Further on, we will describe the phenomena which accompany the principal fermentations.

SUGAR OF MILK.

This substance is met with in the milk of the mammiferous animals, and even in that of the carnivorous animals, kept exclusively on meat regimen.

To extract it, the milk is treated with an acid which coagulates the casein; the liquor, which holds in solution the sugar of milk, is filtered, and from this, by properly evaporating, crystals of sugar of milk are obtained. The liquor is often clarified with animal charcoal. In Switzerland, sugar of milk is prepared by the evaporation of the whey, after the cream and cheesy portion which are used in the manufacture of Gruyere cheese are separated.

Pure sugar of milk presents itself in prismatic crystals. That which is found in commerce is generally in crystalline masses, compact, and semitransparent. It has a sweet and agreeable taste; it is to its presence that milk owes its sweet taste.

The fermentation of sugar of milk constitutes one of the most curious points in its history; the results vary according to the nature and state of the ferment. Thus, when fresh milk is raised to a temperature of 40° C., the casein contained in the milk acts as an alcoholic ferment, and transforms the sugar of milk into alcohol and carbonic acid. It is probable that the sugar contained in the milk is transformed into glucose before fermenting. But if the milk is exposed for a time to the air, and the casein is allowed to alter itself, this body then acts in an entirely different manner on the sugar of milk; it makes it undergo an isomeric modification, and changes it into lactic acid. Some animal matters, altered in the air, may produce the lactic fermentation in presence of sugar of milk.

At one time, when the price of sugar was very high, sugar of milk was used for adulterating raw sugar. This may be detected by treating the sugar with alcohol at 33°, which only dissolves the sugar, and leaves the sugar of milk.

Glucose, Grape Sugar, Sugar of Amidon.

Glucose exists ready formed in the vegetable organization. It may be extracted from honey; it is found in all acid fruits, and principally in the grape; it is this which forms the white and crystalline substance which covers prunes and figs. It is generally obtained by subjecting neutral bodies, lignine, amidon, the gums, and sugar of milk, to the action of feeble acids. Thus, glucose is found in the animal organization; it exists in the urine of diabetes.

In comparing the composition of glucose with that of amidon and lignine, it is seen that glucose does not differ from the neutral bodies we have named, except in the elements of water. Amidon, lignine, and the gums, in changing into glucose, simply undergo a hydratation.

Glucose separates itself from water in small mammellated crystals; its taste is feebly saccharine. Two and a half parts of glucose sweeten only equal to one part of cane sugar. Alcohol dissolves glucose more readily than cane sugar. Heat, at about 60° C., softens glucose; at 100° C., it is transformed into a yellow deliquescent mass; at 150° C., it caramelizes. It is less soluble in water than sugar cane; it requires for its solution, one and a third times its weight of cold water. It is soluble in alcohol, even when absolute, though in small proportions, and will crystallize from it. Glucose is transformed, under the influence of a ferment, into alcohol, carbonic acid, and water.

MANUFACTURE OF GLUCOSE, AND SYRUP OF FECULA.

Glucose exists ready formed, as we have before said, in most saccharine fruits, and particularly in the grape. In the years 1810, 1811, and 1812, in many parts of France, it was extracted from the white grape in the form of syrup, which was intended as a substitute for cane sugar, the price of which was very high. The acids of the grape were saturated with chalk, and sulphite of lime was added to the *must*, to prevent fermentation; this must, filtered, was rapidly evaporated to 20° of the areometer; it was left to cool and settle for twenty-four hours, to deposit the salts of lime which it might contain; it was then decanted, and subjected to a new evaporation to bring it to 32°. This syrup, brought to the density of 35°, deposited crystals of glucose.

This syrup of grapes, properly prepared, was amber-colored, clear, and agreeable to the taste; though much less saccharine than the syrup of the cane, it was sufficiently so to answer for most domestic purposes, and for use in the hospitals of the south, where this syrup was made in large quantities.

We will recall to mind here, that, by the reaction of diastase on amidon by means of sprouted barley, the brewer produces the quantity of glucose which is necessary for the manufacture of beer.

Glucose is obtained in the arts by making sulphuric acid react on fecula.

Water, at 50° C., holding fecula in suspension, is poured into a covered basin containing water acidulated with a hundredth part of sulphuric acid, and so heated with steam as to obtain a temperature of 100° to 104° C.

This operation ought to be carried on, so that the temperature does not fall, that the reaction of the acid on the fecula should be almost instantaneous, and that starch should not be formed. Ten kilogrammes of sulphuric acid and 1000 kilogrammes of water are used for 500 kilogrammes of fecula.

When all the fecula has been poured into the basin, the liquor should remain clear, and after twenty or twenty-five minutes of ebullition, the conversion of the fecula into glucose is accomplished.

The liquor is drawn off and chalk is gradually added to saturate the sulphuric acid, it is left to repose until the sulphate of lime is precipitated, then decanted and rapidly evaporated to 32°; during this evaporation the liquor deposits sulphate of lime, from which the syrup is cleared by allowing it to settle in reservoirs. This syrup is brought by a rapid evaporation to the density of 45°; on cooling, it forms into a white amorphous mass, which constitutes the sugar of amidon of commerce. In this state it is soapy, and dissolves with difficulty in water. Glucose is obtained in the large way pure and granular by following a process due to M. Fouchard. Instead of evaporating the syrup to 45°, the evaporation is stopped when the liquor marks 30° only; it is then run into casks having only one head, which is pierced with holes stopped with plugs: after some days the crystals of glucose are seen through the liquor; these crystals go on increasing, and in a short time the plugs may be taken out and the molasses drained off.

When the draining is complete, the crystals are removed to a stove furnished with plaster shelves which absorb the syrup; they are then dried by a current of air at 25° C.

Glucose thus granulated is much purer than glucose in mass. In fact, the foreign substances which give to glucose in mass its fat aspect, and frequently bitter taste, are carried off in the molasses.

It is doubtful, however, whether anything can be gained by this method of making glucose to adulterate raw sugars.

Sugar of Diabetes.—The urine of diabetics often contains a considerable quantity of glucose. Diabetics may pass as much as eighteen litres of urine in twenty-four hours; each litre often contains 85 grammes of sugar. To extract this sugar from the urine, it is sufficient to evaporate it in a water-bath, and to treat the residue with boiling alcohol of 94°, which dissolves the glucose. The alcoholic liquor is decolorized with animal charcoal, and brought to a syrupy consistence, when it shortly gives up crystals of glucose which are purified by repeated crystallizations.

CANE SUGAR.

Cane sugar has been known from the earliest times. It exists in sugar-cane, beet root, the sap of the maple tree, pumpkin, corn-stalks, chestnut, horse-chestnut, turnip, carrot, cocoa, and a great number of tropical fruits.

Cane sugar crystallizes in rhomboidal prisms. It is soluble in a third of its weight of cold water, and in every proportion in boiling water. Weak alcohol dissolves it readily, but it is very slightly soluble in anhydrous alcohol. It fuses at about 180° C. Cane sugar is not precipitated from its solution, either by the neutral acetate, or the subacetate of lead; in analysis, this property is made use of to separate it from different organic bodies, which are precipitated by these reagents.

When a concentrated solution of cane sugar is rapidly evaporated, a thick liquid is obtained, which forms in mass when poured on a cold body; this is called *barley sugar.* This sugar is transparent and amorphous; but when kept for some time in the air, or in bottles hermetically sealed, it becomes opaque and crystalline, and passes back to the state of common sugar. This crystallization is prevented by the addition of a small quantity of vinegar.

Sugar becomes phosphorescent by striking it; when it is

rubbed for a long time by a hard body, it acquires a disagreeable taste.

Water acts upon sugar; under the influence of heat, it hydrates, and is transformed into glucose. This fact is important in the manufacture of cane or beet sugar. When the syrups are maintained at a prolonged ebullition, they are transformed for the most part into glucose.

The acids also change cane sugar into glucose. Cane sugar is rapidly carbonized by heat; at 215° C., it changes into *caramel.*

Cane sugar does not appear to ferment directly, as has been shown by M. Dubrunfaut; but under the influence of a ferment, is first transformed into glucose. It is this last sugar which, in presence of a ferment, gives rise to water, carbonic acid, and alcohol. Sugar, influenced by different ferments, may undergo four different kinds of fermentation.

1st, In presence of the yeast of beer, it changes into carbonic acid and alcohol. This transformation constitutes *alcoholic fermentation.*

2d, If yeast of beer, first boiled in water, is poured into a solution of sugar, it produces, according to the observations of M. Desfosses, a peculiar fermentation called *viscous fermentation;* the juice then changes into a neutral substance, which renders the water viscous, and which appears to be represented in its composition by carbon and water. The viscous matter is almost always accompanied with mannite.

3d, A great number of azotized organic substances, such as albumen, fibrine, and casein, which have undergone in the air the commencement of an alteration, may cause cane sugar to undergo an isomeric modification, and to change into lactic acid. This transformation constitutes the *lactic fermentation.*

4th, Finally, sugar, in presence of ferments altered in the air, undergoes the *butyric fermentation.*

BEET-ROOT SUGAR.

All the beet-roots contain sugar; but generally (and especially in France), it is extracted from the *white* or *Silesian beet-root.* It is this which gives the purest and most dense juice, and that which is the most easily worked. The density of this juice is ordinarily from 6° to 7°. The *red beet-root* is not used, on account of its color. The *field beet-root* is very large, but its juice is watery, and hard to work.

The beet-root, when healthy, seems to contain only crystallizable sugar. The efforts of the manufacturers should be directed to obtain the largest quantity of crystallizable sugar, which reaches as high as 10 per cent.

The loss which is suffered is owing to an alteration of the sugar during the evaporation of the juice.

Beet root of good quality contains from 10 to 12 per cent. of sugar, though but from 4 to 5 per cent. are generally extracted, and rarely as high as 7 per cent. This manufacture then, it is seen, is open to important improvements.

The beet roots, taken from the pit or warehouse, are deprived of their neck and radicles, and then put into a cylindrical washing machine divided into spaces of three or four centimetres, partly plunged into a vessel of water. The cylinder is inclined, so that, by the rotatory motion which is given to it, the beet roots pass its entire length, and are thus cleaned of the dirt.

The beet roots when cleaned are rasped in a cylindrical machine armed with saws. The resulting pulp is then subjected to the action of hydraulic presses: after having been pressed, they still contain 15 to 20 per cent. of juice. They should be pressed in as short a time as possible, as the juice readily ferments.

The extraction of the juice by a methodical washing has often been tried, but has not given good results.

The objection to this process is the introduction of an

additional quantity of water, requiring, of course, a longer time for the evaporation of the juice.

The treatment of the juice comprises six different operations, which succeed each other in the following order: *defecation, first filtration through black, first evaporation, second filtration through black, boiling, crystallization.*

The defecation is intended to remove the acids, albumen and viscous matter, which prevent crystallization and alter the sugar.

The boilers for defecating are composed of round-bottomed, cylindrical vessels; they have a double bottom of copper for receiving high-pressure steam.

The juice is rapidly raised to the temperature of 60° C. 50 grammes of lime to the hectolitre of juice are then added: the lime should be slaked and dissolved or suspended in water.

The liquor is raised to boiling; a scum is formed, which collects in a more or less consistent form on the surface of the liquid. Another portion of the foreign matters sinks to the bottom.

The lime, reacting on the ammonaical salts, disengages ammonia. The liquor is allowed to settle, and is then run upon the bone-black filters.

This operation is done with *Dumont's filters*, or filters with black in grains; it consists of a copper box, having a false bottom pierced with holes: a moist cloth is spread over this, and the filter is filled with granular black well packed, and over this is placed a cloth covered with a plate pierced with holes. The use of this filter is a great improvement in the manufacture of beet sugar. The juice, in passing through the Dumont filter, is clarified and decolorized.

Of late, the copper boxes have been replaced by tanks of iron, which will contain 40 hectolitres of juice.

The filtered and decolored juice is then subjected to a first evaporation. This was formerly done with an open fire. But in this way a great proportion of the sugar was altered.

For some time the evaporation was effected with low steam, but by this method the evaporation was slow, and much of the sugar was transformed into glucose; now high steam is used. The evaporation is generally done in open pans, though in many cases, to hasten the evaporation, it is done in vacuo.

When the juice is reduced by evaporation, it is a second time passed over Dumont's filter to decolorize it anew, and to free it from matter deposited during the evaporation, and the second evaporation is then proceeded with.

This is done in the open air, or better, in vacuo, which is produced by an air-pump, or by the condensation of the vapor (which escapes from the sugar) in passing through cylindrical refrigerators.

These cylinders are generally cooled by the juice, which is thus heated and undergoes a first evaporation.

Then, in order to separate the particles which are in suspension in the liquid, it is first filtered through *Taylor's filters.* These filters are formed of large bags 50 centimetres wide and 1 metre long, which are enclosed in a linen sheath 18 centimetres wide, to keep the bag in folds.

The liquor is then passed through Dumont's filters to decolorize it; after this comes the boiling, which ought never to be done with a direct fire.

Boiling in vacuo is preferred. This is done between 112° and 115° C., and the syrup should mark 43° on the areometer of Baumé.

The degree of evaporation is judged of, by taking on a skimmer a drop of syrup and pressing it between the forefinger and thumb, and then rapidly separating them: the syrup forms a sort of string, the length of the thread and the form of it at the point where it breaks, show the degree of concentration: this is called the *string test.* The *bubble test* is performed by blowing on some syrup held in a spoon; if the boiling is done, the cool air of the breath detaches a large number of bubbles, which solidify in the form of feathers.

After the syrup is concentrated, it is drawn off into a vessel called a cooler, where, as the temperature abates, the crystallization manifests itself, by forming a crust on the surface, which is broken up and stirred through the mass, causing a general and uniform crystallization.

When the syrup has lost its transparency in consequence of the formation of crystals, it is poured into large conical moulds of earthen-ware, or of galvanized iron with stoppers, or into large rectangular vessels.

After several days, when the crystallization is complete, the moulds or forms are placed on pots, or better, on metallic troughs; the stoppers are then taken out, in order that the uncrystallized syrup may drain off, and when the dripping has ceased, the sugar is taken out and sent to the refiners under the name of *raw sugar.*

The syrup which has dripped out still contains crystallizable sugar, but the evaporation which it should undergo requires great precautions.

This syrup, deprived of its sugar by repeated boiling and crystallization, and brought to the density of 35°, is sold as *molasses* to the distillers.

When a whiter sugar is required, the sugar undergoes a process, before being sent to the refinery, called *liquoring;* this consists in washing the sugar with syrups sufficiently dense not to dissolve the sugar but to carry off the foreign bodies which it contains.

The upper surface of the mould is loosened up, and the liquor (that is syrup of 33°), is poured over it; this operation is repeated until the sugar becomes sufficiently white and drained.

REFINING OF SUGAR.

Cane and beet root sugar in general cannot be put into the market till they are refined, that is, purified, from the foreign matters which they contain, the weight of which in general, including moisture, amounts to 10 or 15 per cent. of the weight of the sugar.

These foreign matters are water, sand, earth, organic remains, coloring-matters, and those producing smell, uncrystallizable sugar, and some salts having for base lime, potash, soda, magnesia, and ammonia.

Water always forms the greater part of the foreign matters contained in raw sugar; it is seldom that all the other substances combined amount to more than 3 or 4 hundredths.

The first operation in refining is the melting of the sugar; this is done in a large boiler (blow up) heated with steam: the syrup is then clarified by adding 3 or 4 per cent. of fine black, and a small quantity of albuminous matter (blood, or the white of egg). This liquor is then boiled by steam circulating through pipes. The albumen, in coagulating, collects together the black, and completely clarifies the syrup.

The decolorized syrup is passed into steam-pans, provided with apparatus for producing a vacuum, in order that it may be rapidly evaporated; it passes from these pans into metallic crystallizers.

When the crystals begin to form, they are detached from the sides of the crystallizer with a large wooden stirrer, without breaking them: this operation is gone over three times, when it is filled out into moulds, which are placed in apartments, the temperature of which is 20° to 30° C. The syrup soon becomes covered with a crystalline crust, which is repeatedly broken up by wooden tools.

After 15 to 20 hours of rest, the moulds are removed into

another apartment to drain, they are placed on a false floor perforated with holes, and drain into troughs lined with zinc, from which the syrup is conducted to a common reservoir.

When the draining is finished, which is known by knocking out the loaves, the first claying is performed.

Claying is based on the same principle as the purification of saltpetre by means of water saturated with pure nitrate of potash. It consists in treating the sugar still impure, with a syrup of pure sugar, so concentrated that it will dissolve no more sugar, and that it operates only on the foreign matters contained in the raw sugar.

The claying only differs from *liquoring* in a peculiar disposition of the vessels in which the two operations are performed.

In claying, the sugar in the mould is covered with a layer of clay saturated with water, of about a centimetre in thickness. This clay gradually gives up the water which it contains, which dissolves the layer of powdered sugar on the mould. The syrup which results runs through the mass, pushing before it the molasses which colored the raw sugar. A single claying is not sufficient to whiten the raw sugar, from the beet; it must be repeated three or four times. When the clay, nearly dry, can be lifted off in one piece, the purification is terminated.

The first claying lasts about ten days. It is performed in an apartment not heated. The second claying lasts seven or eight days. These two clayings are sufficient when the refining operations have been properly done. Care should be taken that the clay used contains neither lime nor sulphur, nor sulphate of iron.

After being kept for some days exposed to the air in the moulds, the loaves are placed in stoves to be dried. The syrups which have drained out during these different operations are again boiled, producing sugars of a second quality, called *bastards* and *lumps*.

In some sugar-houses, the raw sugar, before being sent to the refineries, is subjected to the operation of *liquoring*, which consists in washing it directly with liquor, that is to say, with a syrup marking 33° on the areometer. This liquoring is generally done in rectangular boxes, the bottom of which is a metallic cloth, and in this the sugar in powder or small crystals receives the liquor more or less white.

It is thus brought to a state of decoloration and purity, from which very white loaves of sugar may be prepared, by re-dissolving the products and filtering them anew over coarse black, before boiling again.*

---

* The process of refining, as practised in the best refineries of the United States, has for its object to free the raw sugars brought from Louisiana and the West Indies from the impurities which they contain, thus rendering them fit for consumption. The sugar is first dissolved in a large open iron vessel called a *blow-up*, by means of steam and hot water; it is then, while in this blow-up, clarified by means of albumen and fine bone-black and some lime-water. This mixture, after having been thoroughly agitated, is then skimmed to remove the impurities which float on the surface; this being done, it is run through bag-filters, which retain the grosser impurities, preparatory to its being filtered in the char cisterns. The char cisterns are large, deep, iron vessels, packed with *bone-black* or *char*. In these, this solution of sugar (*the liquor*) becomes decolorized and thoroughly clarified; after having percolated and filtered through the char, this liquor should be perfectly limpid, colorless, and brilliant, when it will be ready for concentration in the vacuum-pan. In the pan it is boiled by steam at a low temperature, and so concentrated as to be ready for crystallization. At this period of the process, the mass, which is in a semifluid state, is run off into a large iron vessel called a *cooler*. This vessel is provided with a double bottom supplied with steam, so as, when properly regulated, to prevent the too sudden cooling of this mass of crystallizing sugar.

Here the crystallization of the sugar goes on, and the mass is constantly stirred with wooden stirrers, so as to equalize the crystallization, or *graining*, throughout. From the coolers, at the proper time, the sugar is filled out into moulds for completing its crystallization

Sucre Royal.—This name is given to sugar highly refined, perfectly white, and in brilliant crystals, which is obtained by subjecting loaf-sugar of a fine quality to a new clarification with the white of egg, to black, and two clayings.

Stamped Sugar.—To obtain stamp loaves, brass moulds are filled with lump sugar in powder and slightly moist, the sugar is packed by three blows repeated on the base of the form; the loaf then has sufficient body to be knocked out and put into the stove.

This sugar is far from having the quality of refined loaves, for in making it sugar of the second quality is used.

---

and solidification, to be then drained and *liquored.* The moulds are conical vessels made of iron painted, having an orifice at the tip to admit of the draining. These orifices are fitted with plugs before the moulds are filled, and when ready for draining the plugs are removed, and the fluid, uncrystallized portion of the sugar drips out into suitable vessels, over which the moulds are placed. The sugar in the moulds is then liquored.

The principle of this process of liquoring is as follows: the liquor, being a *saturated solution of refined sugar*, is poured over the surface of the sugar in the moulds, (which has been first loosened up and levelled) and in percolating or descending through the sugar, drives before it the molasses or colored syrup, which coats the grains of sugar: this liquor, being already saturated with sugar, does not in its passage dissolve any of the sugar, but simply acts by displacing the dark-colored syrup which envelopes the grains. This liquoring is repeated until the sugar becomes of the required whiteness, and then, after draining sufficiently, it is knocked out of the moulds, to be shaved, or ground, or crushed, as may be required, or it is dried in an oven to form loaf-sugar—loaf and crushed sugar are the more highly refined qualities.

Syrup is a solution of sugar in water. Molasses, properly speaking, is a solution of uncrystallizable saccharine matter in water, containing also impurities. Syrup from the sugar-houses is generally a mixture of syrup and molasses, and is constituted of the last drainings from the moulds. The first drainings or drips are mostly boiled again to make the yellow sugars.

It gives solutions which are colored, so that it has not the same value in the market as refined loaf-sugar.

Sugar-Candy. — The white, light colored, and brown sugar-candy which is exposed for sale, is obtained from syrups of different shades: the syrups should be first, second, and third proofs.

The syrups are then poured into vessels of copper, of the contents of 10 to 12 litres, and in the form of a cone or truncated pyramid; strings are extended across these vessels for the attachment of the crystals, and those which form on the strings are preferred to those adhering to the sides of the vessel. These vessels, placed first in a stove where they are filled, are there subjected to a temperature of 75° C. for eight days, causing the evaporation of the syrup and formation of the crystals. These crystallizers are withdrawn, the crust broken on one side, and the mother-water decanted, when the crystals are placed to drain; the vessels are placed for a short time in boiling water, and the sugar-candy is then detached by a light blow, which knocks it out in the shape of the vessel. The sugar is then dried for twenty-four hours in a stove.

The syrups resulting from this draining, unless too highly colored, as those from brown candy, may be boiled again, to form more candy. If they are too much colored, the syrups are brought to the point, where the sugar has lost nearly the whole of its water of crystallization; the vessel is then taken from the fire and briskly agitated with a wooden spatula, and thus is obtained a brown sandy sugar, used in making common chocolate.

Barley Sugar, Apple Sugar. — Sugar sold under these names, has for a long time been in commerce containing neither barley nor apple.

When first prepared, this sugar is of a light brown color and transparent. Some days after its preparation, it turns and becomes opaque. The point to which the syrup is reduced by boiling is judged of, by first dipping the finger in water, and then in the syrup, and replacing it rapidly in cold water. If it has attained the proper point, it should detach itself from the finger readily, and easily break between the teeth. It is then poured on a stone slightly greased, and as soon as it can be worked with the hand, it should be kneaded to prevent crystallization; it is cut into pieces and rolled on a stone into sticks of a uniform size.

Sugar sold as *sucre de pomme*, is prepared with whiter sugar than that used for barley sugar.

ALCOHOLIC FERMENTATION—PANIFICATION.

There are some organic nitrogenous substances, of an albuminous nature, which rapidly alter when exposed to moist air. These bodies possess the curious property of involving in their decomposition, other organic bodies with which they are placed in contact. This decomposition of an organic substance under the influence of another body which only acts by its presence, has received the name of *fermentation*. The azotized body which causes the decomposition, is called *ferment*.

The ferments are produced in a great number of circumstances. The name *alcoholic fermentation* is given to the transformation of sugar, under the influence of a ferment, into alcohol and carbonic acid. When grape juice is exposed to the air, the albuminous matter which it holds in solution soon becomes changed into a ferment, which then causes the transformation of the sugar into alcohol, and carbonic acid. It is known, in fact, that the juice of fruits soon becomes foaming, and that their saccharine taste disappears completely, to be replaced by an alcoholic taste. It is thus that wine is produced; and in a general way, all the alcoholic drinks. The best alcoholic ferment, that which most readily changes sugar into alcohol and carbonic acid, is the *yeast of beer*. This body is produced in the manufacture of beer.

Bread-making, or at least the operation intended to make bread light and porous, also depends on alcoholic fermentation.

Bread is made with flour and water, carefully kneaded, put to ferment, and then properly baked. Flour is a mixture of amidon and a soft, elastic, nitrogenous, and very nutritive substance, called *gluten*. Gluten plays a great part in panification; it is this which renders dough light, and allows of its *rising* during the baking.

The *leaven* is the substance added to the dough before baking, and which causes the puffing up and lightness of the mass. It is known to all that unleavened bread is heavy and compact. This leaven is sometimes made of dough left to sour and ferment, and sometimes of the *yeast of beer*, of which we have previously spoken.

Thus, bread is made of a mixture of flour, water, and leaven. It is easy to comprehend now what takes place in panification.

When the dough is left in a warm place, it *rises*, and becomes porous. At this moment, a small quantity of sugar which existed in the flour, or was produced there, undergoes, in presence of a ferment, a true alcoholic fermentation; alcohol and carbonic acid are developed. This last gas not being able to escape, on account of the viscidity of the dough, rises it, and produces the small cavities which are found in well-prepared bread. When the bread is put to bake, a part of the water of the dough is disengaged, the mass solidifies, and remains full of small cavities which are produced by the disengagement of the carbonic acid, and which make the bread white, light, and easy of digestion.

Alcoholic fermentation can only be produced in certain conditions; thus, it does not manifest itself, as Gay-Lussac has proved, except under the influence of atmospheric air. A temperature of 20° to 25° C., favors fermentation, while a temperature of 70° to 80° C. coagulates the ferment, and immediately arrests its action on the sugar. A liquid, then, may be prevented from fermenting, either by keeping it from the contact of the air, or by carrying it to the boiling-point.

Some bodies, by their presence, arrest fermentation; such are concentrated alcohol, essence of turpentine, and creosote: these bodies are called antiseptics.

ALCOHOL.

Alcohol is the product of the fermentation of sugars or saccharine liquids; it is obtained by the distillation of wine, cider, beer, and all liquors which have undergone alcoholic fermentation. The alcohol resulting from the distillation of fermented liquids, without the addition of any drying-matter, always retains some water which must be abstracted from it. In laboratories, to prepare anhydrous or *absolute* alcohol, alcohol of commerce is put into a glass retort, three-fourths filled with pieces of quicklime of the size of a hazel-nut. The alcohol should entirely cover the lime; after twenty-four hours of contact, the retort is placed in a water-bath, the water of which is made to boil, and the distillation is kept up till no more alcohol comes over. This liquid, distilled a second and a third time on fresh quantities of caustic lime, may be considered as absolutely anhydrous.

It has long been remarked that alcohol, kept in porous vessels or in bladders, after a time becomes concentrated. The water by *endosmose* passes more readily through the membranes than alcohol; but this latter liquid cannot thus be brought beyond 95° to 98°. Besides, it takes up from the bladder a fat substance, and gives to it a disagreeable taste and smell.

There are apparatus for continuous distillation, in which the alcohol successively undergoes several rectifications, which bring it to an almost complete anhydrous state.

Alcohol is a colorless liquid, very fluid, of a burning and caustic taste; when concentrated, it is poisonous. The injection of anhydrous alcohol into the veins, causes immediate death; in this case, it undoubtedly acts by coagulating the albumen of the blood; it is lighter than water. Alcohol boils at 78° C. It cannot be solidified by cold. Alcohol is soluble in water in all proportions. It is very inflammable; its flame is yellow, and often deposits carbon; weak alcohol burns with a blue flame.

Alcohol is transformed into acetic acid in presence of oxygen, under the influence of black of platinum, or the ferments.

Some salts which are soluble, or even insoluble in alcohol, communicate to its flame different colors, which serve to characterise these salts; thus its flame is colored purple by the salts of strontium, red by salts of lime, and green by salts of copper.

Alcohol possesses an undoubted affinity for water; a very sensible elevation of temperature is recognized in the mixture of these two liquids. Alcohol absorbs moisture from the air, and becomes rapidly aqueous. Most organic bodies which are plunged in alcohol, lose a part of their water, and may then be preserved without alteration.

The name *brandy* is given to mixtures of alcohol and water, formed of about equal parts of each.

Spirits of commerce, is alcohol containing less water than brandy. The strength of spirits is estimated nearly always by the quantity of real alcohol which it contains; this is not, however, the case with brandy; its value is not always proportional to the quantity of alcohol which it contains: it often depends upon its growth and age. The *spirits* of commerce was for a long time tested by throwing some on gunpowder, and then setting it on fire; when the powder took fire, the spirit was judged of good quality. This proof was not attended with exactness.

To subject spirit to the *Holland proof*, it was put in a bottle, and rapidly shaken; if it *beads*, it then marks 19° on the areometer of Cartier. It is to this spirit that all those of commerce are referred. It has become a commercial type, and contains about one-half its volume of water.

The name *three-sixths*, is given to a spirit, 3 volumes of which, with 3 of water, give a spirit which beads. In France, the legal alcoholometer is that of Gay-Lussac; it enables us to appreciate directly the quantity of absolute alcohol which a liquor contains. The experiment should be made at 15° C; if the liquor is not at this temperature, it is easily brought there by heating it with the hand. In fact, Gay-Lussac has made tables of correction, which, by means of the alcoholometer, enable us to determine the strength of an alcoholic liquor, taken at different temperatures.

The principle of the graduation of this instrument is very simple. The alcoholometer, placed in absolute alcohol, sinks to the point where 100° is marked; placed in pure distilled water, it stops at a point which is the zero. This scale is then divided into 100 parts by the aid of mixtures of water and alcohol in known proportions. It must not be forgotten that this instrument indicates the relations of volume, and not of weight. A spirit which marks 50°, contains 50 per cent. of pure alcohol.

The areometer (test liquor) of Cartier, is also used in commerce; in this instrument, distilled water marks 10°, and anhydrous alcohol 44°.

Alcohol has many important uses. It is used in laboratories as a dissolvent and combustible; it is used in perfumery for dissolving essential oils, which are insoluble in water. Cologne-water is a solution of several essential oils in alcohol. When cologne-water is diluted with water, it becomes turbid, because the essential oils which are insoluble in water, are precipitated. The most simple recipe for making cologne-water, consists in pouring into a litre of alcohol, at 33°, twenty-four drops of the following essential oils: orange, cedrat, citron, bergamot, and rosemary.

Alcohol is used in preparing the fulminating powders,

called *fulminates of silver and mercury*. The fulminate of mercury is used in the manufacture of *percussion priming*, or *caps*. To make the fulminate of mercury, 1 part of mercury is dissolved in 12 parts of nitric acid at 34°, and to this solution are added 11 parts of alcohol at 36°; the mixture is heated in a water-bath, and shortly there is precipitated a white crystalline powder, which is the fulminate of mercury. The fulminate of silver is prepared in the same way, only the solution of silver in nitric acid is substituted for that of mercury.

The fulminates are very dangerous to handle; friction between two hard bodies, or a slight elevation of temperature, are sufficient to explode them. These explosions are very powerful, and may give rise to the most serious results.

Ordinary percussion caps are made of a small capsule of copper, in the bottom of which is introduced a mixture of fulminate of mercury and gunpowder; each cap contains about 16 milligrammes of the fulminate.

Alcohol distilled with the acids produces compound ethers, such as acetic, nitric, muriatic, &c., which are remarkable for their agreeable odor, and are sometimes used in medicine.

ETHER.

This compound is indifferently called *ether*, *sulphuric ether*, *normal ether*.

Ether is obtained by making sulphuric acid react by the aid of heat on alcohol.

The operation is easily performed by the apparatus of Scottman, pharmaceutist of Berlin.

This apparatus is composed, 1st, of a large glass tubulated retort, which is placed in a sand-bath to the depth of the mixture within the retort. 2d, of a tube which separates the retort from the condenser. 3d, of a matrass which receives the end of the tube, having its neck connected by a cork to a condenser. (The serpentine of an ordinary alembic may be advantageously used). Near the retort is a large bottle filled with alcohol, with a tubular on one side at the bottom, from which a tube passes through the tubular of the retort into the mixture of alcohol and sulphuric acid. This bottle is to be placed above the retort.

The apparatus being arranged, the alcohol and sulphuric acid are mixed in an earthen vessel, using the precaution to introduce the alcohol first, and gradually to pour on the acid while it is constantly agitated. This mixture, when it has become cool, is introduced into the retort, and made to boil: a cock, fitted to the glass tube, is then opened, and the alcohol is made to flow into the retort continually, so that during the operation it may replace that portion of the product which has been distilled.

According to M. Souberain, the proportions of alcohol and acid which are best suited to the preparation of ether are as follows:

Alcohol at 32° alcoholometer. . .	70 parts.
Concentrated sulphuric acid. . . .	100 "

In this reaction the sulphuric acid dehydrates the alcohol and transforms it into ether.

Ether is a colorless liquid, very fluid; its odor is strong and its taste burning. It is entirely neutral.

It is lighter than water, and boils, according to Gay-Lussac, at 35·6° C.

Ether in evaporating produces a great degree of cold; the temperature falls many degrees below zero. Ether may be solidified by cold. The vapor of ether is heavy; it frequently happens, that in falling on inflamed bodies or lighted coals, it takes fire, and causes serious accidents. Ether in burning produces a beautiful white, fuliginous flame, much more illuminating than that of alcohol.

Vapor of ether mixed with oxygen or the air, forms an explosive mixture.

Ether is a valuable solvent in analysis; it is useful in preparing a great number of organic bases, for purifying the fat bodies, for dissolving caoutchouc, &c.

Ether is used in medicine, chiefly as an anti-spasmodic. Applied to the forehead it is useful, from the cold it produces by its evaporation, in the treatment of headaches.

The history of ether is connected, as is well known, with one of the most beautiful chemical applications which have been made in the art of medicine. When inspired, it destroys sensibility, and allows the surgeon to perform operations, often without pain to the patient.

An ethereal liquid called *chloroform*, is successfully used with this intention.

MANUFACTURE OF WINE.

Wine is the product of the fermentation of the must or juice of the grape. This juice, when first expressed, is saccharine; it soon loses this taste, to contract another which is acid and alcoholic.

The quality of the grape necessarily influences that of the wine. All kinds of grape, though uniting the proper conditions of exposure and maturation, are not susceptible of producing a wine fit to drink, and which will keep. It has been vainly attempted in years of abundance to convert the best chasselas grapes into wine; but they only furnish a flat wine, which is apt to turn sour.

The grape for making wine should come from an exposure, where it receives the action of an elevated and regular temperature; a mixed soil, rather silicious and gravelly than clayey, appears to be one of the conditions of good quality for the grape. The vine should have but little stable manure. The branches laid in the ground, called *layers* (provins), to obtain new stocks, ought alone, so far as possible, to be manured; the rest of the vine should be dressed with earth.

Finally, long experience has shown that wine produced from grapes, the yield of old vines, is better than that from young vines. Thus, in vine-growing countries, the proprietors preserve the old vines, and do not give them up till they are past yielding crops.

We should remark that these general observations only apply to those growths in which the quality rather than the quantity is sought.

The maturity of the grape is also a condition essential to the quality of the wine. This condition is accomplished most generally in wine-growing countries, which are in the south of France, especially when the vintage takes place in the month of September, and when the crop is not gathered too early.

The crop is gathered as far as possible in a dry time, and when there is sun; the grapes collected in baskets, are then emptied into casks placed at the vines. In some places, the grapes in the casks are crushed with a stick having three teeth at its extremity, and the grape is added till these vessels are full; in this way, fewer casks are required.

This process is not practicable in the vineyards when the weather is warm, as, when the grape is crushed, fermentation takes place more readily. This observation is important where black grapes are used for making white wine; because the alcohol produced by the fermentation, would react on the coloring-matter of the skin, and color the wine.

Manufacture of White Wine.—When the grapes are brought to the cellar or vault, the crushing is completed. If the grapes are black, they are raked on a wooden frame placed over a wooden or stone tub; this frame is furnished with a net made of stout cord, having the meshes arranged so that only the seed of the grape will pass through. When white grapes are used, the casks are emptied into the tub, and the crushing of the grapes is completed either with the three-toothed fork, or by having them trodden by a man who goes into the tub; while this is doing, a cock is opened through which the must flows out into casks. Over the orifice of the cock within the tub, is placed a basket or grating to intercept the skins and seed.

When the must is done flowing, the skins, &c., are taken to a press. There is a difference between unpressed wine (*mère goutte*), and that which comes from the press: the last is considered inferior in quality; it is very commonly kept to be mixed with the must from black grapes too highly colored.

If it is desired to preserve the wine without color, it must be kept as much as possible from the air; the casks, if they are new, ought to be limed with milk of lime at 40°, and then carefully rinsed out; if old casks are used, they should first be fumed with sulphur: the object of these two operations is to purify the casks from all cause of coloration.

The must, introduced into the cask, there undergoes the alcoholic fermentation which manifests itself by the disengagement of carbonic acid gas, and by the formation of a white scum, holding in suspension a substance of the nature of yeast. The casks should be so filled, that the gas and scum may escape without the loss of liquid: they should not be bunged till the wine begins to clarify; this should be done with the hand, so that the bung may be thrown out should the disengagement of gas manifest itself.

When the fermentation of the must is completed, those white wines which are intended for immediate use are drawn off, while those intended to be kept are not drawn off till after the first frost. The casks should be bunged accurately, and the leakage refilled.

This operation is one of the conditions essential to the preservation of all kinds of wine; it requires great care during the first weeks of its preparation, but after a time it suffices every month.

When there is not enough of good wine to fill up the cask, washed and non-effervescing river-sand is sometimes substituted. The diminution of volume of the wine during its preservation depends particularly on the more or less perfect ventilation of the cellars or vaults.

White wines sometimes suffer from a disease called the *grease* (viscous fermentation), occasioned by an azotized substance which M. François called *gliädine.*

The white wines which contain this azotized substance cannot be converted into sparkling wines till the gliädine has been precipitated by a certain proportion of tannin, which, according to M. François, is 1 gramme or ½ gramme to the litre.

Manufacture of Red Wine.—The black grape is brought from the vineyard to the cellar, crushed or entire. For wines of quality, they are sometimes kept in this last state, for in some places, where the ripening is unequal, the grapes are picked or sorted.

When the grape is crushed, as we have above described for the white grape, the stems are taken off.

The wine-growers differ as to the time when the stems are to be taken off; this is sometimes done before and sometimes after fermentation.

The grape, crushed and properly trodden, is put into a tub, the capacity of which varies from 30 to 45 hectolitres, where gradually it loses its saccharine taste. A bubbling up is heard, the stones, seeds, and skins collect on the upper surface, where they form a sort of crust, called *chapeau,* which preserves the liquid from contact of the air. In some wine countries, this fermentation goes on in open vessels; in others, the tub is covered with a lid, with a tube for the escape of the carbonic acid gas; finally, in some places, a false lid with holes is used, fastened down below the level of the must: this lid keeps the crust (chapeau) constantly immersed, and in contact with the fermenting liquid, which by means of the alcohol dissolves the coloring matter of the skins.

It is of use to keep the crust from contact with the air; for, when the fermentation is slow, acetic acid may be formed, which injures the quality of the wine. When the fermentation is sensibly abated, the crust is made to descend into the tub, to be submitted to the action of the fermented liquid.

The fermentation takes place more or less slowly, according to the temperature. When it is feared that the slowness of the fermentation will injure the quality of the wine, it is often made active by heating a portion of the must to raise the temperature of the mass.

When the liquor becomes of a red color, they then proceed to *tun it*.

This operation is effected as for white wine, by separating the unpressed from the pressed wine, at least for the best wines; the tunning and the preparation of the casks require the same care. The bungs should be put in more firmly than for white wine, for the fermentation is for the most part finished in the tubs. The replenishing from leakage should always be exactly done, and in general the first drawing off should not be done till after winter.

Sparkling Wines. — The most sought after are those of Champagne. The sparkling wines of Burgundy and of Touraine are of fine quality, but they have not the bouquet nor the lightness of Champagne wine.

Carbonic acid gas may be compressed into the white wines, rendering them sparkling in the same way as in gaseous waters; but the foam is temporary, and not like that proceeding from the reaction of the elements of the must on themselves.

Champagne wine is prepared from the black grape. None of the precautions we have above pointed out for preserving choice wines should be neglected. Thus, the grapes should be carefully sorted, they should be pressed with great celerity, and only the unpressed juice used for the white sparkling wines. The wine from the press, which is already colored during this operation, is converted into rose-colored wine by the addition of a suitable quantity of turnsol, where the rose tint is not sufficiently pronounced.

The must intended for the white Mousseux is treated as above described for the manufacture of white wines. At the time it is put into the tun, a litre of eau de vie de Cognac is added to every hundred litres of must. It is generally in December, in a cold dry time, that the first racking and clarifying is performed. The clarification is done with 16 grammes of isinglass for 200 bottles; the isinglass is dissolved in a sufficient quantity of white wine. The wine is racked off a second and a third time in January and February. The third racking is followed by a second clarification. Finally, in April, the wine is placed in bottles which ought to contain a greater or less quantity of a saturated solution of sugar-candy in white wine. The taste of the customer is almost always suited in this matter.

The wine thus bottled, and corked with the best quality of corks, compressed strongly into the necks of the bottles and kept there with an iron wire, is left to itself for eight to ten months. The fermentation continues in the bottles on their sides, the carbonic acid is dissolved in the wine, and a deposit of ferment is formed which collects in the lower part of the bottle.

It is necessary that this deposit should be removed by an operation called *degorgement* (clearing, cleansing).

To do this, a workman turns over the bottle, giving it a rotatory motion, so as to bring the deposit from the centre of the bottle on to the cork, and then places the bottles on shelves with holes, so that the deposit may be entirely removed. When the wine has become perfectly clear, which takes place at the end of from fifteen to twenty days, he takes up the bottle, neck down, and cuts the wire; the cork escapes, entirely covered over with ferment, as well also as a small portion of wine, which is promptly replaced by clear wine or syrup. The bottle is then newly-corked, tied, and dipped in pitch, and at the end of five or six months, is ready for consumption.

For some time past, tin-foil has been substituted for pitch; this protects the corks and wire from the dampness of the cellars.

Vins de Liqueur—Sweet Wines.—These liqueurs contain a certain quantity of alcohol, and are at the same time saccharine. Very sweet grapes are used in the preparation of these wines, in which the ferment is not in sufficient quantity proportionably for the sugar; there is then added to the must a portion of this same must concentrated by evaporation. Sometimes, to obtain liqueurs, the fermentation is stopped by the addition of some alcohol.

It is in this way are prepared the liqueurs of Spain, Portugal, Italy, Tokay, and Hungary.

Diseases of Wines.—Wines are subject to diseases known under the names of *pousse*, *graisse*, *acidity*, *bitterness*, *flowers*.

La pousse is a second fermentation, which is developed in the cask, and which turns the wine bitter. The remedy for this disease is to rack off the wine into a cask previously fumigated with sulphur; the sulphurous acid gas which fills the cask has the property of stopping the fermentation. When the wine becomes viscid, (graisseux) some tannin is added to it.

If the wine naturally contains an excess of tartaric acid, it may be turned into a bitartrate, but slightly soluble, by the addition of a small quantity of neutral tartrate of potassa; but when the wine turns sour, it is difficult to cure this disease.

The red wines, particularly those of Burgundy, often become bitter with age. This disease sometimes disappears by the addition of a little alcohol, but it is better to add to it new wine of the same quality. In all cases it is necessary, after these additions, to leave the wine for several months to complete repose.

Finally, wines are sometimes covered with a species of white mushroom, finely divided, called *flowers*. This disease is thought to be the result of an elevation of the temperature of the liquid, which may be arrested by cooling it. The flowers are seen especially in casks or bottles badly stopped, and consequently are owing to the action of the air on the wine. To purify the wine of these *flowers*, it suffices to fill the cask completely; the flowers come to the surface, and by means of a jar given to the cask, they disappear. These rarely reappear when the conditions above recommended of replenishing leakage, and stopping, are fully carried out.

In cold seasons, when the sugar is not developed in the grape in sufficient quantity to give quality to the wine, some sugar is added to the must at the time of fermentation.

In Lower Burgundy, cane sugar is generally used; in localities near the manufactories of glucose, this saccharine material is preferred.

According to Gay-Lussac, wine, on an average, may be said to be composed in weight of 8 to 10 parts of alcohol, of 85 to 90 of water, and 2 to 5 of a residue, formed of coloring and extractive matter, of tartar and other salts, with lime, alumine, etc., as bases, of ferment, and sometimes of sugar, which has not been destroyed by fermentation.

Alcohol is the essential principle of wine: its proportion varies in general from 5 to 17 hundredths in volumes, according to climate, soil, culture, and chiefly temperature at the time of the ripening of the grape.

The vines of the South are more spirituous than those of the North, and in the same locality, they differ one year with another. We give here a table made up principally from the analyses of Gay-Lussac, which represents the quantity of alcohol in volumes, contained in the principal wines.

Table of the quantity of Alcohol, contained in certain Wines and Spirituous Drinks.

Wine.	Alcohol p. 100.	Wine.	Alcohol p. 100.
Bagnouls	17·0	Chateau Destournel	9·0
Grenache	16·0	Brannes Mouton	9·0
Jurançon white	15·2	Léoville	9·1
" red	13·7	Grand-Larose-Kirwen	9·8
Saint Georges	15·0	Cantenac	9·2
Malaga	15·1	Giscours	9·1
Madeira, very old	16·0	Lalagune	9·3
Cyprus	15·1	Therme Cantenac	9·1
Fine wine of the South	13·0	Tronquoi-Lalande	9·9
Common "	9.8	Saint Estéphe	9·7
Vauvert	13·3	Phelan	9·2
Frontignac	11·8	Tokay	9.1
Hermitage, red	11·3	Volney	11·0
Côte Rôtie	11·3	Champagne	11·6
Sauterne, white	15·0	Mâcon	10·0
Bomme "	12·2	Vins du Cher	8·7
St. Pierre du Mont	11·5	Angers (coteaux)	12·9
Barzac, white, first growth	14·7	Saumur	9·9
" second "	12·6	Vins de l'Ouest	10·0
" third "	12·1	White wines of la Vendée	8·8
Poudensac, white, first growth	13·7	Wachenheim (Rhine)	11·9
" second "	13·0	Forst	11·5
" third "	12·1	Sherwiller (Bas Rhin)	11·0
Chateau Lafitte	8·7	Westhoffen	10·0
" Margaux	8·7	Molsheim	9·2
" Latour	9·7	Rosheim	8·6
" Haut Brion	9·0	Barr	6.9

Wine.	Alcohol p. 100	Wine.	Alcohol p. 100.
Egersheim	6·0	Perry	6·7
Châtillon (near Paris)	7.5	Burton ale	8·2
Verrieres (4 leagues from Paris)	6.2	Edinburgh ale	5·7
Wine, for retail at Paris	8·8	London porter	3·9
Vin de la Societé ænophile	9·3	Strasburg beer (old)	3·9
" id. bottled	10·5	New beer	3·0
Vins de lies pressées, Paris	7·6	Red beer of Lille	2·9
Cider, strong	9·1	White beer	2·9
" weak	4·8	Paris beer	1·9

ESSENTIAL OILS.

The name *essential oils*, is given to the oily and volatile products, which are found in the aromatic vegetables.

Essential oils often exist ready formed in vegetables. It is, in fact, well known, that a piece of lemon or orange peel, on compression, immediately throws out a volatile and very inflammable essential oil; but sometimes the volatile oils do not pre-exist in plants: they do not form except at the moment when the plants are placed in contact with water, as is the case with oil of almonds and mustard.

The essential oils are extracted by different methods. They are generally obtained by distillation; the fragrant plant is put into an alembic, and covered with water. This addition of water fills the double end of preventing the vegetable from carbonizing in the alembic, and facilitating the distillation of the oil, which carries over with it the vapor of water.

Some essential oils which alter by heat, may be obtained more easily by passing into the alembic a stream of hydrogen or carbonic acid, which carries over the essential oil.

When an essential oil boils only at a high temperature, it is often useful to raise the point of ebullition of the water. Common salt is for this purpose put into the alembic. By continuing the distillation until the water no longer has an odor, an essential oil is obtained, which is either heavier or lighter than water. Often the essence remains in solution in the water, and forms distilled aromatic waters. Generally, the essential oils are collected in a peculiar receiver, called *Florentine receiver*, which preserves the oil, and lets the distilled water flow off.

The essential oil which is in solution in the water, may be separated by saturating the water with common salt, when the essence swims on the surface, forming an oily stratum. The essential oil may likewise be separated from the water by agitating it with ether, and then the essence is obtained by distillation.

When an essential oil easily alters, solvents, such as ether or the fat oils, may be used to extract it. In this way are obtained the fragrant principles of the flowers of the linden, jessamine, &c.

When vegetables contain a large quantity of essential oil, it is obtained from them by compression. The essences obtained by the processes above pointed out, are never pure; they generally contain in solution solid bodies, called *stéaroptènes*. Some essential oils, such as those of lavender and valerian, are saturated with camphor.

The essential oils are rarely colorless; they are mostly yellow; their color increases when they remain exposed to the air.

Their boiling-point varies from 140° to 200° C. Though volatile, they are often decomposed by ebullition. Their density is variable. They are generally known as oils heavier, and oils lighter, than water; those most dense are generally the most volatile. An essential oil, dropped on a sheet of white paper, makes a spot on it like that formed by a fat body, but when the paper is heated, the spot produced by the essential oil disappears, while that made by a fixed oil, remains. Water sometimes dissolves the essential oils in moderate proportions, and forms aromatic waters, known in pharmacy as *distilled waters*. The essential oils are generally soluble in alcohol, ether, and the fat oils. The volatile oils absorb oxygen slowly; becoming transformed into resins or acids. Some give rise to acetic acid. In this case, the oxygen not only attaches itself to the molecule of the essential oil, but often causes the combustion of part of its elements to form water and carbonic acid.

According to Theodore de Saussure, the oil of aniseed absorbs in two years, 150 times its volume of oxygen, and produces 56 volumes of carbonic acid.

Like results have also been obtained from other oils, such as those of lavender and citron.

The essential oils dissolve sulphur and phosphorus, and give up these bodies, in the form of crystals, when they are evaporated.

Nitric acid often acts energetically on the essential oils, and sometimes causes their inflammation.

The essential oils are used in medicine as aromatics: they are also used in compounding some varnishes, to dissolve the resins, and for taking out spots.

ESSENCE OF TURPENTINE.

The essence of turpentine is obtained, by distilling the resinous principle which exudes from the pine. The substance which volatilizes, is the essence of turpentine, and a resin remains in the distilling apparatus, which is *colophane.* The essence of turpentine is colorless, and very fluid; its odor is strong and characteristic, its taste is sharp and burning; it boils at 156° C., it is inflammable, and burns with a smoky flame; exposed to the air, the essence of turpentine gradually hardens, and then becomes completely resinous.

The essence of turpentine very readily dissolves the resins. It is most used in oil painting, and in the manufacture of varnishes.

RESINS.

The resinous substances are of great importance in the arts. They are abundant in vegetation; but their chemical study still leaves much to be known. They are generally extracted, by making incisions in certain trees, from which the mixture of resin and essential oil flows out. The separation of these two bodies is effected by distilling the natural resin, with the direct fire, or with water.

The resins are all insoluble in water, and soluble in warm alcohol; their alcoholic solution mixed with water becomes milky, and the resin separates from it. Some, as copal resin, are insoluble in alcohol. Most resins are soluble in ether: some, however, as resin of jalap, are not soluble in it. The resins, in general, dissolve in the fixed and volatile oils. Generally they are uncrystallizable, though some resins crystallize.

When they are heated, they soften, fuse, and give off by distillation carburets of hydrogen, solid, liquid and gaseous.

They are all combustible; their flame is not brilliant, is very smoky, and leaves a carbonaceous deposit.

The alkalies often dissolve the resins, which act in this case like feeble acids. Nitric acid oxidizes them with energy.

The resins form with bases, salts called *resinates* and improperly *soaps of resin.*

The soaps of resin make a froth or lather in water, like soaps formed with fat bodies, but are not precipitated by salt water, as are ordinary soaps.

VARNISH.

The resinous substances, the principal properties of which we have just described, are generally used in the manufacture of varnishes.

A varnish may be considered as a solution of one or more resinous materials in a volatile liquid, or in one having the power of drying in the air. The quality of a varnish in general depends on the hardness of the resin dissolved. We shall here give a list of the principal bodies which enter into the composition of varnishes.

Dissolving liquids.	Solid bodies.	Coloring matters.
Oil of poppy,	Copal,	Gamboge,
" linseed,	Amber,	Dragon-blood,
" turpentine,	Mastic,	Aloes,
" rosemary,	Sandarac,	Saffron.
Alcohol,	Lac,	
Ether,	Benzoin,	
Pyroxylic spirit, or spirit of wood,	Colophony,	
Acetone.	Arcanson, or dried resin of pine,	
	Caoutchouc.	

Some resins may be used in the composition of varnishes without preparation; but others, as lac and copal, require a previous preparation, to render them soluble in alcohol and ether.

Good varnishes should have the following characters:

1st. After desiccation, they should remain brilliant, without presenting a dull or greasy look.

2d. They should adhere intimately to the surface of bodies, and should not scale off, even after a long time.

3d. Their desiccation should be as rapid as possible, without diminishing their hardness.

The name of *fat varnishes*, is given to those which contain a fat drying oil. In fat varnishes of good quality, copal or amber is generally used.

Alcoholic varnishes may be polished, but they have not in general the same hardness as the turpentine varnishes. The alcohol, in fact, evaporates more rapidly than turpentine, leaving, as a residue, the resinous substance pure, while the essence of turpentine, oxidizing in the air, forms a resinous layer which fixes the resins.

The essence of turpentine applied to an object, may of itself alone produce a sort of varnish. The dryness of an alcoholic varnish, is often corrected by the addition to it of oily substances, or even soft resins.

CAOUTCHOUC.

This substance is also known as *Gum elastic.* It is extracted in S. America from the *hevea guianensis*, or the *jatropha elastica.* Transverse incisions are made in the trunks of these trees, from which flows out a milky juice, which holds in suspension about 31 per cent. of caoutchouc. Other vegetable juices, such as those of the euphorbias, asclepias, poppy, lettuce, also contain caoutchouc.

Caoutchouc was described first in 1751, by La Condamine. The study of its properties has been principally made by Faraday.

Caoutchouc is found in commerce in the form of flasks, generally brown, sometimes smooth, sometimes pressed with various designs, which are made by applying the milky juice of plants to bottles of earth, serving as moulds. These flasks are generally dried in smoke, which colors them: the mould is then broken, leaving the caoutchouc in the flask form. It is often met with also in thick plates: the natural juice has for some time been sent to Europe in bottles: this juice is boiled, and the albumen it contains, coagulating, envelopes also the caoutchouc.

Caoutchouc is transparent and colorless. Its recently cut surfaces, when put together at once coalesce. This property is used in making the tubes which are so much used in chemical laboratories.

Caoutchouc is altered by concentrated sulphuric and nitric acids; but it resists the action of other acids, and also that of chlorine; potash even in concentrated solution does not alter it. It does not conduct electricity. When exposed to a temperature of 0° C., it hardens firmly, but it resumes its original suppleness when heated.

Caoutchouc is insoluble in water, and soluble in pure ether; alcohol precipitates it from this solution; placed in the cold in contact with the oil of petroleum, it first increases in volume, and by boiling it is completely dissolved. It is also soluble in several empyreumatic oils, in the essential oils, fat oils, and the sulphuret of carbon. It fuses at about 120° C., giving rise to an oily substance; it appears to undergo in this case a simple isomeric modification.

When caoutchouc is distilled, it is transformed into different carburets of hydrogen.

The water-proof cloths are generally prepared by placing between two cloths a thin layer of caoutchouc, first dissolved in pure essence of turpentine. The caoutchouc is applied as a coating, which should be in a pasty condition, so as not to spot the stuff.

Caoutchouc is used for rubbing out pencil-marks on paper. It forms part of the composition called *marine glue*, a mixture remarkable for the strong adhesion it causes between two pieces of wood. This glue consists of a solution of caoutchouc in the essential oil of tar, to which gum-lac is added. It is used at a temperature of about 120° C., in the construction of made masts, for repairing masts at sea, yards, &c. Caoutchouc is used for making such surgical apparatus as require suppleness and flexibility; but its principal use consists in the preparation of waterproof cloths, overshoes, suspenders, and other articles of dress requiring elasticity.

Its solvents used in the arts are ether, sulphuret of carbon, essence of coal tar, different carburets of hydrogen, and particularly those obtained by the distillation of caoutchouc itself. The solvent most used is essence of turpentine well rectified.

GUTTA PERCHA.

Gutta Percha, a substance brought from China, has been for some time known in commerce: this presents a great analogy with caoutchouc. Gutta percha often resembles clippings of leather or horn: it is whitish, coriaceous, hard, and flexible; it becomes soft and elastic when heated. It may, in a measure, be kneaded in boiling water. It is lighter than water; when distilled it decomposes, giving oils which are very inflammable. It is insoluble in water and alcohol, soluble in sulphuret of carbon; ether swells it up and dissolves it slowly; it resists the action of alkaline solutions and chlorhydric acid. Concentrated sulphuric acid carbonizes it with difficulty; nitric acid transforms it into a yellow resinous substance.

Gutta percha, freed from the foreign substances which it nearly always contains, such as resins, a peculiar acid, etc., exhibited to M. Souberain a composition approaching closely to that of caoutchouc. This body may then be considered a solid carburet of hydrogen, comparable to caoutchouc.

Gutta percha is used for making whip-handles, riding-whips, belts for machinery, etc.

FAT BODIES.

Before the remarkable works of M. Chevreul on fat bodies, these substances were but little known. It was well known that many fat bodies would produce soaps or plasters when treated with potash, soda, or oxide of lead.

Scheele had shown, in the products of the action of the oxide of lead on oils, the existence of a saccharine soluble substance, which he designated the *sweet principle of oils;* but the theory of saponification was entirely unknown.

About the year 1813, M. Chevreul published a series of works on the neutral fat bodies, and on the products formed in saponification, which threw great light on this question.

He showed that the fat matters known as *oil*, *grease*, *suet*, were formed, with a few exceptions, by a mixture of immediate principles, which he described under the names of *stearine*, *margarine*, *oleine ;* that these immediate principles doubled themselves, under the influence of alkalies, into the sweet principle of oils, or *glycerine*, and peculiar fat acids; that thus stearine produced glycerine and *oleic acid;* that margarine doubled itself into glycerine and *margaric acid;* and he remarked that if in saponification mixtures of different acids are formed, it is because the neutral bodies subjected to the action of bases were themselves mixtures of margarine, oleine, butyrine, etc.

From the very beginning of his labors, M. Chevreul had assimilated the oils and fat to the ethers; he had shown that saponification could take place in vacuo, without disengagement as without absorption of gas, and that it consisted solely in the fixation of the elements of water upon the fat matter, which then doubles itself into glycerine and fat acid.

He did not restrict himself to make known the general phenomena of saponification: he also described with the greatest care, the properties of most neutral or acid fat bodies. The late researches which have been made on fat bodies, only confirm the accuracy of M. Chevreul's labors.

STEARINE.

Stearine exists in nearly all solid fat, and in many vegetable oils. Its proportion in fat bodies increases with their consistence, and as their point of fusion is high. It is generally extracted from mutton-suet. Stearine is white, very combustible, without odor or taste, and fusible at 62° C. It is insoluble in water. Boiling alcohol dissolves about the seventh of its weight, and deposits a great part of it on cooling. It is much more soluble in boiling ether; but this liquid, when cold, holds in solution but a small proportion of it.

The bases and particularly potash, soda and lime, decompose stearine with water, and prolonged ebullition. This reaction, known as *saponification*, produces hydrated glycerine, and an alkaline stearate.

It would be desirable that stearine could be easily extracted from suet, for from its great resemblance to wax, it is believed that it would answer for the manufacture of candles; it has been proposed with this view, to treat suet with the essence of turpentine, which dissolves the oleine, and leaves the stearine, which then could be compressed. But this operation has not yet been executed in the large way with economy.

MARGARINE.

Margarine is found in human fat, in olive oil, and many other fat bodies; it is generally mixed with oleine and stearine; it is often found combined with oleine. It is obtained by treating human fat with boiling alcohol; margarine precipitates in micaceous scales; it is purified by repeated crystallizations. It resembles stearine, but differs in its point of fusion, which is 47° C.

Margarine may, like stearine, saponify under the influence of alkalies and metallic oxides, transforming itself into glycerine and margaric acid.

OLEINE.

Oleine exists in oil and fats in variable proportions; it predominates in oils, and is not so abundant in solid fat.

Oleine is slightly yellow; it is discolored by the direct light of the sun; it remains liquid even at a temperature of 0° C. It absorbs the oxygen of the air, giving off carbonic acid; in this case, it in part becomes resinous. This property is one of the inconveniences of its use in watchmaking, requiring the oils of a watch to be changed after a time.

Oleine, like the two preceding fat bodies, saponifies with the alkalies, becoming transformed into glycerine and a liquid fat acid, called by M. Chevreul, *oleic acid.*

Oleine, mixed in different proportions with margarine and stearine, forms a great part of the fat bodies of vegetable or animal origin.

STEARIC, MARGARIC AND OLEIC ACIDS.

Stearic acid is produced when stearine is saponified with a base, potash for example; stearate of potash and glycerine are formed; these two bodies are soluble in water: on pouring an acid into the solution, stearic acid, which is insoluble, precipitates.

Stearic acid is white; it crystallises by fusion in white needles; it is insoluble in water, soluble in all proportions in alcohol and ether. It melts at 70° C.; it is combustible, and burns with a brilliant flame; it forms a great part of the stearic candles. It combines with bases; the stearates of potash, soda and ammonia, are the only stearates soluble in water. Margaric acid presents a great analogy of properties with stearic acid; it is produced in the saponification of margarine: this acid is more fusible than stearic acid; it melts at 60° C.

As to oleic acid which proceeds from the saponification of oleine, it differs essentially from the two preceding acids, for it is liquid at the ordinary temperature. The oleates of potash, soda and ammonia, are soluble in water, and uncrystallizable. The other oleates are insoluble.

GLYCERINE.

Glycerine, or the *sweet principle of oils,* is, as we have said above, a product from the saponification of most fat bodies. It is liquid, colorless, very soluble in water, uncrystallizable; its taste is very sweet; it is not, however, a sugar, for it does not ferment. It is easily prepared, by saponifying a fat body, oil for example, by the oxide of lead; oil is a mixture of oleine and margarine; under the influence of oxide of lead, these two bodies double themselves into margarate and oleate of lead, which are insoluble in water, and into glycerine, which, on the contrary, remains in solution: this liquor often contains oxide of lead, which is precipitated by sulphuretted hydrogen; this is filtered to separate the sulphuret of lead, and by a suitable concentration the glycerine is obtained pure.

GENERAL PROPERTIES OF THE NEUTRAL FAT BODIES.

The three substances which constitute most fat bodies, whether of vegetable or animal origin, are *stearine*, *margarine*, and *oleine*. It is understood then, that the vegetable oils and animal fat should have a great number of common properties. Mostly, the small quantity of coloring, sapid, or odorous matters which they contain, hardly modify their chemical properties.

The fat bodies of vegetable seeds are generally enclosed in the cellules: heat is not sufficient to break these cellules; this must be done by trituration; the seeds are then pressed, and often the action of heat is added.

In some cases, the fat matter contained in the seeds is displaced, by subjecting them to the action of water; it is seldom that solvents are used for extracting vegetable oils.

The oils have a very variable consistence: some remain liquid in a low temperature; others, like olive oil, always solidify in winter; while some have the consistence of grease; such are the butters of palm, cocoa, etc.

The oils are lighter than water.

Air acts differently on them; some dry rapidly, and are called *drying oils;* others, on the contrary, are called *non-drying oils*, and become resinous slowly.

Oils absorb oxygen when drying. Theodore de Saussure, who has observed these phenomena, noted, that this absorption takes place in an irregular manner, and is favored by heat. Thus an oil of nuts, which, in six months, had only absorbed but twenty to thirty times its volume of oxygen, absorbed 80 volumes of oxygen in the following month.

When an oil oxidizes in the air, its temperature rises rapidly, and fires have been produced by the spontaneous inflammation of oils in oxidizing. These combustions are more likely to take place when the oil is divided by certain organic matters, such as cotton.

Oils, when kept in cellars, may absorb a sufficient quantity of oxygen to render the atmosphere unfit for respiration.

It is generally admitted that oils are formed by the mixture of margarine with oleine.

Oils exposed to the air, become acid and contract a disagreeable odor; this is called *rancidity*. The oil in this case undergoes a sort of fermentation, which is due to the presence of nitrogenous bodies which it holds in solution. The rancidity may be prevented, to a certain extent, by heating the oils; the heat coagulates the animal matter and suspends its action.

OLIVE OIL.

This oil principally comes from Provençe, Italy, Spain, and the coast of Africa. The olives are gathered some time before they are ripe; they are crushed and pressed cold; this first operation gives *fine oil* or *virgin oil.* To extract more oil from the compressed olives, they are acted on by hot water, by which is obtained oil of the second quality. There is also in commerce another oil of inferior quality, obtained by fermenting the olives whole and the residue which is left after compression.

In the extraction of the olive oil, care must be taken that the olives be pressed as soon as crushed, for if they are left for some little time, there is a disposition in the mass to ferment, which is a great injury to the quality of the oil.

Olive oil begins to congeal at a few degrees below 0° C.: it readily becomes rancid, caused by the alteration of a substance which it holds in solution and which gives it its agreeable taste. It is rapidly solidified by the nitrate of the protoxide of mercury. To test it, it is rubbed with the twelfth of its weight of this salt: this shows the presence of $\frac{1}{10}$th, and often of $\frac{1}{20}$th of the oil of poppies in olive oil. This test is sufficient, for below $\frac{1}{10}$th the fraud is not injurious. The oil of olives, cold, compressed, and repeatedly treated with ether, deposits in confused crystals a white substance, the point of fusion of which cannot be carried beyond 20° C. This substance is a true combination of oleine and margarine, which saponification transforms into a mixture of oleic acid, margaric acid, and glycerine.

LINSEED OIL.

This oil is extracted from the seeds of common flax, which furnish a little more than one-fifth of their weight of it. Its color is yellow, and its odor feeble. It is one of the most difficult oils to congeal with cold: it does not solidify till 15° to 20° below 0° C. It dissolves in 40 parts of cold alcohol, 5 parts of boiling alcohol, and in 1·6 parts of ether.

Linseed oil is one of the most drying of oils. This property, so useful for the manufacture of varnish, and oil colours, becomes much more pronounced in linseed oil, which has been boiled for some hours with litharge, or the binoxide of manganese.

The desiccation of linseed oil is due to an absorption of oxygen: after exposure to the air for some years, it becomes completely resinous.

Linseed oil dried with litharge, is used for making printers' ink, black varnish for leather, &c.

Consistence may be given to it, and to some extent, the elasticity of caoutchouc; and it may be used for preparing sounds, and many other surgical instruments.

Linseed oil, exposed to a high temperature, forms a thick glue, like turpentine. By boiling this residue several hours with water, acidulated with nitric acid, a substance is obtained of the consistence of plaster, which hardens on exposure to the air. This matter softens, without however melting, with the heat of boiling water, and acquires great elasticity; it has a strong resemblance to ordinary gum-elastic. It swells considerably in ether, and is dissolved in the essence of turpentine, and sulphuret of carbon; a concentrated solution of potash hardens it; it dissolves on the contrary, in very dilute alkaline solutions, and reproduces with the acids, a body like caoutchouc.

TALLOW, OR SUET.

The different fats of the herbaceous animals are designated by this term. The masses of fat taken out by the butchers called *suif en branches*, contain *suet* proper, and animal membranes, which are skimmed off when the suet is melted.

There are two different processes for melting the tallow; the first is by rendering, and consists simply in heating it; in this case, the animal membrane shrivels up, and the fat drips out. The residue which has drained out, is called *tallow in cakes.*

The second process is described by M. d'Arcet, and is called the *process with acid.* The crude tallow is heated with sulphuric acid, which dissolves or disintegrates the membrane. The suet thus obtained is of fine quality, but there is this very great objection to it, the sulphuric acid seems to determine the separation of the oleine from the suet; for when prepared with acid the suet is granular, and allows the oleine readily to drain out.

Suet is considered as a mixture of different fat bodies, which are, stearine, margarine, oleine, and besides, a neutral substance, very abundant in the fat of the goat, which M. Chevreul calls *hircine.* Suet gives by saponification, oleic, stearic, margaric and hircic acids. The proportion of this last is extremely feeble. As the solid acids are predominant in the products of the saponification of suet, this fat body is almost always used in the manufacture of stearic candles.

BUTTER.

The composition of butter seems to be very complex. M. Chevreul has shown that this body contains five neutral substances, which are: *oleine, margarine, butyrine, caprine,* and *caproine.*

This fat body, treated with the alkalies, saponifies and is transformed into oleic, margaric, butyric, capric and caproic acids.

Capric, caproic, and butyric acids are volatile, and may be separated from the oleic and margaric acids by distillation.

Butter is one of the richest of the fat bodies in margarine, but it does not appear to contain stearine. Rancid butter is used sometimes for making candles.

BEESWAX.

Wax is found in commerce in two different forms. When taken from the bees and melted, without any purification, a wax of a yellow color is obtained, the point of fusion of which is 62° or 63°C.

This wax, exposed to air and light, loses its color; it then forms white wax, the point of fusion of which is between 64° and 65°C.

It is well known that wax is chiefly used for making

candles, that it is originally yellow, and that to bleach it, it is exposed to the air and light in thin layers.

It has been attempted to use chlorine for bleaching wax. This does bleach it, but it always retains the chlorine, which on combustion produces muriatic acid.

SOAPS.

It appears from the beautiful observations of M. Chevreul that *saponification* is an operation, which has for result to transform, under the influence of bases, the fat neutral bodies into glycerine and fat acids, which unite with the base used to make the soap.

Soaps then are true salts formed by the combination of fat acids with the metallic oxides. Those which are most frequently met with in commerce, have for a base, soda or potash, and for acids, the stearic, margaric, oleic, and palmitic. The great resemblance which the different fat acids have to each other is followed out in their combinations with bases, that is to say, in the different soaps. The consistence of soaps is great in proportion as the point of fusion of the fat bodies with which they are prepared is elevated. Soda forms harder soaps, other things being equal, than potash.

The soaps of soda, potash, and ammonia, are alone soluble in water: all the others are insoluble in this liquid, and may be obtained by double decomposition, as are the other salts which are insoluble in water. It is thus that if the oleo-margarate of potash or soda (soap) is poured into nitrate of lime or copper, a precipitate is obtained which consists of oleo-margarate of lime or of copper. Soaps with alkaline bases are soluble in alcohol and ether; the metallic soaps, properly called, with the exception of the soaps of copper, protoxide of iron and manganese, are insoluble in these liquids. The fat oils and the essence of turpentine will dissolve these last soaps.

The acids decompose soaps uniting with their bases and

eliminating the fat acids which swim on the surface of the aqueous solutions in which they are insoluble. The soluble soaps are prepared either by directly uniting the fat acids with potash and soda, or by treating oils, fat, and the suets, with boiling alkaline solutions. This operation, which constitutes *saponification*, requires much time and patience for its completion. In laboratories, soap of soda or potash is prepared by boiling in a porcelain capsule 100 parts of a fat neutral body (olive oil, suet, lard, etc.) with 20 to 25 parts of caustic potash, or soda, and 200 to 250 parts of water. The mixture should be constantly stirred with a glass rod; the water, in proportion as it evaporates, is replaced by distilled boiling water. It is known that the saponification is terminated, when a small quantity of the material, tried with pure water, is entirely dissolved without leaving any trace of fat matter.

It is also known that the soap is well prepared, when chlorhydric and sulphuric acids separate from it a fat acid entirely soluble in alcohol.

We have said that the insoluble soaps may be obtained by a double decomposition; but those having for base the protoxide of lead, lime, barytes, and strontian, may also be prepared in the same way as the alkaline soaps: that is to say, by boiling these oxides directly with water and the fat neutral bodies. Thus are formed insoluble soaps, and the water retains the glycerine.

Simple lead plaster is always prepared in the shops, by boiling a mixture of oil, litharge and water.

Soaps are known in commerce as *hard* and *soft soaps*. the soft soaps are always made with potash for base, with the oils of hempseed, linseed, and colza or rapeseed, etc.

Hard soaps have soda for base, and are made with olive oil, suet, different fats, etc., or with fat bodies which contain a great quantity of solid matter.

Soaps may be coagulated by a great number of alkaline

salts; we will mention principally the carbonates of potash and soda, chloride of sodium, sulphate of soda, chlorhydrate of ammonia, etc. This property is used to advantage in the manufacture of soaps. When it is judged, from certain physical characters, that the fat matter is completely saponified, in order to free it from an excess of alkali, with which it is always sticky, sea-salt is thrown into the pan, which causes a separation of the soap and alkaline water.

Sometimes, at the moment the soap is about to solidify, a soap of ferruginous alumine, colored green, is added, forming the *marbling* or *mottling* of soaps.

The marbling of soap is of advantage in showing the proportion of water which it contains, for this operation is not practicable if the soap contains more than 30 per cent. of water.

In the preparation of dark soap and some toilet soaps, the alkali is made to react on the fat body; but the excess of alkali is not removed, as in the preceding case, by adding salt, for the soaps thus made always contain a great excess of alkali. The proportion of water is also very variable.

Soft soaps have a much stronger alkaline reaction than hard soaps; they dissolve more rapidly in water, which, in some cases, is an advantage; they contain, independently of the oleate and margarate of potash, an excess of alkali, chlorides, sulphates, and glycerine set free during the saponification. They are much used in the north of France, and in Belgium and Holland; they are used not only for washing, but in the fulling and scouring of woollen stuffs, and generally give a disagreeable odor to linen. When soft soap has been made with hempseed oil, it has, naturally, a greenish color; mostly it is yellow, and to give it the green color which is required in the market, a small quantity of indigo is added to the mass.

Toilet soaps owe their perfume to the essential oils which are introduced in minute quantities into the mass at the moment of pouring out. Sometimes it proceeds directly from the aroma contained naturally in the oils used in their manufacture. Such are particularly soaps made with palm and olive oil. Tallow, butter, and fish oil, on the contrary, give a disagreeable and characteristic odor to the soaps made with them. This property is due to the presence of hircine, butyrine, *phocénine* (oil from the dolphin or fish).

Transparent soaps, colorless, or of various colors, are found in commerce. To prepare them, soaps of tallow well dried are generally dissolved in alcohol. The warm and limpid mass is poured into moulds; it does not become transparent till after desiccation.

Pearly Soap, or Almond Cream, is obtained by cooling slowly a soap with potash for base, and pounding it strongly in a mortar when cold.

Some resins, particularly common resins (colophane), form, with alkaline bases, salts, which have an analogy with soaps; thus, sometimes in England, a considerable quantity of resin is used in the composition of common soaps.

MANUFACTURE OF STEARIC CANDLES.

The manufacture of stearic candles is the result of the works of M. Chevreul on fat bodies. This manufacture, now so important, originated in Paris, from whence it has gradually extended into the principal towns of France and other countries. Stearic candles have now almost entirely displaced wax and spermaceti candles, the price of which is much higher. This new application of chemistry to the arts and domestic economy is regarded as one of the most useful which has been made this century.

Up to this time, it has been impossible practically to extract stearine from tallow and other fat neutral bodies, either by pressure or by the aid of solvents; but when these matters are saponified, and the soap thus produced is decomposed by sulphuric acid, the margaric, stearic, and oleic acids, which result from this decomposition, subjected to pressure, separate into two parts, of which one is liquid, that is oleic acid, while the other, formed of stearic and margaric acids, constitutes a solid and white mass, which is used in the preparation of stearic candles. The oleic acid which is produced in the manufacture of stearic candles, is used for making soap.

M. Chevreul having shown that the alkaline oxides and alkaline earths, particularly lime, caused the saponification of fat matters, two ingenious manufacturers, Messrs. de Milly and Motard, first took the idea of applying lime to the extraction practically of the fat acids from tallow. They showed that this oxide, the cost of which is so little, saponified fat bodies more promptly than potash and soda, because the lime mixes itself intimately with fat matters.

COLORING MATTERS.

Coloring matters are found indiscriminately in all the organs of living beings; they are found in the roots of plants, as the coloring matters of orchanet, turmeric, madder, etc. They may be extracted also from the stems of vegetables, as from sanders or sandal wood, Campeachy, Brazil wood and quercitron; the leaves, the flowers, the fruit, or the seed of vegetables, are rich in coloring matters. Whole insects are used as dyeing substances; such are cochineal and kermes (a species of cochineal). It is also known that some liquids of the animal organization, such as the blood and bile, are highly colored. Coloring matters are seldom found isolated; it is often even difficult to separate them from their combinations. They do not always pre exist in the vegetable organization, and are formed only when organic bodies are submitted to fermentation or the influence of oxidizing forces. Some coloring matters form only in the presence of ammonia.

The most common coloring matters are yellow, red or

green; those which belong to the first two classes are, in general, easy to isolate. It is not the same with the green substances. The green coloring matter of leaves, which has been called *chlorophyll*, and which plays so important a part in the vegetable organisation, is not yet known in a state of absolute purity. Coloring matters often have a pungent and saccharine taste; they are inodorous; there are some of them which crystallise easily, as those which are extracted from indigo and madder; they are often volatile; they should always be distilled with precaution, for a temperature of 150°C will often decompose them.

All the coloring matters are decomposed by light; under the influence of this agent they absorb oxygen; combustion commences and they become decolorized. Coloring matters are said to be of *good or bad dye*, according to the rapidity with which they decolorize in the light.

M. Chevreul, to whom science is indebted for numerous works on coloring matters, has determined the different circumstances which influence their decoloration. Coloring matters are often soluble in water; some are only soluble in ether, alcohol, or the essential oils; the presence of acids sometimes facilitates their solution; others dissolve on the contrary with facility in the alkalies.

The different chemical agents in general modify the tint of coloring matters; this property is often taken advantage of in dyeing, and even in chemical analysis. We will call to mind here, that alkalimetry and the analysis of borax are based on the different colors produced in the tincture of turnsole by weak acids, like boracic or carbonic acids, or when it is modified by an energetic acid, like sulphuric acid.

Concentrated bases decompose all coloring matters; but when they are diluted, they only modify them, making them take the shades inclining in a great degree to the dye. In some cases, coloring matters originate under the influence of the bases and oxygen; we shall here cite tannin, which according to the experiments of M. Chevreul, forms coloring matters of a reddish brown when submitted to the influence of oxygen and an alkali.

The metallic oxides in some cases contract true combinations with coloring matters which then behave as acids; the preparation of *lakes* is especially founded on this property, for they are combinations of coloring matters with alumine, oxide of tin, etc. Some salts act on coloring matters through their acid, or even through their base; they combine with them and fix them in the cloth. This property serves as a base for the application of the mordants.

Powdered carbon and animal black absorb the coloring matters, but do not destroy them. If for example a decoction of Brazil-wood is decolorized by animal black, the absorbed coloring matter may be withdrawn by subjecting the animal black to the influence of a liquor slightly alkaline.

No coloring matter is known which resists the action of chlorine; under the influence of this energetic agent, the coloring matters undergo a true combustion, losing hydrogen which is transformed into muriatic acid, and in some cases taking chlorine. Under the influence of moist chlorine, the water is decomposed, muriatic acid is formed, and the oxygen of the water goes to the coloring matter to decompose it; in this case, the action of the chlorine may be compared to that of oxygen.

Since the beautiful experiments of Berthollet, cloths intended for dyeing are in most cases bleached by chlorine.

Sulphurous acid exercises an action on coloring matters of advantage in the arts; this agent, in fact, generally destroys the coloring matters without altering the cloth. Sulphurous acid is especially used for bleaching silk and wool. It often acts by taking off the oxygen; it often acts also, by uniting with the coloring matter, forming with it a colorless combination. It is admitted, also, that the sulphurous acid decomposes the water, seizing upon its oxygen, while the hydrogen of the water, uniting with the coloring matter, forms a hydrate, which is colorless. It is well known that fruit-stains may be taken out of linen by using sulphurous acid; in this case, care must be taken to use plenty of water to dissolve out thoroughly the sulphuric acid produced, which, in the end, would destroy the fibres of the cloth.

Some bodies, greedy of oxygen, decolorize by producing on the coloring matters a slight modification, so that these substances may immediately resume their primitive color, when they are exposed to the air. We will here name, as decolorizing agents, hydrogen, sulphuric acid, sulphydrate of ammonia, the alkaline sulphurets, the protoxide of iron, etc. It is supposed that the preceding bodies, which have a great affinity for oxygen, decompose water, absorb the oxygen, while the hydrogen of the water combines with the coloring matter. Some chemists, however, think that the coloring matters, under the influence of the above bodies, undergo a true disoxygenation.

We will here quote a very remarkable experiment, reported by M. Persoz, which gives a glimpse into one of the most important functions of vegetables. When a balsamine is plunged into a solution of a coloring matter, it decolorizes it by means of its roots; a colorless fluid is then seen circulating in the capillaries of the plant; but when the liquid arises at the petals of the flower, it immediately resumes its first color. The roots of a plant may then be considered as forming an apparatus of reduction, while the petals of the flowers bring about true combustions.

To extract from a vegetable the colorless principles, which may then be colored in the air, the dyeing material is dissolved, and the coloring substance is engaged in combination with the oxide of lead; there is thus formed a true salt, insoluble in water. This salt is decomposed by sulphuric acid, which, in seizing upon the oxide of lead, deoxidizes, at the same time, the coloring matter; the liquor then, by evaporation, deposits colorless crystals, which, under the influence of oxygen or oxydizing bodies, reproduce the color of the dyeing substances, from which they were extracted.

Cloths absorb the colored bodies. *The art of dyeing* is founded on this principle: a species of combination takes place between the coloring matter and the cloth, which differs from the chemical combinations, for it is not in a definite proportion. Coloring matters have not an equal affinity for cloth. Cloths which readily absorb the coloring principles, are, in general, of an animal nature, such as wool and silks; cotton, hemp, and flax, appear to have less affinity for coloring matters.

As to the composition of coloring matters, it is somewhat variable; some are formed only of carbon, oxygen, and hydrogen; others contain azote. There are some coloring substances which may be represented by carbon and water.

GENERAL PRINCIPLES OF DYEING.

The art of dyeing has for its end to fix coloring substances in materials, which are generally of wool, silk, cotton, hemp, linen, the skins of animals, paper, wood, etc. In order that the coloring matters should combine with tissues, it is indispensable that these substances be liquid; and further, that the tissue be purified of the foreign matters which color it, and which besides prevent the coloring matter from being fixed; thus, before dyeing a tissue, it should be *bleached.*

There are some coloring matters which combine immediately with the tissue without the aid of an intermediate body. Such are indigo, saffron, archil, etc.; but others, as madder and woad, give a color which water immediately takes out. The name *mordant* is given to all bodies which, having at the same time an affinity both for the tissue and the coloring matter, may be used as an intermediary to fix the colored substances. The principal mordants are alumine, oxides of iron, tin, alum, the acetates of iron, alumine, and the chlorides of tin.

When the coloring substances are to keep their primitive colors on the tissues, mordants not colored are used, such as those having alumine and oxide of tin for base. When the color is to be modified, the oxides of iron and copper are used as mordants. The use of mordants is made at variable temperatures, according to the nature of the tissue. The mordanting of silk is done at the ordinary temperature, wool at the boiling temperature, linen and cotton at between 30° and 40° C. To mordant, the tissue is first dipped in the mordant, and then into the dyeing bath. Sometimes the mordant is mixed with the coloring matter.

The quantity of coloring matter which fixes itself in the tissue is in proportion to the concentration of the solution of the mordant. The dyers make use of this property to obtain different tints with the same coloring matter.

The dyers' baths are prepared by methods which vary with the nature of the coloring substances to be used.

Generally the colored woods are divided into shavings, or are reduced into powder; they are treated with boiling water, and sometimes with alkaline waters.

The temperature varies with the nature of the stuff, and that of the coloring matter. Indigo and carthame are used cold in dyeing. When warm, the dyeing is more homogeneous. In general, the textile materials dye better in threads and flocks, than when they are converted into cloth. To dye the threads, sticks are passed through the hank, which is made to turn on the sticks till the color reaches its proper shade. This manœuvre is termed *lissage*.

Stuffs are dyed by means of a wheel placed on the edge of the boiler; one end of the piece is put over the wheel, while the other is plunged into the bath. By moving the wheel, the stuff may be dipped in and drawn out at will. On taking the stuff out of the baths, it is rinsed in an abundance of water. This operation is called *scouring* (*degorgeage*). Sometimes a first coloring is given to stuff, and afterwards a second. In dyeing this is called *donner un pied.*

The operations of dyeing above spoken of, are intended to give a uniform tint to the tissue. In the manufacture of prints, different colors, on the contrary, are applied to the cloth, so as to produce designs. We shall here mention only the principles of the ingenious processes used in this manufacture.

1st. A mixture of the mordant and coloring substances is made, which is thickened with gum, starch, or farina, and by means of a cylinder, on which the designs are hollowed out, only those parts of the stuff which are to be colored are printed. This mode of printing is particularly applicable to silks and woollen goods.

2d. The mordants, properly thickened, are applied on certain parts of the stuff, where the coloring matter is to be fixed; it is then passed through the coloring bath. The coloring matters combine strongly only with those parts which are covered with the mordant. By simply washing in running water, the coloring matter is removed where it is not fixed by the mordant. Thus are obtained colored designs on white ground. This mode of printing is particularly applicable to calicoes.

3d. To obtain white designs on colored grounds, the parts which ought to remain white are covered with substances called *reserves.* The stuff is then colored by the ordinary methods. Then, in taking off the reserves, white designs on colored grounds are obtained. These designs may then be colored with a tint different from that of the ground.

4th. To remove in places a color applied equally over a stuff, substances, called *discharges*, are used, which may be either chlorine, oxalic acid, salts of tin, or those of which the composition varies with the nature of the coloring matter which it is intended to remove. When certain parts of a stuff have become white, they may be colored, and thus designs may be obtained, presenting various tints.

ANIMAL CHEMISTRY.

Animal chemistry consists in the study of bodies taken from the animal organisation, and also in the chemical and physiological examination of the principal liquids and tissues which constitute animals. We will first examine the properties of the principal bodies which exist in the animal organisation.

FIBRINE.

Fibrine is the substance which is found in suspension in the blood, and which gives it the property of coagulating. It is that which constitutes, in great part, the solid substance of the muscles. It is there traversed by prolongations of vessels, by arteries, nerves, and aponeuroses, from which it is difficult to free it.

When the blood is drawn from the vessels which contain it, it separates into two parts; the one forms a species of jelly, called *clot*, while the other is liquid, and constitutes the *serum*. The fibrine remains altogether in the clot; it is that which retains, as it were, in net-work, the red globules. To separate the fibrine from the clot, it is sliced into thin pieces, which are broken down, and then placed on a cloth and subjected to the action of a jet of cold water; the globules are carried off, while the fibrine remains on the cloth, in the form of white and elastic filaments. Fibrine may also be obtained by stirring the blood with a bunch of twigs, to the extremities of which long filaments of impure fibrine soon begin to attach themselves. At first, this fibrine is washed in plenty of water, to free it from the soluble principles of the blood which it may contain; it is then dried, and afterwards washed with alcohol and ether, which take off the fatty matters. It is then treated with weak acids, and afterwards with distilled water. Fibrine thus purified is white, completely insoluble in water, alcohol and ether. When it is burned, it always leaves 2 to 3 per cent. of ashes.

Fibrine dried in a stove becomes horny, grey, and opaque: heated to 300°C it decomposes, giving off ammoniacal products and leaving a brilliant and voluminous char.

The acids act upon fibrine, and produce with it a white and gelatinous mass.

Nitric acid unites with fibrine and colors it yellow. Very weak muriatic acid, only containing 0·694. gram. of acid to 1 litre of water, after some hours of contact in the cold transforms the fibrine into a transparent jelly, which dissolves in pure water. This solution coagulates by heat; it is precipitated by tannin, by the cyanoferride of potassium, and by acids.

Dilute muriatic acid dissolves fibrine with still more facility when it is mixed with a few drops of gastric juice. These properties explain, according to Messrs. Bouchardat and Sandras, the rapidity with which fibrine dissolves in the stomach.

Fibrine placed in contact with acetic acid, produces a colorless and transparent jelly, which is soluble in boiling water.

Fibrine dissolves in potash, even dilute; the acids precipitate it, but in this case it is altered.

ALBUMEN.

Albumen exists abundantly in organized beings. It exists in the animal and vegetable organisation. It is found in a state of solution in some liquids of the animal organisation, such as the blood, or the white of egg.

When a solution of albumen is exposed to a temperature of 65°C., it becomes opaline, and if carried to a temperature of 75°C., the albumen then coagulates completely. In coagulating by heat, albumen collects in a sort of netting all particles which are in suspension in the liquid; in this way it is used for clarifying different liquids.

The coagulation of the albumen under the influence of heat, is incomplete when the solution is very dilute. Thus a liquor which is formed of 1 part of albumen and 10 parts of water does not coagulate by heat, but becomes simply opaline. When albumen is evaporated at a temperature which does not reach its point of coagulation, a gummy and transparent mass is obtained, which may be entirely redissolved in water.

Alcohol causes the complete precipitation of albumen.

Albumen coagulated by alcohol is found to be in the same state as albumen coagulated by heat. Nearly all the acids precipitate albumen white, with the exception of trihydrated phosphoric acid and acetic acid. This last acid turns into jelly the concentrated solutions of albumine.

Nitric acid, is of all acids, that which most readily coagulates albumine; this property enables us to recognise the presence of albumine in the liquids of the animal organisation.

Nearly all the metallic salts are precipitated by albumen: we will name particularly corrosive sublimate, which forms in the solutions of albumine a white precipitate, insoluble in water, which does not act on the animal economy. For this reason albumen is considered the best antidote for corrosive sublimate. A vegetable albumen is found in a great number of plants, which has been for a long time compared with animal albumen; which, in fact, has the same properties, and coagulates at the same temperature. The identity of animal and vegetable albumen has been demonstrated analytically, by M. Mulder. This fact is of great importance for physiology; it proves that vegetables contain, ready formed, some of the principles which are found in the animal organization.

CASEINE.

Caseine is an albuminous substance, which exists in milk. To obtain it in a state of purity, the milk may be first treated with a certain quantity of sulphuric acid; a precipitate is formed, which is washed in plenty of water, and then subjected, cold, to the action of carbonate of soda, which dissolves the caseine; the solution is exposed to a temperature of 20° C., in order that the butter may completely separate. The solution is then precipitated by sulphuric acid. The precipitate is washed until the water used is no longer acid. The caseine still retains traces of sulphuric acid, which are removed by carbonate of soda; the caseine is then treated with alcohol and ether, which dissolve the fat matters which it may retain.

Caseine thus obtained is white, slightly soluble in water, and insoluble in alcohol; it is soluble in the alkalies. The acids cause its precipitation, and then tend to combine with it. It feebly reddens litmus-paper.

Phosphoric acid is the only acid which does not cause the coagulation of caseine.

The solutions of caseine in the acids or alkalies, subjected to evaporation, become covered with a white pellicle, like to that which is formed when milk is evaporated.

Caseine is precipitated from its solution by rennet.

GELATINE.

The skin, bones, cartilages, etc., give up to boiling water, a substance which has received the name of *gelatine.*

Gelatine, which in the arts is known as isinglass, (fish-glue) is colorless and transparent when pure; it is remarkable for its great coherence; it is inodorous, insipid, neutral with colored reagents. When heated, it first fuses and then decomposes, giving off a disagreeable odor of burnt horn.

Gelatine does not dissolve in cold water, and only softens in it; in this case it is hydrated, and may take as much as six times its weight of water. Under the influence of boiling water, gelatine dissolves, and when cold, turns the water into jelly; a liquid which contains but one hundredth of gelatine, forms a jelly in cooling; this jelly alters rapidly under the influence of heat, and becomes acid. The solution of tannin completely precipitates gelatine; the tanning of hides is founded upon the affinity of tannin for gelatinous substances.

BLOOD.

The blood of man, mammiferous animals, and birds, is an alkaline liquid, of a brown red or scarlet red color, slightly thick and viscous, of a specific gravity much greater than that of water. It has a saline, repulsive taste, of a mawkish, peculiar odor characteristic with some animals. Its temperature is the same as that of the body.

Alkalinity of the Blood.—Rouelle demonstrated in 1776, that the alkalinity of the blood is due to soda. This property is essential to the blood, in order that it may serve for the accomplishment of the phenomena of life. This liquid has never been seen to present any other reaction with man or living animals; with these last, the sanguine fluid can never be made acid with the aid of direct injections; life ceases long before this result takes place.

The *color of the blood* presents some variations in the animals of the lower degrees in the zoological scale. It is of a pure deep red with reptiles, and of a bluish color with fishes. Among the non-vertebrated animals, leeches alone have red blood. It is colorless with certain mollusca, of a slightly shaded and milky blue in others. With insects, the blood of the dorsal vessel is transparent, or is of a greenish tint in some; in others, it is yellow or orange, or a deep brown.

With man and warm-blooded animals, there are two kinds of blood distinguished by color, into, 1st, *arterial*, of a vermilion red; 2d, *venous*, of a deep brown. This difference of color does not take place till after birth, when respiration is effected. During *intra uterine life*, the blood is the same color in the arterial and venous system, and the color which it presents is of an intermediate tint between the venous and arterial blood of the adult.

Specific gravity of Blood.—The blood of warm-blooded animals is denser and more viscous than that of cold. The density and viscosity of the blood may vary to a certain extent, caused by the difference of aliment, hæmorrhages, sanguine emissions, etc.

Further, the different portions of blood from the same bleeding may be of different densities. These circumstances explain the different numbers given by observers. However, at 15°, the specific weight of the sanguine fluid of the adult only varies from 1.050 to 1.058. This density is in general greater with man than with woman.

The phenomena of the circulation are more easy and compatible with a state of health in proportion as the blood is dense. When this fluid loses its viscosity, and becomes more aqueous, it is imbibed into the tissues, and circulates with difficulty in the capillaries, as has been proved by M. Magendie.

The *odor of the blood* is characteristic in each animal species, according to some observers, and it is more marked with the male than the female. This special odor is developed and exalted when the blood is treated with sulphuric acid. This character, which M. Barruel has attempted to apply in medical jurisprudence, is generally thought to be of little value.

The *heat of the blood*, taken in the heart, is, with men and mammiferous animals, about 38° to 40° C. With birds, it is 4° or 5° higher. Exercise and digestion increase the heat of the blood; rest and abstinence diminish it. The experiments of J. Davy, Becquerel and Breschet, Mayer and Saissy, tend to show that the heat of the arterial blood in the left side of the heart is higher by $1\frac{1}{8}$° than that of the venous blood.

Microscopic examination of the Blood.—When the circulation in the web of the foot of a frog, or that in the membrane of the eye of a living bat, is observed under the microscope, it is seen that the blood in movement in the living body is a colorless fluid, in which swim the peculiar corpuscules called the *globules of the blood.* The liquid in which these corpuscules are held in suspension during life, is a solution of albumine, of fibrine, and of salt, called *liquor sanguinis.* These globules, invisible to the naked eye on account of their tenuity, give to the blood its characteristic red color. Independently of these bodies, the sanguine fluid may also contain in suspension globules of fat and corpuscules of lymph and chyle.

The *globules of blood* exist in the blood of all the vertebrated animals. With man, and most mammiferous animals, they are circular, flattened in form of discs, and swelled out on their edges. With birds and reptiles, they are elliptical, and also flattened. There are two sorts of globules of the blood; one, not colored, are much the most numerous, demi-transparent, and of a yellow color when they are isolated;

they are of a red color when many of them are united together; the other sort are colorless and much smaller.

The diameter of the colored circular globules of the blood varies in different kinds of animals. With man it is the $\frac{1}{120}$ of a millimeter. In animals with elliptical globules, there are likewise relative differences of volume. With all animals, the globules of blood, whatever be their form and volume, present themselves in the form of corpuscles, smooth, flexible, and elastic, which gives them the power of being able to circulate freely, and gliding one on the other, and of temporarily elongating themselves to traverse those capillaries, which are smaller than their ordinary diameter. When the blood circulates in the vessels, the corpuscles of the blood appear simple and homogeneous; but, as soon as they quit the living body, a central spot is perceived on them, and on their two faces is seen a convexity or swelling out, which corresponds to it. This is what is called the nucleus (noyau) of the sanguine globules.

In looking at its physical constitution, the globule of the blood is composed of an exterior envelope, which encloses the nucleus, and of the coloring matter.

When the globules of the blood are preserved in the serum, or in an albuminous liquid, they alter but slowly; but if water is added, a phenomenon of endosmose is produced, in virtue of which the water penetrating the envelope of the globule distends it, and gives it a spherical form. In this state it is seen that the interior nucleus becomes more and more apparent in proportion as the envelope grows pale, and as the coloring matter spreads in the liquid.

Coagulation of the Blood.—When the blood is extracted from the living vessels, and left in repose, it soon undergoes a change, in virtue of which it separates into a limpid liquid of a yellowish green color, and into a solid reddish mass, which retains the sanguine globules. This change constitutes the phenomenon of the coagulation of the blood. The solid part forms the sanguine *clot*, and the liquid portion is called *serum*.

The coagulation of the blood commences to take place soon after its escape from the blood-vessels. It commences habitually at the expiration of five or six minutes; and at the end of ten to twelve hours, it is complete. When this phenomenon is examined with care, the blood is first seen to become thick, and to assume the consistence of a soft jelly; then on the surface of this blood, a clear citrine liquid, the serum, is seen to ooze out, which is, as it were, expressed from the clot.

The *serum* is a liquid slightly viscid, of a greenish yellow, or yellowish red color, due, according to some authors, to small quantities of hematosine and of the biliary pigment held in solution. During digestion, the serum contracts a milky appearance proceeding from the particles of fat which have been carried to it by the chyle. It has a saline and somewhat mawkish taste. It holds in solution albumen and salts, and constantly has an alkaline reaction on the red paper of turnsole. When heated to 76° C., it coagulates without disengaging gas.

The *clot* is a red mass of the consistence of a firm jelly, which may be penetrated by the finger. The clot is constituted of fibres which hold in their interstices the sanguine globules, and there is imbibed in it some serum. Its surface, exposed to the air, is of a clear red white; its interior is of a red verging on brown. It is heavier than serum. The clot is constantly contracted, because the fibrine, in solidifying, contracts at the same time; the expulsion and separation of the serum is due to this circumstance.

The fibrine and sanguine corpuscles are found distributed in an inverse manner, at different heights in the clot. Thus the inferior parts of the clot are very rich in globules and very poor in fibrine; the superior parts on the contrary are very rich in fibrine and very poor in globules. This disposition, in these two elements, is very well explained by their different densities respectively. Before the coagulation, the fibrine, on account of its less density, tends to rise to the surface, while the globules, whose specific gravity is much greater, tend to fall to the bottom. The solidification of the clot comes on and seizes the fibrine, and the globules direct themselves in an opposite direction. It may even happen that the sanguine corpuscles precipitate themselves before the coagulation, and that a layer of fibrine more or less thick coagulates on the surface without taking up any of them. In this case, which may depend on the fibrine rising more rapidly, or on the globules descending more rapidly, a pellicle more or less thick is formed on the surface of the clot, deprived of globules, to which is given the name of *coat* of the blood.

The proportion in weight of serum and clot presents numerous variations, which depend either on the particular state of the organism, or on the more or less energetic contraction of the fibrine, which thus drives out of its cellules more or less considerable quantities of serum. However, Berzelius asserts that the serum constitutes about $\frac{3}{4}$ of the weight of the blood, while the clot, still moist and not pressed, would form $\frac{1}{4}$ of this weight.

The blood of all animals does not coagulate equally well; it is established that of all blood, that of birds coagulates with most rapidity, while that of fishes and of reptiles coagulates slowest. It is the same with the blood of hibernating animals during their sleep. With non-vertebrated animals the coagulation is very imperfect, and is even denied by some authors.

Circumstances which influence the coagulation of the blood.—The cause which determines the coagulation of the blood is unknown. It takes place in a vessel exposed to the air or placed in the vacuum of an air-pump. It is effected equally in oxygen, carbonic acid, and hydrogen. We will, however, mention causes, some of which retard and others of which accelerate coagulation.

The coagulation of the fibrine is prevented by the sulphate of soda, chloride of sodium, nitrate of potash, chloride of potassium, acetate of potash, and borax, provided these substances are added in the proportion of 30 grammes to 180 grammes of blood. According to Hamburger, the carbonates and acetates would prevent coagulation, whatever might be their degree of concentration, while the sulphates in concentrated solution would retard it, and in a state of dilute solution would favor it. The same thing appears to take place with the tartrates and borates. M. Magendie has shown that the dilute mineral acids prevent the coagulation of the blood, at the same time that they thicken it, and give it an oily appearance. The nitrates of strychnine, morphine, and nicotine, also prevent the coagulation of the blood, according to M. Magendie. Hunter saw the same effect produced by a solution of opium.

Temperature exercises an undoubted influence on the coagulation of the blood. Cold retards it and arrests it in certain cases. Thus blood just drawn from a vein and exposed to great cold, freezes without coagulation, then again becomes liquid with heat, and afterwards coagulates like fresh blood. Heat is then necessary to bring about coagulation; a temperature of 38° to 40° C., equal to that of the living body, is that which is most favorable to it.

The coagulation of the blood is again retarded by its contact with the animal membranes or tissues; it can be explained, from this cause, why blood, infiltrated into the cellular tissue, remains fluid a long time, sometimes many weeks before coagulating. The coagulation of the blood is accelerated by heat, and also by the action of a galvanic current. In dry air it is more rapid than in moist air, without doubt, because the evaporation of the water renders the blood more coagulable.

The most simple physiological notions lead us to suppose that the blood is not a liquid chemically identical in all parts of the body. This remark however is relative only to venous blood, and ought not to be applied to arterial blood. In fact, the arterial blood which goes out from the lungs passes into the left side of the heart, and circulates in the aortic system, to be distributed in all the tissues and all the organs of the body. In this passage, it is subjected to no cause capable of changing its composition; it is only in going through the capillary vessels of each organ that it is modified and becomes venous, as it is called. Therefore, it is evident that this denomination does not express a chemical change which would be every where the same, and it is clear that the venous blood which has passed through the kidney, after having furnished materials for the urine, ought to differ from the venous blood which has traversed the pancreas after having furnished the elements of the pancreatic juice. This diversity in the composition of venous blood is a thing perfectly proved at this time by analysis. The composition of the blood offers some differences in the extreme ages of life. In the new-born infant from two weeks to five months old, the proportion of water increases, and that of globules diminishes. From five months to forty years, the proportion of water diminishes, and the proportion of globules increases. From forty to seventy years, the proportion of water increases anew, and that of the globules diminishes. The quantity of albumen does not sensibly vary in the blood, considered in infancy, mature age, or youth. The blood also presents differences according to the age, and the individual temperament and constitution.

With individuals of a sanguine temperament, strong and robust, the blood encloses a much larger proportion of globules. With individuals of a lymphatic temperament, the blood is poorer in solid matters, and especially in globules.

MILK.

Milk is a liquid constantly alkaline, white, and opaque, secreted by the mammary glands. The milk of the cow is formed of 87 per cent. of water, which holds in solution 2 to 3 hundredths of *caseine*, 3 to 4 hundredths of sugar of milk, several alkaline and calcareous salts, principally phosphate of lime. Milk, besides, holds in suspension 4 hundredths of butter. It is this fat body which, in the form of globules, renders milk opaline. When milk is kept in a cool place, a layer of *cream*, which is produced by the globules of butter, forms on its surface. If the milk is briskly agitated at a temperature slightly elevated, the globules of the fat body collect together and form *butter*. When milk is left for some time in the air, it rapidly becomes acid; it is the sugar of milk which changes into lactic acid, and this acid causes the precipitation of the caseine.

It is then said that the milk *curdles*. This coagulum is used for making *cheese*. Milk may be prevented from coagulating by putting into it a small quantity of bicarbonate of potash or soda. The liquid part of the curdled milk is called *whey;* it holds in solution lactic acid, alkaline and calcareous salts, and sugar of milk.

MUSCULAR FLESH.

Muscular flesh consists principally of fibrine, but it contains besides different bodies which are interposed between the fibres of the muscles, such as fat, blood, etc.

BONES.

The bones of animals are essentially formed by an inorganic matter, which consists for the most part of a mixture of phosphate and carbonate of lime, and of an organic substance, which by boiling in water is transformed into gelatine. The bones of man generally contain —

Organic matter	28
Phosphate of lime	58
Carbonate of lime	7
Different salts	7

BILE.

The bile is a liquid secreted by the liver; in man and in a great number of animals, it is collected in an especial reservoir, the gall-bladder, from which it pours into that part of the intestine which immediately connects with the stomach.

The bile is yellow, and sometimes green; it is soluble in water in every proportion; its taste, at first bitter, often leaves behind a sweetish taste. It is often alkaline, and sometimes neutral.

The chemical composition of the bile appears to be very complicated; by using different reagents, a great number of different bodies may be extracted from it. The bile, however, may be considered to be an aqueous liquid containing different salts, some organic matters of an albuminous nature, and principally a sort of soap with soda for base.

With man, biliary concretions frequently exist, which lodge in the gall-bladder; these calculi are formed by a fat crystalline substance, which is called *cholesterine.*

ELEMENTARY NOTIONS ON RESPIRATION AND NUTRITION.

The waste which the animal organisation suffers is repaired by two functions, which are *digestion* and *respiration.* These functions are, besides, required for the growth of all the parts of the animal organisation.

DIGESTION.

Digestion is an act essentially chemical, which has for result the assimilation of the aliments and their transformation into blood. The aliments are first divided in the mouth by the *teeth*, and mixed with a liquid secreted by the *salivary glands*, which is the *saliva*. The action of the saliva is both physical and chemical. It is in general alkaline; it facilitates the division of the aliments, and brings about certain chemical reactions, as the transformation into sugar of the neutral organic substances, such as amidon.

The aliments having undergone the action of the saliva, take the name of *alimentary bole*, pass, by the mechanism of *deglutition*, into an elastic tube called the *œsophagus*, and thence into the *stomach*. The mucous membrane of the stomach secretes an acid liquid, the *gastric juice*, the chemical role of which is very important.

The gastric juice acts chemically on the aliments, *digests them;* it modifies and dissolves nitrogenous substances, does not act on the fat bodies, and determines the hydratation of neutral bodies. The aliments remain some time in the stomach, and are then by a series of its contractions taken into the *duodenum*, where they are wet with the *pancreatic juice*. This, as M. Bernard has shown, mixes with fat bodies, and renders them absorbable; it moreover transforms the amidon into sugar.

The duodenum receives still another alkaline liquid, which is the *bile*. The bile is secreted by the liver; it is retained for a time in the gall-bladder, and then comes into the duodenum by a special duct called the *biliary duct*.

The alimentary substances, after having received the influence of the saliva, the gastric juice, the pancreatic juice, and the bile, constitutes the *chyme*, and pass into the *small intestine*.

The small intestine is a canal, always narrow, the development of which varies with the nature of the aliments it is to receive.

In man, the intestines are from six to seven times the length of the body.

In carnivorous animals, they are much shorter. In the lion, they are about three times the length of the body.

In the herbivorous animals, they are always much developed. The intestines of the sheep are twenty-eight times the length of the body. The chyme is gradually pushed by a vermicular motion from the small intestine; a multitude of vessels are inserted into this intestine, called *chiliferous vessels*, which come in, as it were, to suck (*sucer*) the small intestine, and to extract from the chyme a white and sometimes rose-colored liquid, called *chyle.*

This chyle comes by the capillary vessels into the venous system, and restores to the blood that which it has given up to the different organs during circulation.

The chyme having lost the chyle, that is to say, the reparative and nutritive part of the aliments becomes more and more solid, and penetrates into the large intestine, and finally the non-nutritive parts are rejected by the *cœcum.*

RESPIRATION.

We have just seen how the blood, which is continually wasting by the phenomenon of assimilation, and which gives some of its substance to each organ which it nourishes, may be regenerated by the act of nutrition.

The blood holds in solution oxygen, which is changed into carbonic acid by this same act of assimilation. This carbonic acid would be hurtful to assimilation if the blood did not carry it off.

Thus the blood is for animals a purifier as well as restorer.

Respiration has for its object to give to the blood, oxygen which is useful to the phenomena of assimilation, and at the same time to take away from the blood the carbonic acid with which it is charged. The blood arrives in the right ventricle of the heart; the contractions of the heart drive it into the pulmonary artery, and thence into the lungs.

The lungs are spongy; they receive in one direction air by

the trachea, in another venous blood by the pulmonary artery.

In the lungs, the blood dissolves oxygen, loses its carbonic acid, becomes of a red color and changes into arterial blood, which, arriving at the left ventricle, is thrown into the arterial system.

The blood produces the phenomena of nutrition by placing itself in relation with the organs; the oxygen changes into carbonic acid, and the blood from being red becomes black, that is to say, is transformed into venous blood, which is mixed with chyle, then brought back into the lungs to receive anew the action of the atmosphere. Animals absorb air by the lungs or by the skin: thus two kinds of respiration are known, *cutaneous* and *pulmonary* respiration.

With warm-blooded animals, pulmonary respiration is much more active than the other respiration. By enclosing for some days an animal in an impermeable sack filled with air, so that his head may be out, the composition of the air is found not to be sensibly altered.

In animals with cold blood, the cutaneous respiration is, on the contrary, very active, and may even sometimes replace the pulmonary respiration. Thus frogs live for many days after their lungs have been removed.

Salamanders live entire months after their heads have been cut off.

The ensemble of the cutaneous and pulmonary functions is called *perspiration.*

The most simple experiments show that carbonic acid is given off in the act of respiration; thus common air contains about $\frac{3}{10000}$ths of carbonic acid, while the air expired contains as much as $\frac{4}{100}$ths.

It was for a long time thought that this species of combustion took place in the lungs, that is, this transformation of oxygen into carbonic acid. Now, it is admitted that the air in the lungs is only dissolved in the blood, and displaces the carbonic acid which is there found, and that it is then, when in the circulation, and in consequence of the phenomena of assimilation, that the oxygen changes into carbonic acid.

A complete and very important work on the phenomena of *perspiration* is due to MM. Regnault and Reiset. We will here quote some of the conclusions of their work.

The quantity of oxygen which an animal takes from the air is not all found in the carbonic acid he expires; so that a part of the oxygen is absorbed by the animal.

The diet has an influence on this phenomenon. When an animal is fed on meat, he absorbs a great deal of oxygen.

When he is fed on vegetables or bread, he exhales a quantity of carbonic acid which contains more oxygen than was furnished by the air; in this case the excess of oxygen comes from the food. The animal heat evidently proceeds from the phenomena of combustion, which are produced in the living body; but it cannot be admitted that the heat produced is equal to that which would result from the active combustion in oxygen, of the carbon which is found in the carbonic acid expired, or of the hydrogen which may have been burnt. There are evidently absorptions and disengagements of heat taking place in the body, the results of which can never probably be subjected to calculation.

CHEMICAL PHENOMENA OF VEGETATION.

It is well known that a seed placed in certain conditions of heat and moisture soon sprouts.

In the act of germination the seed takes nothing from the soil; for numerous experiments show that a seed will readily germinate in flowers of sulphur, pure sand, cotton, sponge, etc. These different substances, it is evident, can give nothing to the plant. The atmosphere, on the contrary, interposes in the phenomena of germination, with one of its elements, which is oxygen; and in causing seed to germinate in the atmosphere, it is seen that the volume of the gas does not vary, but that the oxygen is transformed into carbonic acid. Germination is impossible in nitrogen, hydrogen, or carbonic acid.

The seed germinates with the assistance of the air, and without taking any thing from the soil.

The immediate principles which constitute vegetables and animals being principally formed of carbon, oxygen, hydrogen, and nitrogen, it is important to study the mode of assimilation of these different elements. According to the observations of MM. de Saussure and Boussingault, seeds properly moistened with distilled water and put to germinate in powdered brick, which does not contain traces of organic bodies, have without the assistance of soil gone through all the phases of vegetation, from the first development of the seed to the fructification. It is then well shown from this that a vegetable *may* develope itself without the influence of the soil; it is then from the air and from the water that the elements which constitute organic bodies are derived, that is to say, oxygen, hydrogen, carbon, and azote or nitrogen.

The carbon which exists in organic substances evidently proceeds from the carbonic acid of the air. The experiments of Priestley, Sennebier, Ingen-Housz, and Theodore de Saussure, show that plants decompose carbonic acid under solar influence, that they assimilate the carbon, and that they reject in part the oxygen existing in this carbonic acid. This beautiful observation explains not only the mode of assimilation of the carbon by vegetables, but it demonstrates also that the composition of the air should remain invariable; and in fact, the oxygen of the air constantly tends to change itself into carbonic acid in the act of respiration or in the phenomena of combustion; this carbonic acid is decomposed by the vegetation, its oxygen is regenerated, and the composition of the air is thus found to be brought back to its normal state.

The hydrogen which exists in organic substances, proceeds evidently from the water which is fixed integrally, or which, in some cases, may be decomposed during the act of vegetation.

The oxygen contained in these organic bodies proceeds from the air and water. It is important that the air should have access to the roots of plants. This explains the necessity of those labors and operations which are given to the earth in its cultivation. When the roots of a tree penetrate into stagnant water, which does not contain oxygen, but carbonic acid in solution, it is seen that the tree soon dies.

As to the nitrogen, it may either come from the atmosphere or from manures.

It results from the very important experiments made by Boussingault, that some plants, as clover, take a considerable quantity of nitrogen from the air, so that these plants, cultivated without manure, are often themselves valuable manure; in practising what is called *enfouissement en vert* (the ploughing in of green crops), nitrogen taken from the air is thus introduced into the soil. The cereals, and principally wheat and oats, do not take a sensible quantity of nitrogen from the air, and consequently take it from manures; so that these plants require the employment of highly azotised manures.

When it is proposed to ameliorate a soil which is not of a good quality, it is best to commence by cultivating plants which take nitrogen from the atmosphere, and to feed stock, which, by their droppings, make manure useful for the cultivation of the cereals.

Nitrogen is not probably assimilated by the plants in a free state; it is generally thought that it proceeds from the decomposition of ammoniacal salts.

AGRICULTURAL CHEMISTRY.

Agriculture has for its object to produce *economically* a quantity of vegetation much more considerable than that which has been entrusted to the earth by seeding.

The means employed are: cultivation, manures and often the plants themselves, when they draw the elements of fertilization from the air, which they then deposit in the soil.

It is very important for the farmer to know how to appreciate the value of these means; it is not till he has acquired this knowledge, that he can divide his farm in such a way as to suit the different crops, according to the nature and condition of the soil, and according as the species of plants improve or wear out the soil by their growth. It is then only that he can direct judiciously the use of the manures which he has at command. It is this division of a farm for the cultivation of different crops which is called *assolement.* Assolement is *triennial* when the land is divided so as to produce three different crops in three years.

1st. Potatoes. 2d. Wheat. 3d. Clover.

Or, as it is the case in many countries—

1st. Fallow. 2d. Wheat. 3d. Oats.

It is quadrennial, quinquennial, etc., when four or five different kinds of cultivation succeed each other before the same crop returns.

The operation is called *rotation;* a culture is *alternate* when a plant succeeds one of a different nature, without ever leaving the soil unproductive.

FARM LAND.

The soil is to be considered, as destined to support the plants, to permit the development of their roots, to transmit to them some of the principles which it contains, to absorb the hygrometric water in the atmosphere, gases that the plants may assimilate them, to receive, to retain, or let flow out, as it is more or less permeable, the water necessary for vegetation; to retain manures till their elements are ready to be transmitted to the plants; to receive the heat necessary for the development of the plants, and then to give it up by means of its power of conduction.

It is seldom a soil is found which reunites all the conditions necessary for the accomplishment of these different functions. The agriculturist must thus study the soil which he is about to work, in order to appreciate its qualities, and to supply by cultivation the manure and improvements which it requires.

The elements of arable soil are, in general, sand, clay, lime, and an organic substance known under the name of *humus*.

To these matters may also be added, though in smaller proportion, magnesian salts, metallic oxides, and phosphates.

What is called sand, is an aggregation of silex, alumine, and calcareous salts, colored by humus and oxide of iron; in some, the silex is in large proportion.

Clay, which is also an aggregation of the substances which exist in sand, is moreover remarkable for the large quantity of alumine which it contains.

By calcareous soil, is understood one which, among other elements, contains a white, chalky substance, effervescent with the acids, composed of carbonic acid and lime. Finally, humus cannot be considered as originally making an essential part of the soil, as do sand, clay, and lime; for it has probably been carried there in the manure, by ameliorating plants, and in cleared lands, by the leaves of trees which pre-existed there.

Agriculturists call land *light* or sandy when silex predominates, and *heavy* or clayey when there is in it a large proportion of alumine.

This division of soil into sandy, clayey, and calcareous, is not, however, absolute. There are intermediate soils proceeding from a mixture of sand and clay in different proportions. The qualities of arable land depend almost always upon a natural mixture of these different soils in proper proportions, or on modifications, which are produced in them by the hand of man.

SANDY SOIL.

The base of sandy soil is silex (silicic acid), a white pulverulent, tasteless substance, which rarely exists pure in them. It is almost always colored by the oxide of iron or humus, and is mixed with clay and lime. Silex, in consequence of its granular form, has but little retentive power for water, and the roots of plants cultivated in sandy soils soon become dry; the easy circulation of air between its particles increases the dryness still more. Sandy soils are not fit to absorb and preserve the fertilizing principles of the atmosphere and of manures; thus they require a greater quantity of manure than other soils; and plants which derive most of their nourishment from the air, are best suited to these soils. Soils which are too sandy, are but little esteemed. But when these soils are properly mixed with calcareous and clayey soils, or with humus, they constitute soil of good quality, light, and easily worked in all seasons, allowing the air and water to circulate freely, so as to keep the roots of the plants fresh; they are permeable to atmospheric influences, preserving the manure a sufficiently long time for the plants to assimilate the results of their decomposition, and are able, with the aid of manure and alternate culture, to produce every kind of crop.

CLAYEY SOIL.

Clay owes its principal qualities to the alumine which it contains. It is known that this base shrivels the tongue, absorbs water, and retains it with persistence. It forms paste with it; and constitutes, by its mixture with lime, a substance known as *marl*, which is much valued by agriculturists.

Claycy soils are the opposite of sandy soils. They absorb water abundantly; the sort of paste which is the result dries with difficulty. This paste is, in a measure, impermeable. Thus, land which contains too large a proportion of clay, is sometimes wet; it is cold, and can hardly be worked before spring. This kind of soil often presents great difficulties to cultivators; the plough goes through it with difficulty; it is raised up in great clods, which are an impediment to seeding.

These soils, in drying, are furrowed with cracks, and when they contain clay in excess, refuse all kinds of cultivation, and can only be used for the manufacture of earthenware, bricks, and tiles.

It is, however, among the clay soils that the best land is found, where the properties of the alumine are modified by those of silex, lime, metallic oxides, and humus. This variety of land is remarkable for its blackish brown color, and its readiness for improvement. It has, besides, the property of retaining the water of imbibition in a proportion suitable for keeping the plants in a state of freshness favorable for their development, of absorbing the fertilizing matters, liquid and gaseous, of manures, and of preventing their dispersion.

CALCAREOUS SOIL.

The name calcareous soil is given to that in which the carbonate of lime predominates. The soils of this species, which are often very extended, have a whitish aspect, like clay and pure sand; they refuse all sort of fertilization, and are only fit for culture when mixed with other soil. They are then easy to cultivate, as is seen in Champagne and in the south of France. They are, however, but little esteemed, because they dry with facility, and, of their cold temperature, which is in great part due to their white color.

The calcareous matter, however, produces excellent effects in agriculture. It often acts mechanically, by dividing soils which are too clayey; and further, on account of its porosity, it absorbs the gases of the atmosphere, and may determine the production of nitrates.

It is also on account of this porosity that it absorbs the liquid and gaseous emanations of manures, and retains water better than the sandy soils, without the inconveniences which clayey soils present. The calcareous matter may also replace sand for dividing soils which are too clayey, and ameliorate soil for grain; it furnishes to plants whose growth is rapid, the lime which is suitable for them.

VEGETABLE SOIL.

There exists in arable soils a black organic substance to which they in part owe their color. This substance being mixed with different kinds of soils constitutes *humus, vegetable earth.* This has been added to the soil by a natural covering up (enfouissement), as in the forests of America, or covering up of the roots, leaves, parts of vegetables or manures, by tilling.

We will now speak of the modifications which soil may be made to undergo, when its constitution is not of a nature to fulfil the object of the farmer, or when repeated croppings have worn it out. This part of the science of agriculture is to be considered under two different heads.

1st. When it is desired to modify the nature of a soil which is too sandy, and to render it more fit for absorbing atmospheric emanations, or to hold the water which flows through it too readily.

2d. When it is proposed to render a soil which is too clayey, more light, less moist and compact, and to increase its absorbing powers.

These modifications can only be brought about by the

introduction of elements having properties opposed to those of the element which is in excess in the soil.

Thus in too sandy a soil, lime or clay must be introduced and nearly always both. Soil too heavy, on the contrary, should have sand and lime. These mechanical means of modifying soil are called *mineral manures.*

When arable land contains but small quantities of humus and has lost its fertility, and plants have been grown there which draw their support from the soil and not from the atmosphere, the soil is exhausted, and azotised organic substances known as manures must be added.

MINERAL MANURES.

The mineral manures are lime and marl. Lime absorbs water from the atmosphere and becomes a hydrate; it also combines with the carbonic acid of the air and of manures, to form carbonates, which always preserve the same state of tenuity as the hydrate. It decomposes ammoniacal salts, and thus furnishes to the plants ammonia or its elements. It reacts on the elements of the air, to produce nitrates which are found in marked quantities in many plants. It interposes itself between the particles of soil too abundant in silex or alumine. Finally, by its mixture with earth and different organic substances, it forms a fertilizing manure, called *compost*, very highly appreciated in many localities. Several kinds of lime are known in the arts; fat lime, poor lime; hydraulic lime, and magnesian lime. Fat lime, which is considered the purest, because it comes from a carbonate which contains but little foreign matter, is that which is used in agriculture. Its use requires some precautions. It must be borne in mind that most of its properties useful in agriculture are developed when in a state of great division. To attain these results, lime freshly calcined should be used, or it should be preserved from moisture. Without this precaution the expense of transport would be unnecessarily increased, for the hydrate of lime contains a fifth part of water. Besides, the spreading of lime should be avoided in damp weather, for then the lime cakes, and its separation or division is much more difficult; it absorbs carbonic acid less promptly, it preserves its causticity a longer time, and in this state it exercises a baneful influence on the roots of plants. It is generally towards the end of summer, when the ground is dry, that lime should be spread. For some crops, such as beets and potatoes, the lime is put on at the time of planting.

To incorporate fresh lime into the soil, it is put in heaps of about 25 to 30 litres, 5 to 6 metres apart. As soon as it is reduced to the state of impalpable powder by hydratation, it is spread as regularly as possible with shovels, a harrow is then passed over the ground, which is afterwards worked.

Very variable quantities of lime are spread over soils of somewhat similar qualities.

Thus while in France, on an average, 3 to 5 hectolitres of lime are spread to the hectare, in Germany 8 to 10, in England the quantity is as high as 100 to 600 hectolitres. When the worthless plants of sandy soils make their appearance, it is a sign that land requires lime. When it is desirable to introduce the calcareous element into a soil which is deprived of it, or which contains but a small proportion of it, porous or fragile chalk may be used with advantage, but in this case it should be exposed to the influence of the autumn rains and frosts, to bring it to a state of proper disintegration.

MARLING OR MANURING WITH MARL.

It is seldom the agriculturist is obliged to have recourse to the use of lime alone to improve the soil he is working. There exists in many subsoils, at a greater or less depth, a mixture of chalk and clay, known under the name of *marl*, which is one of the most powerful improvers of soil. According to M. de Gasparin, all efforts to imitate this natural geological production have been fruitless; and the products obtained are entirely different in properties from those of natural marl.

Carbonate of lime exists in marls in the proportion of from 15 to 90 per cent. In some marls, clay is replaced by sand, whence the designations *clayey* and *sandy marls.* It is important to find out by analysis of the soil to be marled, and by that of the marl, the proportions of the elements of each; for it would be a useless expense to introduce into a calcareous, clayey, or sandy soil, a marl which would only bring with it elements which the soil already possessed in sufficient quantity.

Of whatever nature may be the marl intended for the modification of a soil, it ought to have for its principal qualities that of being sensible to atmospheric influences, and of pulverizing readily.

The same care is required in the use of marl as in that of lime. It should be placed in small heaps at equal distances. It should not be spread till completely pulverized, which often requires the alternate action of the heat of summer and of the frost of winter. The water then penetrates it, breaks it up, and divides it. The effect of marling is incontestable; the grain crops have been seen to be doubled, and almost tripled, by marl; but its use should be limited by the quantity of lime taken off by the crops, and by its richness in lime.

PLASTER.

Gypsum is formed of 80 parts of sulphate of lime, and 20 parts of water. In this state it is called *crude plaster.* It is called *calcined plaster* when the water of crystallization has been carried off by calcining. It may be used in agriculture in either state. The calcined plaster, however, is more easily reduced to powder, and its separation among the plants is more easy than that of crude plaster.

Though in some localities the plaster is applied at the time of working the ground in the autumn, it is more generally used by sprinkling it over the plants in the morning, while they are still wet.

It is an acknowledged fact in agriculture, that plaster should only be used on soils properly manured; and that it is without effect on thin and poor soils.

The quantity of plaster used per hectare of ground is from 200 to 2000 kilogrammes. These quantities appear to be limited only by the price of the plaster. Its efficacy on plants is not general. It suits leguminous plants, clover, sainfoin, lucerne. It has but little effect on natural meadows, hoed crops, and grain.

USE OF SALT (CHLORIDE OF SODIUM) IN AGRICULTURE.

The effects of salt have been differently appreciated by agriculturists. Some consider it hurtful; others, on the contrary, think that in some localities it may be used with considerable advantage.

It appears, however, to have good effects when spread on clayey or marled soils. It acts either as an excitant, or by furnishing, to soils without potash, soda, which is necessary to their growth.

The quantity of salt which ought to be used depends on the nature of the soil and that of the subsoil. If the subsoil is permeable, it may be that salt produces no effect, and that it is carried off by the rain.

DRAINAGE.

We have pointed out the inconveniences which a very clayey soil offers for cultivation. These inconveniences, sufficiently serious for the tillable soil, are not less so, when under this there is found a clayey subsoil, which prevents the percolation of rain, thus rendering the soil wet, cold, and marshy. This circumstance often interferes with the culture of good soil. Meadows of this kind produce a sour, hard, and dry grass, often mixed with plants unfit for feeding animals, and which are produced on account of the wetness of the soil alone.

To prevent the effects of this impermeability, drains filled with pebble-stones are made in the subsoil, to carry off the excess of water in the arable soil, and thus take the place of a permeable soil. This process is practised in England and Scotland, under the name of *drainage.*

But it often happens that the interstices of the pebbles fill with earth, partly obstructing them, so that the escape of water takes place but partially.

A modern substitute for this mode of draining has been put in practice, as follows: Trenches are dug 1 metre 32 centimeters in depth, distant from each other 10 to 12 metres in some places, and in others from 16 to 18 metres; in these are placed earthenware pipes 33 centimeters long, the diameter of which varies from 2 to 3 centimeters; these pipes are fitted one within the other, and at the depth at which they are placed, they cannot be reached by the plough; the water drains into the pipes at the joints, and flows off in a continuous manner when the soil is wet. Drainage produced such advantageous results in England, that Parliament, in voting 75 millions of francs for the encouragement of agriculture, recommended that a part of this sum be appropriated for drainage. A like measure, in France, might recover for agriculture lands yet uncultivated.

MANURE.

Whatever be the nature of the mineral fertilizers which have been used for fertilizing arable land, they soon become insufficient, inasmuch as they give to plants only a part of the substances necessary for their development; vegetables will grow in a soil thus improved, but they are weak and stunted:

their cultivation rewards but imperfectly the labor and expense of the cultivator. To re-establish a suitable fertility in a soil, there must be introduced into it, that mixture of animal and vegetable matters, to which the name of manure has been given, which carries into the soil with carbon, oxygen, hydrogen, and nitrogen, the phosphates, carbonates, nitrates, sulphates, silicates, chlorides, etc., which the plants assimilate to themselves, and which are proper for their development.

It must not be forgotten that the principal object of manures is to restore to the earth that which vegetation takes away each year. The principal manures are the tops of turnips, cabbage, hemp, the pressed skins of olives and other fruits, residue from starch factories, blood, the flesh of dead animals, urine, fecal matters, guano, char from refineries; but of all manures, the best is stable manure.

Sometimes salts are used with advantage: these are called *inorganic manures*, such are ashes of vegetables, ammoniacal salts, phosphate of lime, plaster, sea-salt, etc.

The production of manure is an essential part of agricultural chemistry, so much so, says M. Boussingault, that the intelligence of a cultivator may be known by the care he gives to it.

To obtain manure for the farm, litter is spread under the animals: for the best manure, this should be, as far as possible, straw from the cereals in sufficient quantity to absorb the animal dejections.

This is easily done for horses, but for horned cattle it is difficult, especially in the spring and summer, when the excrements of these beasts are very liquid. It is then best to make trenches or gutters in the stables to carry off to reservoirs the liquid dejections which the litter does not absorb. Sheep stables do not require this arrangement, for on account of the solidity of the excrements of these animals the litter is sufficient to absorb the urine.

Manure is removed in different ways from horse and cattle stables. In some places, as in cavalry quarters, the litter is removed daily; after having taken up the manure and separated it, the litter which is not too much altered is used again; in other places this is only done once or twice a week, while some again, every day, put fresh litter on the old manure, and this goes on till the accumulation is so great in the stables as to require removal.

If only the manure is kept in view, this last method is the best; the litter broken up by the tramping of the animals, is more thoroughly impregnated with their dejections, and will more readily undergo an ulterior disorganization; but the health of the animals must also be considered; and it is evident that in those stables where this is practised, the air becomes vitiated by a considerable disengagement of ammonia.

The manure taken from the stables, is carried to trenches, where it soon heats, entering into fermentation, and gives off carbonate of ammonia. This fermentation should be moderated, for if it is too rapid, the azote is dissipated and lost, and this is the essential element of fertilisation.

The fermentation is retarded by spreading new manure daily on the heap, to maintain in the mass a certain degree of moisture, which prevents the temperature from rising too rapidly, and acts also in absorbing the volatile principles of the fermentation.

The manure ought not to be stirred, as is done in some countries, to hasten the decomposition. In this way a considerable quantity of ammonia is lost.

Precautions to prevent loss in the fermentation of manures are always useful. Gazzeri proved in fact that in four months, the manure of the horse loses more than half of the dry matter which it contained before putrefaction. M. Payen found also that the manure-heap, after it had ceased to ferment, had lost two-thirds of its primitive azote.

It might be supposed, from these observations, that it would be of advantage to the cultivator, to transport the manure directly from the stables to the soil, without allowing it to ferment in trenches ; but is it quite certain that this fermentation, the inconveniences of which may be prevented, is not necessary to reduce the organic parts of the manure to a proper state of assimilation ?

Spreading manure fresh from the stables cannot be practised except in the small way. The daily transport of fresh manure over the ground of a large farm is very expensive, while the transport of the manure of trenches, whose volume is diminished one-half, is done in a convenient manner in the winter, when labor is less dear, when the teams have no other occupation, and when the land, almost bare, may be easily improved. The questions as to the employment of fresh or fermented manure, are not yet resolved. However, the most distinguished agriculturists appear really to give the preference to manure which has not yet undergone fermentation.

PLOUGHING IN CROPS.

Though it is the interest of the agriculturist to stimulate the act of vegetation by animalised manures, which carry to the plants the quantity of azote necessary for their development, there may be circumstances when this is impossible. Thus it may happen that a farmer taking possession, receives neglected lands, without fodder for the nourishment of animals, and that he has no means of procuring manure. In this case he is obliged to get the azote from plants, which he cannot get elsewhere ; it is by *ploughing in green crops* that he arrives at this result.

This mode of manuring, to be fruitful, should be practised with judgment. Thus the soil being worn out, it is important not to plant on it vegetables which it must support, but those which will draw from the atmosphere the water, the carbon, and the azote necessary for their development. For this purpose, leguminous plants are generally used, such as lupine (peas), beans, and vetches; also, spergula, or spurry, tobacco, buckwheat, *madia sativa*, and turnips.

It is not only by their hygrometric qualities, and because they carry to the soil azotised organic matters, that plants ploughed in improve vegetation, we must also take into consideration the saline matters which they contain, the presence of which explains the efficacy of ploughing in all the marine plants of the alga family, which are collected on the coasts of Normandy, Britain, Scotland, and Ireland, in contributing to the fertility of the soil.

CULTURE OF TREFOIL.

The cultivation of trefoil, so far from wearing out the soil, returns to it, by the decay of its roots, nearly all that which it took away.

This plant, which is sown in the spring like a cereal, developes itself in one year, in such a manner that in the following year it takes the place of the fallow by giving two or three crops of fodder, the last of which may be ploughed in to increase the fertility of the soil. This crop, giving to the farmer the means of raising a great many cattle, is a true source of manure.

The introduction of trefoil into farm culture was simultaneous with a system of alternate cropping, which is calculated so as to bring back to the soil plants, the culture of which will not be hurtful to the succeeding crops. It is a happy result brought about by the judicious classification of plants which exhaust the soil by the action of their roots, and of those which enrich it by the faculty they possess of drawing their sustenance from the atmosphere.

We ought, however, to observe with M. Boussingault, that it must not be expected that trefoil can be profitably introduced in a triennial rotation. It should be returned every fourth or fifth year; this requires the introduction in the rotation, which would then be quadrennial or quinquennial, of hoed plants, which clean the soil.

EMPLOYMENT OF AZOTISED SALTS.

After having shown that manures have the property of carrying azote into the soil, an element indispensable to the growth of vegetables, it has been thought that azote might be presented to the plants in another form, and in a state more easy to assimilate, by using nitrates or ammoniacal salts.

Practical agriculture is already on this course, and it is only necessary to draw the attention of cultivators to the experiments of men whose scientific authority is enhanced by the experience of the farm. Although the results obtained are already satisfactory, we cannot too often repeat that these trials are not sufficiently numerous and generally conclusive, and that they have not yet sufficient authority, to justify their use in the large way. Agriculturists should experiment themselves with caution, first on a small scale, taking account of all the circumstances and all the results.

It must be recollected, that at present, there is nothing more economical and certain than stable manure, which in decomposing slowly as vegetation progresses, gives successively to the plants, salts, ammoniacal vapors, and carbonic acid, all of which are useful to their growth.

INDEX.

THE END.

www.ingramcontent.com/pod-product-compliance
Lightning Source LLC
LaVergne TN
LVHW020932110826
845150LV00004B/838

* 9 7 8 1 4 2 5 5 5 2 6 5 7 *